Studies in Inherited Metabolic Disease
Lipoproteins; Ethical Issues

Studies in Inherited Metabolic Disease
Lipoproteins
Ethical Issues

Proceedings of the 25th Annual
Symposium of the SSIEM,
Sheffield, UK,
September 1987

The combined supplements of *Journal of Inherited Metabolic Disease* Volume 11 (1988)

edited by R. J. Pollitt,
R. A. Harkness and G. M. Addison

SPRINGER-SCIENCE+BUSINESS MEDIA, B.V.

ISBN 978-94-010-7059-1   ISBN 978-94-009-1259-5 (eBook)
DOI 10.1007/978-94-009-1259-5

Copyright

Contents

Author Index   ix

Title Index   xi

Preface – Inherited Metabolic Disease 25 years on   1

The biochemistry of lipoproteins
A. M. Salter and D. N. Brindley   4

Clinical consequences of hyperlipidaemia
G. R. Thompson   18

Lipase deficiencies
H. Greten and F. U. Beil   29

The use of recombinant DNA techniques for the diagnosis of familial hypercholesterolaemia
S. Humphries, R. Taylor, M. Jennah and M. Seed   33

Familial LCAT deficiency and fish eye disease
N. McIntyre   45

Biochemical, genetic and metabolic studies of hyperapobetalipoproteinaemia
P. O. Kwiterovich, Jr   57

Apolipoprotein polymorphism and multifactorial hyperlipidaemia
G. Utermann   74

A neonatal screening approach to the detection of familial hypercholesterolaemia and family-based coronary prevention
D. E. L. Wilcken, B. L. Blades and N. P. B. Dudman   87

The paediatric lipid clinic in Birmingham
M. Tarlow, A. Green, D. Worthington and E. Buchanan   91

Recent advances in cystic fibrosis
M. A. McPherson   94

Symposium: the ethics of antenatal diagnosis and the termination of pregnancy
Introduction
D. P. Brenton and J. W. Seakins   110

Antenatal diagnosis and the termination of pregnancy – what the churches have to say
Rev. N. R. Holtam   111

Ethics and clinical practice
R. Gillon   120

Discussion   125

Short Communications
Preface to Short Communications   131

Free Communications   132

An erroneous apolipoprotein E-3 band in high density lipoprotein fractions
A. V. Rawlings and T Deegan   135

Histochemical abnormalities in liver and jejunal biopsies from a case of cholesterol ester storage disease
A. Lageron and J. Polonovski   139

Cholesteryl ester storage disease: risk factors for atherosclerosis in a 15-year-old boy
R. Longhi, C. Vergani, R. Valsasina, E. Riva, C. Galluzzo, C. Agostoni and M. Giovannini   143

Cholesteryl ester storage disease with secondary lecithin cholesterol acyl transferase deficiency
  S. Van Erum, D. Gnat, C. Finne, D. Blum, C. Vanhelleput, E. Vamos and F. Vertongen   146

A treatable familial neuromyopathy with vitamin E deficiency, normal absorption, and evidence of increased consumption of vitamin E
  A. Kohlschütter, C. Hübner, W. Jansen and S. G. Lindner   149

Fat malabsorption, vitamin E deficiency, scoliosis and cataracts
  R. D. Griffiths, C. J. Taylor, D. M. Isherwood and M. J. Jackson   153

Familial high-density lipoprotein deficiency (Tangier disease): the third Italan case
  G. Bracco, G. Dotti, F. Levis, E. David, G. Saracco, M. Rizzetto and G. Verme   155

Failure of taurine to improve fat absorption in cystic fibrosis
  G. N. Thompson   158

Peroxisomes and peroxisomal functions in hyperpipecolic acidaemia
  R. J. A. Wanders, C. W. T. van Roermund, M. J. A. van Wijland, R. B. H. Schutgens, J. M. Tager, H. van den Bosch and G. H. Thomas   161

Bile acid analyses in 'pseudo-Zellweger' syndrome: clues to the defect in peroxisomal β-oxidation
  P. T. Clayton, B. D. Lake, M. Hjelm, J. B. P. Stephenson, G. T. N. Besley, R. J. A. Wanders, A. W. Schram, J. M. Tager, R. B. H. Schutgens and A. M. Lawson   165

Adrenomyeloneurodystrophy with late cerebral involvement and evidence of a multiple autoimmune disorder
  A. Federico, M. T. Dotti, P. Annunziata, U. Bonuccelli, G. Fenzi, G. Ciacci, A. Malandrini, G. Meucci and G. C. Guazzi   169

X-linked adrenoleukodystrophy: identification of the primary defect at the level of a deficient peroxisomal very-long-chain fatty acyl-CoA synthetase using a newly developed method for the isolation of peroxisomes from skin fibroblasts
  R. J. A. Wanders, C. W. T. van Roermund, M. J. A. van Wijland, R. B. H. Schutgens, A. W. Schram, J. M. Tager, H. van den Bosch and C. Schalkwijk   173

Infanto-juvenile encephaloneuropathy and pigmentary retinopathy in a girl, associated with congenital adrenal insufficiency and altered plasma medium chain fatty acid levels
  A. Federico, G. Barachini, M. T. Dotti, L. Ibba, A. Malandrini, G. Ciacci, M. Meloni, S. Palmeri, A. Pompella and G. C. Guazzi   178

Familial hypoketotic hypoglycaemia associated with peripheral neuropathy, pigmentary retinopathy and C6–C14 hydroxydicarboxylic aciduria. A new defect in fatty acid oxidation?
  B. T. Poll-The, J. P. Bonnefont, H. Ogier, C. Charpentier, A. Pelet, J. M. le Fur, C. Jakobs, R. M. Kok, M. Duran, P. Divry, J. Scotto and J. M. Saudubray   183

A new type of mitochondrial encephalomyopathy with stroke-like episodes due to cytochrome oxidase deficiency
  P. Maertens, R. Richardson, F. Bastian, J. P. Williams and F. Hommes   186

Cytochrome c oxidase deficiency in three patients with Leigh's disease
  M. Di Rocco, E. Veneselli, M. O. Ciccone, A. Taccone, M. Stroppiano and F. Cottafava   189

Histochemical, ultrastructural and biochemical study of muscle mitochondria in Leber's hereditary optic atrophy
  A. Federico, L. Manneschi, M. Meloni, C. Alessandrini, A. M. Bardelli, M. T. Dotti and P. Sabatelli   193

Morphometric and biochemical study of muscle mitochondria in adult chronic progressive external ophthalmoplegia (ACPEO)
  A. Federico, L. Manneschi, P. Sabatelli, M. T. Dotti, G. Ciacci, L. Ibba and R. Gerli   198

Cytochrome c oxidase: organ-specific isoenzymes and deficiencies
  K. M. C. Sinjoro, T. B. M. Hakvoort, A. O. Miujsers, A. W. Schram and J. M. Tager   202

Enzymatic heterogeneity in primary hyperoxaluria type 1 (hepatic peroxisomal alanine:glyoxylate aminotransferase deficiency)
  C. J. Danpure and P. R. Jennings   205

Diversity in residual alanine glyoxylate aminotransferase activity in hyperoxaluria type I: correlation with pyridoxine-responsiveness
  R. J. A. Wanders, C. W. T. van Roermund, S. Jurriaans, R. B. H. Schutgens, J. M. Tager, H. van den Bosch, E. D. Wolff, H. Przyrembel, R. Berger, F. G. Schaaphok, W. Reitsma and W. H. J. van Luyk   208

The subcellular metabolism of glyoxylate in primary hyperoxaluria type I: the relationship between glycine production and oxalate overproduction
  G. N. Thompson, P. Purkiss and C. J. Danpure   212

Primary hyperoxaluria and L-glyceric aciduria in the cat
  W. F. Blakemore, M. F. Heath, M. J. Bennett, C. H. Cromby and R. J. Pollitt   215

Clinical effects of serine medication in non-ketotic hyperglycinaemia due to deficiency of P-protein of the glycine cleavage complex
  F. A. Wijburg, C. J. de Groot, R. B. H. Schutgens, P. G. Barth and K. Tada   218

The use of phenylpropionic acid as a loading test for medium-chain acyl-CoA dehydrogenase deficiency
  J. W. T. Seakins and G. Rumsby   221

Odd-numbered long-chain fatty acid contents in erythrocyte membrane phospholipids in patients with an impaired propionate utilization
  U. Wendel, E. Diekmann and M. D. Laryea   225

Mevalonic aciduria: pathobiochemical effects of mevalonate kinase deficiency on cholesterol metabolism in intact fibroblasts
  G. Hoffmann, K. M. Gibson, W. L. Nyhan and L. Sweetman   229

A patient with mevalonic aciduria presenting with hepatosplenomegaly, congenital anaemia, thrombocytopaenia and leukocytosis
  J. B. C. de Klerk, M. Duran, L. Dorland, H. A. A. Brouwers, L. Bruinvis and D. Ketting   233

A closer look at the eye in homocystinuria – a screened population
  J. P. Burke, M. O'Keefe, R. Bowell and E. R. Naughten   237

Peptiduria presumably caused by aminopeptidase-P deficiency. A new inborn error of metabolism
  N. Blau, A. Niederwieser and D. H. Shmerling   240

Early morning urine galactitol levels in relation to galactose intake: a possible method of monitoring the diet in galactokinase deficiency
  J. T. Allen, J. B. Holton, A. C. Lennox and I. C. Hodges   243

Cataracts in children with classical galactosaemia and in their parents
  J. P. Burke, M. O'Keefe, R. Bowell and E. R. Naughten   246

A patient with severe type of epimerase deficiency galactosaemia
  I. B. Sardharwalla, J. E. Wraith, C. Bridge, B. Fowler and S. A. Roberts   249

Branching enzyme in erythrocytes. Detection of type IV glycogenosis homozygotes and heterozygotes
  Y. S. Shin, H. Steiguber, P. Klemm, W. Endres, O. Schwab and G. Wolff   252

β-Mannosidosis in two brothers with hearing loss
  L. Dorland, M. Duran, F. E. T. Hoefnagels, J. N. Breg, H. Fabery de Jonge, K. Cransbery, F. J. van Sprang and O. P. van Diggelen   255

Infantile sialic acid storage disease in two siblings
  A. Cooper, I. B. Sardharwalla, M. Thornley and K. P. Ward   259

Evaluation of lysosomal enzymes in uncultured and cultured chorionic villi and amniocytes
  G. Bartalini, M. A. Margollicci, P. Balestri and A. Fois   263

Author Index

Agostoni, C. 143
Alessandrini, C. 193
Allen, J. T. 243
Annunziata, P. 169
Balestri, P. 263
Baracchini, G. 178
Bardelli, A. M. 193
Bartalini, G. 263
Barth, P. G. 218
Bastian, F. 186
Beil, F. U. 29
Bennet, M. J. 215
Berger, R. 208
Besley, G. T. N. 165
Blades, B. L. 87
Blakemore, W. F. 215
Blau, N. 240
Blum, D. 146
Bonnefont, J. P. 183
Bonuccelli, U. 169
van den Bosch, H. 161, 173, 208
Bowell, R. 237, 246
Bracco, G. 155
Breg, J. N. 255
Brenton, D. P. 110
Bridge, C. 249
Brindley, D. N. 4
Browers, H. A. A. 233
Bruinvis, L. 233
Buchanan, E. 91
Burke, J. P. 237, 246
Charpentier, C. 183
Ciacci, G. 169, 178, 198
Ciccone, M. O. 189
Clayton, P. T. 165
Cooper, A. 259
Cottafava, F. 189
Cransberg, K. 255
Cromby, C. H. 215
Danpure, C. J. 205, 212
David, E. 155
Deegan, T. 135
Di Rocco, M. 189
Diekmann, E. 225
Van Diggelen, O. P. 255
Divry, P. 183
Dorland, L. 233, 255
Dotti, G. 155, 169, 178, 193, 198
Dudman, N. P. B. 87
Duran, M. 183, 233, 255
Endres, W. 252
Van Erum, S. 146
Fabery de Jonge, H. 255

Federico, A. 169, 178, 193, 198
Fenzi, G. 169
Finne, C. 146
Fois, A. 263
Fowler, B. 249
le Fur, J. M. 183
Galluzzo, C. 143
Gerli, R. 198
Gibson, K. M. 229
Gillon, R. 120
Giovannini, M. 143
Gnat, D. 146
Green, A. 91
Greten, H. 29
Griffiths, R. D. 153
de Groot, C. J. 218
Guàzzi, G. C. 169, 178
Hakvoort, T. B. M. 202
Heath, M. F. 215
Hjelm, M. 165
Hodges, I. C. 243
Hoefnagels, F. E. T. 255
Hoffmann, G. 229
Holtam, Rev. N. R. 111
Holton, J. B. 243
Hommes, F. 186
Hübner, C. 149
Humphries, S. 33
Ibba, L. 178, S198
Isherwood, D. M. 153
Jackson, M. J. 153
Jakobs, C. 183
Jansen, W. 149
Jeenah, M. 33
Jennings, P. R. 205
Jurriaans, S. 208
Ketting, D. 233
Klemm, P. 252
de Klerk, J. B. C. 233
Kohlschütter, A. 149
Kok, R. M. 183
Kwiterovich, Jr, P. O. 57
Lageron, A. 139
Lake, B. D. 165
Laryea, M. D. 225
Lawson, A. M. 165
Lennox, A. C. 243
Levis, F. 155
Lindner, S. G. 149
Longhi, R. 143
van Luyk, W. H. J. 208
Maertens, P. 186
Malandrini, A. 169, 178
Manneschi, L. 193, 198

Margollicci, M. A. 263
McIntyre, N. 45
McPherson, M. A. 94
Meloni, M. 178, 193
Meucci, G. 169
Miujsers, A. O. 202
Naughten, E. R. 237, 246
Niederwieser, A. 240
Nyhan, W. L. 229
O'Keefe, M. 237, 246
Ogier, H. 183
Palmeri, S. 178
Pelet, A. 183
Poll-The, B. T. 183
Pollitt, R. J. 215
Polonovski, J. 139
Pompella, A. 178
Przyrembel, H. 208
Purkiss, P. 212
Rawlings, A. V. 135
Reitsma, W. 208
Richardson, R. 186
Riva, E. 143
Rizzetto, M. 155
Roberts, S. A. 249
van Roermund, C. W. T. 161, 173, 208
Rumsby, G. 221
Sabatelli, P. 193, 198
Salter, A. M. 4
Saracco, G. 155
Sardharwalla, I. B. 249, 259
Saudubray, J. M. 183
Schaaphok, F. G. 208
Schalkwijk, C. 173
Schram, A. W. 165, 173, 202
Schutgens, R. B. H. 161, 165, 173, 208, 218
Schwab, O. 252
Scotto, J. 183
Seakins, J. W. 110, 221
Seed, M. 33
Shin, Y. S. 252
Shmerling, D. H. 240
Sinjoro, K. M. C. 202
van Sprang, F. J. 255
Steiguber, H. 252
Stephenson, J. B. P. 165
Stroppiano, M. 189
Sweetman, L. 229
Taccone, A. 189
Tada, K. 218
Tager, J. M. 161, 165, 173,

202, 208
Tarlow, M.   91
Taylor, C. J.   153
Taylor, R.   33
Thomas, G. H.   161
Thompson, G. N.   18, 155, 212
Thornley, M.   259
Utermann, G.   74
Valsasina, R.   143

Vamos, E.   146
Vanhelleput, C.   146
Veneselli, E.   189
Vergani, C.   143
Verme, G.   155
Vertongen, F.   146
Wanders, R. J. A.   161, 165, 173, 208
Ward, K. P.   259
Wendel, W.   225

Wijburg, F. A.   218
van Wijland, M. J. A.   161, 173
Wilchen, D. E. L.   87
Williams, J. P.   186
Wolff, E. D.   208
Wolff, G.   252
Worthington, D.   91
Wraith, J. E.   249

Title Index

Adrenomyeloneurodystrophy with late cerebral involvement and evidence of a multiple autoimmune disorder 169

Antenatal diagnosis and the termination of pregnancy – what the churches have to say 111

Apolipoprotein polymorphism and multifactorial hyperlipidaemia 74

Bile acid analyses in 'pseudo-Zellweger' syndrome: clues to the defect in peroxisomal β-oxidation 165

Biochemical, genetic and metabolic studies of hyperapobetalipoproteinaemia 57

The biochemistry of lipoproteins 4

Branching enzyme in erythrocytes. Detection of type IV glycogenosis homozygotes and heterozygotes 252

Cataracts in children with classical galactosaemia and in their parents 246

Cholesteryl ester storage disease with secondary lecithin cholesterol acyl transferase deficiency 146

Cholesteryl ester storage disease: risk factors for atherosclerosis in a 15-year-old boy 143

Clinical consequences of hyperlipidaemia 18

Clinical effects of serine medication in non-ketotic hyperglycinaemia due to deficiency of P-protein of the glycine cleavage complex 218

A closer look at the eye in homocystinuria – a screened population 237

Cytochrome c oxidase deficiency in three patients with Leigh's disease 189

Cytochrome c oxidase: organ-specific isoenzymes and deficiencies 202

Discussion 125

Diversity in residual alanine glyoxylate aminotransferase activity in hyperoxaluria type I: correlation with pyridoxine-responsiveness 208

Early morning urine galactitol levels in relation to galactose intake: a possible method of monitoring the diet in galactokinase deficiency 243

Enzymatic heterogeneity in primary hyperoxaluria type 1 (hepatic peroxisomal alanine:glyoxylate aminotransferase deficiency) 205

An erroneous apolipoprotein E-3 band in high density lipoprotein fractions 135

Ethics and clinical practice 120

Evaluation of lysosomal enzymes in uncultured and cultured chorionic villi and amniocytes 263

Failure of taurine to improve fat absorption in cystic fibrosis 158

Familial high-density lipoprotein deficiency (Tangier disease): the third Italian case 155

Familial hypoketotic hypoglycaemia associated with peripheral neuropathy, pigmentary retinopathy and C6–C14 hydroxydicarboxylic aciduria. A new defect in fatty acid oxidation? 183

Familial LCAT deficiency and fish eye disease 45

Fat malabsorption, vitamin E deficiency, scoliosis and cataracts 153

Histochemical abnormalities in liver and jejunal biopsies from a case of cholesterol ester storage disease 139

Histochemical, ultrastructural and biochemical study of muscle mitochondria in Leber's hereditary optic atrophy 193

Infantile sialic acid storage disease in two siblings 259

Infanto-juvenile encephaloneuropathy and pigmentary retinopathy in a girl, associated with congenital adrenal insufficiency and altered plasma medium chain fatty acid levels 178

Lipase deficiencies 29

β-Mannosidosis in two brothers with hearing loss 255

Mevalonic aciduria: pathobiochemical effects of mevalonate kinase deficiency on cholesterol metabolism in intact fibroblasts 229

Morphometric and biochemical study of muscle mitochondria in adult chronic progressive external ophthalmoplegia (ACPEO) 198

A neonatal screening approach to the detection of familial hypercholesterolaemia and family-based coronary prevention 87

A new type of mitochondrial encephalomyopathy with stroke-like episodes due to cytochrome oxidase deficiency   186

Odd-numbered long-chain fatty acid contents in erythrocyte membrane phospholipids in patients with an impaired propionate utilization   225

The paediatric lipid clinic in Birmingham   91
A patient with mevalonic aciduria presenting with hepatosplenomegaly, congenital anaemia, thrombocytopaenia and leukocytosis   233
A patient with severe type of epimerase deficiency galactosaemia   249
Peptiduria presumably caused by aminopeptidase-P deficiency. A new inborn error of metabolism   240
Peroxisomes and peroxisomal functions in hyperpipecolic acidaemia   161
Primary hyperoxaluria and L-glyceric aciduria in the cat   215

Recent advances in cystic fibrosis   94

The subcellular metabolism of glyoxylate in primary hyperoxaluria type I: the relationship between glycine production and oxalate overproduction   212
Symposium: the ethics of antenatal diagnosis and the termination of pregnancy. Introduction   110

A treatable familial neuromyopathy with vitamin E deficiency, normal absorption, and evidence of increased consumption of vitamin E   149

The use of phenylpropionic acid as a loading test for medium-chain acyl-CoA dehydrogenase deficiency   221
The use of recombinant DNA techniques for the diagnosis of familial hypercholesterolaemia   33

X-linked adrenoleukodystrophy: identification of the primary defect at the level of a deficient peroxisomal very-long-chain fatty acyl-CoA synthetase using a newly developed method for the isolation of peroxisomes from skin fibroblasts   173

J. Inher. Metab. Dis. 11 Suppl. 1 (1988) 1–3

Preface – Inherited Metabolic Disease 25 Years On

> The interest in neurometabolic conditions is now so widespread that the number of known neurometabolic conditions is ever increasing, and it would be extremely difficult to prepare a complete list of them.
>
> R. S. Illingworth, in Holt, K. S. and Milner, J. (eds.)
> *Neurometabolic Disease in Childhood*
> E. and S. Livingstone, Edinburgh and London, 1964

In 1962 a small group of paediatricians and biochemists, mainly from the North of England, met at the Manchester Royal Infirmary to discuss phenylketonuria. This meeting must have been useful, since the same group then organized a more formal symposium covering a wider range of subjects, held in Sheffield on 31st May 1963. The proceedings of this meeting were published as a small book, *Neurometabolic Disorders in Childhood*. The cover bore the crest of the University of Sheffield and the meeting was referred to simply as "a symposium". The opening line of the Introduction is reproduced above.

The success of the Sheffield meeting led to the formal foundation of the Society for the Study of Inborn Errors of Metabolism (SSIEM) in October 1963. The proceedings of the next symposium, Liverpool, September 1964, the first to be published under the Society's name, were not formally numbered but papers from the 1965 meeting, held in Birmingham, were published as the Proceedings of the 3rd Symposium of the SSIEM. The Sheffield meeting had become, retrospectively, the inaugural symposium of the new society.

Numbering continued on this basis as the subject grew in scope and complexity, the Society's membership steadily increased, and the meetings became two-day and then three-day affairs. The Society rapidly became more international in its membership, with the meeting in Zurich (1968) being an important stage. This international involvement resulted in a further ambiguity in numbering the meetings when the Society agreed to forego its Annual Symposium for 1980 to support the 2nd International Symposium, held in Interlaken. However, the proceedings of this meeting were published as part of the Society's regular series and the numbering carried on accordingly. On this basis the meeting held in Sheffield, 22–25th September 1987, was deemed to be the 25th, the Silver Jubilee Meeting. For the occasion Mr J. Ireland mounted an historical exhibition devoted to the Society and we were honoured by the presence of several participants in the first Sheffield meeting, including the editors of its published proceedings.

The scientific programme had disorders of lipoprotein metabolism as its main topic. This subject was chosen because of its interest to adult physicians as well as to paediatricians, biochemists and those involved in whole population screening, and the way that newer biochemical techniques, particularly DNA studies, are adding a degree of precision to aspects of lipoprotein pathology that have until

Journal of Inherited Metabolic Disease. ISSN 0141–8955. Copyright © SSIEM and MTP Press Limited, Queen Square, Lancaster, UK.

recently been poorly defined. A shorter session was devoted to cystic fibrosis, a disease where recent advances also promise new insights and important practical applications, particularly in the field of prenatal diagnosis.

The ethics of prenatal diagnosis formed the subject of a separate mini-symposium which was particularly noteworthy for the participation of representatives of a number of parents' organisations. Such organisations are an important and growing feature, with aims that variously include social support for families with particular diseases, disseminating information, both to professionals and to the lay public, raising funds for research, and lobbying for improved diagnostic and treatment facilities. They are often a considerable help to individual families with genetic disease and may make the clinician's task in dealing with such families easier. The newsletter of one such organisation, the Research Trust for Metabolic Disease in Children (UK), has, as a special feature, parents' personal accounts of the lives (and deaths) of children affected by various metabolic diseases and the impact of this on their families. These moving and sometimes harrowing articles can remind us, particularly those who work primarily in the laboratory, of the human dimension to our field of study. Moreover, at a time when scientific advances are providing new therapeutic possibilities, some of which are disturbing to conventional morality or religious beliefs, and are coming under increasing political scrutiny, the activities of such organisations must increase public appreciation of the issues involved. Their association with the Society can only enhance our mutual roles and emphasizes that the SSIEM has always aimed to be a multidisciplinary group, bringing together everyone concerned with inherited metabolic conditions, ranging from those engaged in the most esoteric scientific investigation to those caring for affected individuals on the immediate and practical level. Catering for this range of interests, never an easy task, seems, like the compilation of Professor R. S. Illingworth's list, to have become yet more difficult with the passage of time.

The continued prosperity of the Society over the years has owed much to the financial and practical support of a variety of bodies. The crucial role of Scientific Hospital Supplies Ltd. in the early years has frequently been acknowledged and their continued support is gratefully received. Other bodies have contributed towards specific meetings, the symposia this year having been supported by generous grants from the British Heart Foundation, Cystic Fibrosis Research Trust, Duphar Laboratories Ltd., MCP Pharmaceuticals Ltd., and the Nuffield Foundation. The cost of preparing the proceedings of the ethics symposium for publication (pages 110–130 of this Supplement) has been defrayed by grants from the Medical Protection Society and the Medical and Dental Defence Union of Scotland. Above all, though, the Society depends on its members. The willingness of local members to help in the increasingly complex task of organizing the annual meetings is an essential ingredient but their hard work and enthusiasm is always matched by that of the general delegates. Not only are the formal scientific sessions followed closely and the posters scanned with an intense interest, but people from all parts of the world gather in groups and 'talk shop' for hours on end during events labelled 'social' but planned to produce such discussion.

Only a very small proportion of the total information thus exchanged can be captured for the published proceedings. Nevertheless, the two Supplements to this *Journal* (also available together as a hard-backed book) do, over the years, embrace many of the major aspects of the study of inborn errors of metabolism and can, particularly with the Short Communications section, be used as a way into the literature on specific new topics. We hope that with judicious selection of material these supplements will continue to provide, as did the Society's earlier annual publications, a balanced record of the present state of the subject in all its facets, a record of interest to those working in allied fields as well as to the specialist.

R. J. Pollitt
G. M. Addison
R. A. Harkness

The papers listed below were also presented at the meeting. Scripts were not available by the time of publication. 1. Tangier disease and related disorders of apolipoprotein A1. G. Assmann, Munster. 2. Contribution to Ethics Symposium by M. E. Pembrey, London.

J. Inher. Metab. Dis. 11 Suppl. 1 (1988) 4–17

The Biochemistry of Lipoproteins

A. M. SALTER and D. N. BRINDLEY

Department of Biochemistry, University of Nottingham Medical School, Queen's Medical Centre, Nottingham NG7 2UH, UK

Summary: Lipids are transported in the blood in four major classes of lipoproteins. The triacylglycerol-rich lipoproteins are chylomicrons and very-low-density lipoproteins (VLDL) which are produced by the small intestine and liver, respectively. These lipoproteins mainly carry fatty acids to adipose tissue and muscle where the triacylglycerol is hydrolysed by lipoprotein lipase. The resulting particles that remain in the blood are chylomicron remnants and low-density lipoprotein (LDL), respectively. The remnant is taken up by the liver via endocytosis which is mediated by a specific receptor for apolipoprotein E (apoE). LDL, which are rich in cholesterol, can also be taken up by the liver or extrahepatic tissues by a receptor-mediated endocytosis that specifically recognises apoB or apoE. 'Nascent' high-density lipoprotein (HDL) particles are secreted by the liver and intestine and then undergo modification to become HDL_3 and then HDL_2 as they acquire cholesterol ester. They facilitate the reverse transport of cholesterol back to the liver.

Little is known of the hormonal regulation of lipoprotein uptake by the liver. Recently, we have shown that insulin and tri-iodothyronine (T_3) increase the specific binding of LDL to cultured hepatocytes whereas dexamethasone (a synthetic glucocorticoid) has the opposite effect. The changes in binding produced by insulin and dexamethasone are paralleled by alterations in the rate of degradation of apoB. These findings may in part explain the hypercholesterolaemia and increased risk of premature atherosclerosis that can be associated with poorly controlled diabetes or hypothyroidism.

THE PLASMA LIPOPROTEINS

In order to exist in the aqueous environment of the plasma, lipids must be packaged into a stable form. This is achieved by coating the hydrophobic lipids (e.g. triacylglycerol, cholesterol ester) with a surface coat of more amphiphilic compounds. The latter consists of phospholipids (mainly phosphatidylcholine), cholesterol and various proteins referred to as apolipoproteins. Four major classes of lipoprotein exist which differ in both composition and metabolism (Table 1, Figure 1).

Lipoproteins are conventionally described by their density. Thus in fasting human plasma the major classes are very-low-density lipoprotein (VLDL), low-density lipoprotein (LDL) and high-density lipoprotein (HDL). These are characterized

4

Journal of Inherited Metabolic Disease. ISSN 0141–8955. Copyright © SSIEM and MTP Press Limited, Queen Square, Lancaster, UK.

Table 1 Human plasma lipoproteins

Name	Density (d) range	Electrophoretic mobility
Chylomicron	d<1.006	origin
Very-low-density lipoprotein (VLDL)	d<1.006	Pre-β
Intermediate-density lipoprotein (IDL)	1.006<d<1.019	Pre-β–β
Low-density lipoprotein (LDL)	1.019<d<1.063	β
High-density lipoprotein 2 (HDL$_2$)	1.063<d<1.125	α
High-density lipoprotein 3 (HDL$_3$)	1.125<d<1.210	α

by an increasing density which reflects an increased protein/lipid ratio (Table 2). The HDL class can be further separated by ultracentrifugation into HDL$_2$ and HDL$_3$. Another class of lipoprotein, intermediate-density lipoprotein (IDL) can also be found spanning the density range between VLDL and LDL. In the fed state a further class of triacylglycerol-rich lipoprotein, chylomicrons, are also present.

The lipoprotein classes vary not only in the overall content of protein and lipid but also in the specific apolipoproteins (Table 3). With the exception of LDL, which in human beings is generally considered to contain only B100, each class contains a characteristic combination of several apolipoproteins. As can be seen from their molecular weights, the apolipoproteins are a widely diverse group of proteins. It is the presence of particular apolipoproteins on the surface of the lipoprotein that governs the ultimate fate of the particle.

THE METABOLISMS OF CHYLOMICRONS AND VLDL

The major triacylglycerol-carrying lipoproteins are chylomicrons and VLDL. Chylomicrons transport the bulk of dietary fat from the small intestine whereas VLDL is the major transporter from the liver of endogenously produced triacylglycerol (Figure 1). The size and composition of chylomicrons that are secreted from the intestine depend upon the relative rates of lipid and apoprotein synthesis and upon the composition of dietary fat. Thus larger chylomicrons are produced after consumption of high fat loads, at the peak of absorption and when apoprotein synthesis is inhibited (Zilversmit, 1978). The apolipoproteins contained in the newly synthesized chylomicron particle are apoB48, apoAI and apoAIV (Schonfeld *et al.*, 1978). As chylomicrons move from the lymph to the plasma they undergo several changes including the acquisition of apoCs and apoE probably from HDL within the circulation (Havel *et al.*, 1973). The newly acquired apoCII on the chylomicron surface is then available to interact with the enzyme lipoprotein lipase (LPL: E.C. 3.1.1.34) which is located on the capillary endothelium of extrahepatic tissues including skeletal muscle, cardiac muscle and adipose tissue (Bengtsson and Olivecrona, 1980; Jackson *et al.*, 1980). LPL catalyses the hydrolysis of much of the triacylglycerol core of the chylomicron. Unesterified fatty acid and glycerol thus produced are then available to be taken up by the various tissues. The metabolic fate of these depends on the particular cells to which they are delivered.

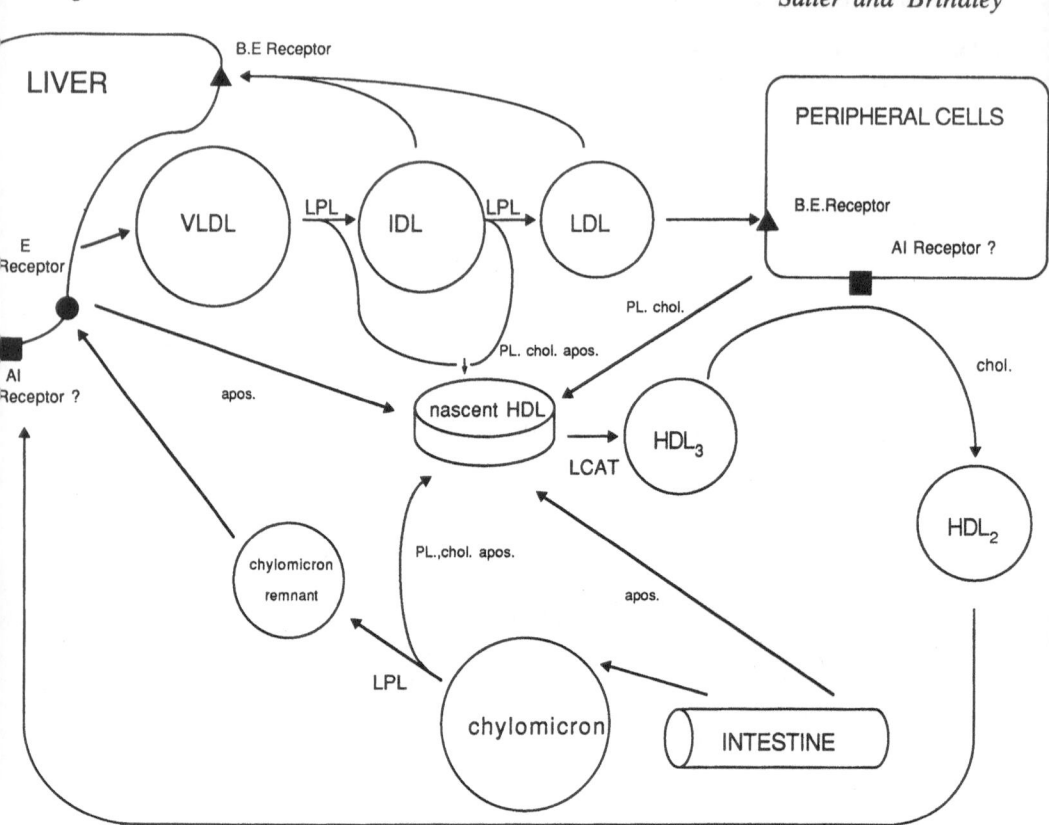

Figure 1 Diagrammatic representation of lipoprotein metabolism. Chylomicrons are secreted by the intestine and on entering the plasma are acted on by the enzyme lipoprotein lipase (LPL). On hydrolysis of the bulk of the triacylglycerol content they are transformed into chylomicron remnants. The remnants are removed from the circulation after interaction with specific apoE receptors on hepatocytes. VLDL is secreted by the liver and again triacylglycerol is hydrolysed by LPL resulting in the formation of IDL. IDL may be removed from the circulation after interaction with B,E receptors on hepatocytes. IDL which is not removed is further metabolized to become LDL which is then removed from the circulation after interaction with B,E receptors either on hepatocytes or on peripheral cells. HDL apolipoproteins are secreted by both the liver and the intestine. These interact with phospholipid (PL), cholesterol (chol) and other apoproteins (apos), shed from the surface of the chylomicrons and VLDL during the action of LPL, to form discoidal 'nascent' HDL. Free cholesterol is esterified by the enzyme lecithin cholesterol acyl transferase (LCAT) and as this moves into the centre of the particles they are transformed into spherical HDL$_3$. HDL$_3$, perhaps through interaction with apoAI receptors on peripheral cells, pick up more cholesterol and are transformed into HDL$_2$. HDL$_2$, through the interaction with an apoAI receptor on hepatocytes, may deliver cholesterol to the liver

Table 2   Percentage composition of the major human plasma lipoproteins

Lipoprotein	Protein	Phospholipid	Cholesterol free	ester	Triacylglycerol
Chylomicron	2	3	3	2	90
VLDL	6	14	16	4	60
IDL	18	22	32	8	20
LDL	21	22	42	8	7
HDL$_2$	44	26	20	5	5
HDL$_3$	52	28	15	3	2

Table 3   The human apolipoproteins

Apo	Molecular weight[a]	Lipoprotein distribution	Function
AI	28 300	chylos, HDL	Activates LCAT receptor recognition
AII	17 400	chylos, HDL	Unknown
AIV	46 000	chylos, VLDL, HDL	Unknown
B48	264 000	chylos	Structural
B100	549 000	VLDL, IDL, LDL	Structural receptor recognition
CI	6 600	chylos, VLDL, IDL, HDL	Unknown
CII	8 800	chylos, VLDL, IDL, HDL	Activates LPL
CIII	9200–9700	chylos, VLDL, IDL, HDL	Inhibits recognition of apoE by receptors Inhibits activation of lipoprotein lipase by apoCII
D	32 000	HDL	Cholesterol ester transfer protein
E	35 000–39 000	chylos, VLDL, IDL, HDL	Receptor recognition

[a]Molecular weights taken from Chapman (1986)

While in muscle fatty acids are readily used as a source of energy, in adipose tissue they are re-esterified and stored within the cell as triacylglycerol. In adipose tissue the activity of functional LPL, and hence the uptake of fatty acids from chylomicrons, depends upon a relatively high availability of insulin and it is normally controlled reciprocally to the activity in muscle tissue (Cryer, 1981).

As triacylglycerol is lost from the chylomicron its volume decreases and much of its surface coat becomes redundant. Phospholipid and some cholesterol together with apoAI, apoAIV and apoCII are shed from the surface of the particle in the form of bilayer discs which are thought to be precursors of plasma HDL (Redgrave and Small, 1979). The resulting particles known as chylomicron remnants are relatively enriched in cholesterol and apoproteins particularly apoB48 and apoE. The loss of apoCII decreases the affinity of LPL for the particle and the remnant leaves the capillary bed. The remmant is ultimately removed from the circulation by the liver. Liver parenchymal cells (hepatocytes) possess specific receptors which recognise the apoE moiety of the remnant (Sherrill *et al.*, 1980; Hui *et al.*, 1981).

VLDL is produced and secreted by hepatocytes. In human beings the newly secreted 'nascent' VLDL particles contain primarily apoproteins B100, E and C (Nestruck *et al.*, 1976; Swift *et al.*, 1980). They are triacylglycerol-rich and contain lesser amounts of phospholipids, free cholesterol and cholesterol esters (Table 2). Following secretion into the plasma the particles are metabolized in a manner analogous to chylomicrons. As such they initially acquire some apoCII from HDL. Most of the triacylglycerol is hydrolysed by LPL and phospholipid, cholesterol and apoproteins are lost from the surface of the particle (Patsch *et al.*, 1978). The resulting remnant particle is normally termed IDL which is relatively enriched in apoB100 and apoE. To varying amounts, depending on the species, IDL can be removed directly from the circulation through the interaction of apoE with the LDL receptor on hepatocytes (Kita *et al.*, 1982; Stalenhoef *et al.*, 1984). Those IDL particles which escape being taken up by the liver are further converted to LDL (Figure 1).

THE METABOLISM OF LDL AND HDL

The major cholesterol-carrying lipoproteins are LDL and HDL. Much interest has focused on these classes in recent years because of their association with the risk of developing atherosclerosis. Elevated plasma concentrations of LDL cholesterol are associated with an increased risk, whereas high plasma HDL cholesterol concentrations indicate a relative protection from the disease. As stated earlier any VLDL particles which are not eventually removed from the circulation in the form of IDL are converted to LDL. The latter are relatively poor in triacylglycerol but rich in cholesterol ester. The only apolipoprotein present is a single molecule of apoB100 (Scott *et al.*, 1987). Compared to chylomicron remnants and IDL, LDL is removed relatively slowly from the circulation. About two thirds of the LDL particles are taken up and metabolized after specific binding to apoB/E receptors on the surface of hepatocytes and extrahepatic cells (Goldstein *et al.*, 1983). The rest of the LDL is cleared by various tissues by so-called 'receptor-independent' mechanisms (Spady *et al.*, 1985). In rabbits, rats and hamsters more than half the total LDL receptors are located in the liver (Carew *et al.*, 1982; Spady *et al.*, 1983).

The binding of LDL by its receptor is followed by internalization of the particle, hydrolysis of the cholesterol ester and degradation of the protein in the lysosomes. In extrahepatic cells the cholesterol released elicits three regulatory responses with respect to cholesterol homeostasis (Goldstein and Brown, 1977):

(1) suppression of hydroxymethylglutaryl -CoA reductase (E.C. 1.1.1.34) activity and thus the inhibition of cholesterol synthesis *de novo*;

(2) the activation of cholesterol acyl transferase (E.C. 2.3.1.26) which enables cholesterol ester to be made and stored; and

(3) the down-regulation of LDL receptors on the cell's surface.

As will be discussed later this last level of regulation may be less pronounced in hepatocytes (Salter *et al.*, 1987a). The importance of the LDL receptors is illustrated by the increased plasma cholesterol and premature atherosclerosis that is observed in patients with familial hypercholesterolaemia. In such patients there is a marked reduction (heterozygotes), or total absence (homozygotes), of functional LDL receptors (Goldstein and Brown, 1977).

Cholesterol delivered to extrahepatic cells via the LDL receptors can be used either for membrane synthesis and repair or in more specialized tissues for steroid hormone synthesis. In the liver cholesterol can be either resecreted in newly formed VLDL or HDL, or secreted in the bile either as free cholesterol or after conversion to bile salts. These latter processes represent the major routes whereby cholesterol is excreted from the body since the reabsorption of cholesterol and bile salts from the intestine is incomplete.

HDL are the most heterogenous class of lipoproteins arising from various sources and containing several subclasses of lipoproteins. HDL_2 and HDL_3 are the major fractions present in most species. Using the analytical ultracentrifuge, Anderson and colleagues (1978) further divided HDL_2 into two fractions HDL_{2b} and HDL_{2a}. HDL_2 can also be subdivided into a major fraction which lacks apoE and a minor one which contains this apolipoprotein (Weisgraber and Mahley, 1980). On cholesterol feeding, the size of this apoE containing pool increases and the particles increase in size and float at lower densities (Mahley and Innerarity, 1983). Such HDL is commonly termed HDL_C.

The different components of HDL arise from various tissues (Figure 1). Intestine and liver are major sources of HDL apolipoproteins (Eisenberg, 1984). Whereas HDL produced by the intestine is rich in apoAI (Green *et al.*, 1978), apoE is the major HDL protein found in liver perfusates (Hamilton *et al.*, 1976). These discoidal particles are composed essentially of apolipoproteins, phospholipid and some free cholesterol. Eisenberg (1984) has challenged whether the particles are secreted directly by the liver or intestine and has suggested that the apolipoproteins, phospholipid and cholesterol become associated in body fluids rather than within the cells before secretion. These lipids could be derived from cell membranes or other lipoproteins or could be the redundant surface components released from chylomicrons and VLDL during the action of LPL. The discoidal HDL precursors are transformed to spherical HDL by the action of the enzyme lecithin–cholesterol acyl transferase (LCAT: E.C. 2.3.1.43) which transfers the unsaturated fatty acid from position 2 of phosphatidylcholine to cholesterol and which is activated by apoAI. The enzyme is believed to circulate in complexes containing apoAI and the cholesterol ester transfer protein apoD which represent a small subfraction of HDL (Fielding and Fielding, 1982). Cholesterol may be derived from other lipoproteins or from peripheral cells and once esterified may be transferred to other lipoproteins. Since cholesterol esters are hydrophobic they would enter the 'core' of the nascent HDL particle which would lead to an expansion of the particle from a flat disc shape to spheres (Hamilton *et al.*, 1976). This initial spherical particle probably represents HDL_3. As the particle continues to accept apolipoprotein and lipid from the degradation of chylomicrons and VLDL, and from the cells of peripheral tissues, it is transformed into HDL_2. The major difference between HDL_2 and HDL_3 is the presence of an extra apoA1 molecule and an approximate doubling of the cholesterol ester content (Eisenberg, 1984).

Many aspects of the degradation of HDL still remain obscure. Unlike LDL, which appears to be degraded as a whole particle, various HDL components appear to be metabolized at different rates and by different tissues. It is generally believed that HDL plays an important role in so-called 'reverse cholesterol' transport. As such it removes excess cholesterol from peripheral tissues and delivers it back to the liver. This function has been suggested to be one of the major reasons for the 'anti-atherogenic' effect of HDL (Miller and Miller, 1975). What proportion of cholesterol HDL itself delivers to the liver and what proportion is delivered via transfer of cholesterol ester to chylomicron remnant, IDL or LDL remains uncertain (Havel, 1985). ApoA1 and cholesterol have been labelled with non-degradable markers and turnover studies have shown a marked dissociation of uptake of protein and lipid in certain tissues of the rat (Glass *et al.*, 1983). The liver and the steroid-producing tissues, adrenal and ovary, all took up a greater proportion of cholesterol ester than apoAI. Thus HDL may be capable of delivering cholesterol to cells without the uptake of the whole particle.

In recent years intensive research has centred around the question of whether the interaction of HDL with cells is mediated by a specific cell surface receptor. While those HDL particles containing apoE are known to interact with both the LDL receptor and the chylomicron remnant receptor (Mahley and Innerarity, 1983), in humans at least, these represent only a small fraction of the total HDL population. Thus the question remains: is there an apoE free-HDL receptor? Further, as HDL can interact with cells in a number of different ways, are there a number of HDL receptors? In tissues which are known to degrade HDL such as liver (Ose *et al.*, 1979; Bachorik *et al.*, 1982; Soltys *et al.*, 1982), steroidogenic tissues (Gwynne and Strauss, 1982) and adipose tissue (Fong *et al.*, 1985, 1987; Salter *et al.*, 1987b) specific high affinity HDL binding sites have been described. In all these tissues the uptake of HDL cholesterol ester apparently exceeds that of apolipoprotein. Work with isolated adipocytes (Fong *et al.*, 1987) strongly suggests that this binding is mediated by interaction with the apoAI of the HDL. Similar studies with hepatocytes (Rifici and Eder, 1984) also suggests this apolipoprotein is important. In cell types were HDL mediates cholesterol efflux such as fibroblasts, smooth muscle cells (Oram *et al.*, 1981; Oram, 1983) and macrophages (Schmitz *et al.*, 1985) specific binding of HDL has also been demonstrated. In macrophages binding has been shown to be mediated by apoAI (Schmitz *et al.*, 1985) while in fibroblasts a role has been suggested for both apoAI and apoAII (Graham and Oram, 1987). One possibility is that the interaction of HDL with all these cell types is mediated by a common receptor and that this is followed by a variety of post-binding events which depend on the cell type involved.

REGULATION OF LIPOPROTEIN METABOLISM BY THE LIVER

From the preceding description of the metabolism of lipoproteins it is clear that the liver plays a central role in regulating plasma concentrations. It secretes VLDL and HDL components. It is also the major site of degradation of chylomicron remnants, IDL, LDL and HDL. As indicated earlier the uptake of these lipoproteins is thought to be mediated via specific cell surface receptors. In recent years we have been interested in the factors which influence the production and catabolism of lipoproteins by the liver. To study this we have been using rat hepatocytes in monolayer culture. Such cells secrete VLDL (Mangiapane and Brindley, 1986) and the rate of secretion can be increased by preincubating the hepatocytes with the synthetic glucocorticoid, dexamethasone. This effect of dexamethasone was found to be antagonized by the addition of insulin. The effect of insulin on its own on VLDL secretion is controversial. While we and others (Durrington *et al.*, 1982; Patsch *et al.*, 1983; Pullinger and Gibbons, 1985; Sparks *et al.*, 1986) have found that insulin inhibits VLDL secretion, other workers found either no effect (Whitton and Hems, 1976; Woodside and Heimberg, 1976) or a stimulation of secretion (Topping and Mayes, 1972; Laker and Mayes, 1984).

In recent years we have also been interested in the uptake of human ^{125}I-LDL by rat hepatocytes. We have described two distinct binding sites for LDL on these cells (Salter *et al.*, 1986). The characteristics of binding to these two sites are listed

Table 4 Binding of human LDL to rat hepatocytes

	Site 1	*Site 2*
LDL displaced by:		
heparin	yes	no
dextran sulphate	yes	no
K_d ($\mu g\,mL^{-1}$)	15	30
Calcium-dependent	yes	partially
Binding inhibited by methylation of apoB	yes	partially

in Table 4. Site 1 appears to be analogous to the 'classical' LDL receptor which was originally described by Goldstein and Brown (1977) on cultured fibroblasts. Site 2 may be part of a distinct system whereby the cells recognize LDL, and which has been demonstrated with monolayer cultures of human hepatocytes of patients with homozygous familial hypercholesterolaemia (Edge *et al.*, 1986; Hoeg *et al.*, 1986) in whom the classical LDL receptor is absent.

We have shown that these two sites differ in the regulation of their expression in response to the lipoprotein composition of the medium in which the cells are cultured (Salter *et al.*, 1987a,d). Binding to Site 1 increased between 20 and 44 h in culture when 10% new-born calf serum was present, whereas binding to Site 2 remained relatively constant. It appeared that the response to serum may be related partly to its HDL content since the addition of human HDL$_3$ to serum-free medium also up-regulated binding to Site 1. In contrast, preincubation of cells with human HDL$_2$ had little effect on binding to either site. This up-regulation by HDL$_3$ has recently been shown to occur in the human hepatocarcinoma cell line HepG2 (Havekes *et al.*, 1986). A possible explanation is that HDL$_3$ promotes cholesterol efflux from the cells and LDL binding increases in response to decreased cellular cholesterol content.

J. Inher. Metab. Dis. 11 (1988)

In marked contrast to other cell types such as fibroblasts (Goldstein and Brown, 1977) we (Salter *et al.*, 1987a,d) and other workers (Havekes *et al.*, 1986; Edge *et al.*, 1986) have found that preincubation of hepatocytes with LDL itself only partly down-regulated subsequent binding to Site 1. This may reflect the role of the liver in removing LDL from the circulation and in the excretion of cholesterol. The complete down-regulation of hepatic LDL receptors in response to high circulating plasma concentrations of LDL would have dramatic pathological implications. Impaired hepatic uptake could result in increased influx of cholesterol into the artery wall and in its uptake by potentially atherogenic 'scavenger' pathways.

Apart from regulation by incubation with lipoprotein fractions, LDL binding to hepatocytes appears to be under hormonal control. We have shown preincubation of hepatocytes with insulin increases and dexamethasone decreases binding of LDL to Site 1 on rat hepatocytes (Salter *et al.*, 1987c). We have also recently found that tri-iodothyronine (T_3) increases binding to Site 1 (Salter *et al.*, 1988). There was some indication of similar effects on binding to Site 2 but these were not as pronounced as the effects on Site 1. Figure 2 shows the effect of the length of incubation with these various hormones on subsequent LDL binding to Site 1. The action of insulin was relatively rapid with significant effects being seen after 1 h and being maintained over the following 24 h. In contrast, T_3 had relatively little effect over the first 6 h with maximum effects being seen between 6 and 24 h. After 1 h incubation with dexamethasone a reproducible stimulation of binding was seen. However, by 12 h we saw a marked decrease in binding. Thus, it appears that while insulin induces a rapid increase in LDL binding, the increase brought about by T_3 and the decrease by dexamethasone are longer-term effects possibly depending upon protein synthesis.

Figure 3 shows the effects of these hormones in combination with each other. Hepatocytes were incubated for 16 h with $10 \, \text{nmol} \, L^{-1}$ concentrations of the indicated hormones. As expected insulin and T_3 alone increased and dexamethasone decreased binding. When insulin and dexamethasone were added together these actions appeared to be antagonistic. In contrast, in the presence of T_3 the action of dexamethasone prevailed. When insulin and T_3 were added together their effects were additive. The changes obtained in the binding of LDL to Site 1 in the presence of insulin and dexamethasone were also paralleled by alterations in the rate of LDL degradation (Salter, Brown, Fisher, Fears and Brindley, unpublished work).

These findings may have important implications for understanding the changes in lipoprotein metabolism that occur in various diseases. Poorly controlled diabetes and metabolic stress are often associated with increased risk of developing premature atherosclerosis. In these conditions the effectiveness of insulin on metabolism is decreased relative to that of the counter-regulatory hormones, cortisol, corticotropin, glucagon, growth hormone and catecholamines. The result of this in terms of LDL uptake by the liver would be a down-regulation in binding due to the increased effectiveness of glucocorticoids. Since the liver is the major site of LDL degradation this may lead to elevated plasma LDL concentrations, confounded by an increased conversion of IDL to LDL. This may result in an increased uptake of LDL by other tissues including the artery wall, resulting in atherosclerosis.

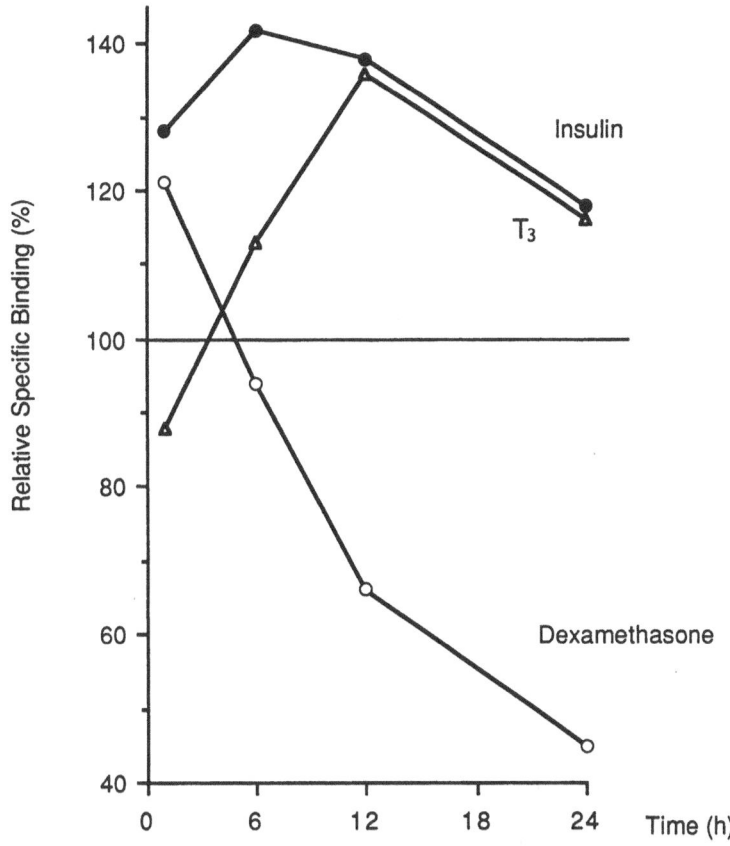

Figure 2 Time course of the actions of insulin, dexamethasone and tri-iodothyronine on LDL binding to hepatocytes. Rat hepatocytes were cultured for 24 h in the presence of 10% new-born calf serum and then for the further indicated period of time in serum-free medium containing $2\,g\,L^{-1}$ of bovine serum albumin and $10\,nmol\,L^{-1}$ of the indicated hormones. The specific binding of ^{125}I-LDL to Site 1 was measured (Salter *et al.*, 1986). Results are the mean±range of duplicate determinations and are expressed relative to binding measured at the indicated times in the absence of any hormone

Hypothroidism is often associated with hypercholesterolaemia and conversely thyrotoxicosis with low plasma cholesterol concentrations (Walton *et al.*, 1965; Heimberg *et al.*, 1985). Our findings suggest that this may be a direct result of the effects of thyroid hormones on the uptake of LDL by the liver. It has recently been suggested that thyromimetics may be used as a treatment for hypercholesterolaemia (Underwood *et al.*, 1986). Our findings support this suggestion.

ACKNOWLEDGEMENTS

Our work in Nottingham was supported by a project grant from the British Heart Foundation and an equipment grant from the Humane Research Trust.

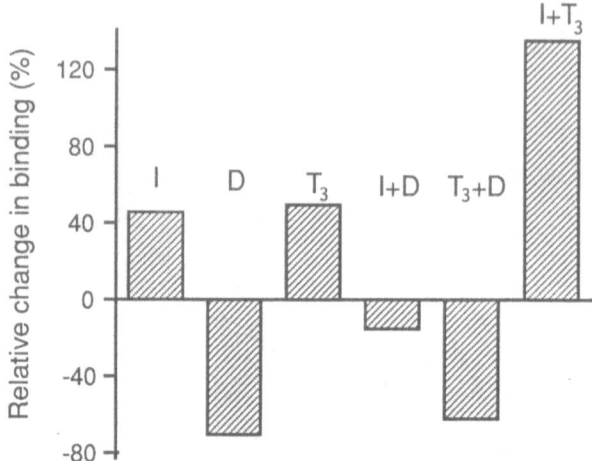

Figure 3 The effect of insulin, dexamethasone and tri-iodothyronine, alone and in combination, on LDL binding to hepatocytes. Rat hepatocytes were cultured for 24 h in the presence of 10% (v/v) new-born calf serum and then for a further 16 h in serum-free medium containing $2 g L^{-1}$ of bovine serum albumin with $10 nmol L^{-1}$ concentrations of insulin (I), dexamethasone (D) and tri-iodothyronine (T_3) as indicated. The specific binding of ^{125}I-LDL to Site 1 was measured (Salter *et al.*, 1986). Results are the mean±range of duplicate determinations and are expressed relative to binding in the absence of added hormone

REFERENCES

Anderson, D. W., Nichols, A. V., Pan, S. S. and Lindgren, F. T. High-density lipoprotein distribution-resolution and determination of three major components in a normal population sample. *Atherosclerosis* 29 (1978) 161–179

Bachorik, P. S., Franklin, F. A., Virgil, D. G. and Kwiterovich, P. O. High affinity uptake and degradation of apolipoprotein E free high-density lipoprotein and low-density lipoprotein in cultured porcine hepatocytes. *Biochemistry* 21 (1982) 5675–5684

Bengtsson, G. and Olivecrona, T. Lipoprotein lipase: Some effects of activator proteins. *Eur. J. Biochem.* 106 (1980) 549–555

Brown, M. S., and Goldstein, J. L. Lipoprotein receptors in the liver. Control signals for cholesterol traffic. *J. Clin. Invest.* 72 (1983) 743–747

Carew, T. E., Pittman, R. C. and Steinberg, D. Tissue sites of degradation of native and reductively methylated [^{14}C]sucrose-labelled low-density lipoprotein in rats. Contribution of receptor-dependent and receptor-independent pathways. *J. Biol. Chem.* 257 (1982) 8001–8008

Chapman, J. M. Comparative analysis of mammalian plasma lipoproteins. In Segrest, J. P. and Albers, J. J. (eds.). *Methods in Enzymology 128. Plasma Lipoproteins. Part A. Preparation, Structure and Molecular Biology*, Academic Press, New York, (1986) 70–147

Cryer, A. Tissue lipoprotein lipase activity and its action in lipoprotein metabolism. *Int. J. Biochem.* 13 (1981) 525–541

Durrington, P. N., Newton, R. S., Weinstein, A. B. and Steinberg, D. Effects of insulin and glucose on very low density lipoprotein secretion by cultured rat hepatocytes. *J. Clin. Invest.* 70 (1982) 63–73

Edge, S. B., Hoeg, J. M., Triche, T., Schneider, P. D. and Brewer, H. B. Cultured human hepatocytes. Evidence for metabolism of low-density lipoproteins by a pathway independent of the classical low-density lipoprotein receptor. *J. Biol. Chem.* 261 (1986) 3800–3806

Eisenberg, S. High-density lipoprotein metabolism. *J. Lipid Res.* 25 (1984) 1017–1058

Fielding, C. J. and Fielding, P. E. Cholesterol transport between cells and body fluids. Role of plasma lipoproteins and the plasma cholesterol esterification system. *Med. Clin. North. Am.* 66 (1982) 363–373

Fong, B. S., Rodrigues, P. O., Salter, A. M., Yip, B. P., Despres, J. P., Gregg, R. E. and Angel, A. Characterization of high-density lipoprotein binding to human adipocyte plasma membranes. *J. Clin. Invest.* 75 (1985) 1804–1812

Fong, B. S., Salter, A. M., Jimenez, J. and Angel, A. The role of apolipoprotein AI and apolipoprotein AII in high-density lipoprotein binding to human adipocyte plasma membranes. *Biochim. Biophys. Acta* 920 (1987) 105–113

Glass, C., Pittman, R. C., Weinstein, D. B. and Steinberg, D. Dissociation of tissue uptake of cholesterol ester from that of apoprotein AI of rat plasma high-density lipoprotein: Selective delivery of cholesterol ester to liver, adrenal, and gonad. *Proc. Natl. Acad. Sci. USA* 80 (1983) 5435–5439

Goldstein, J. L. and Brown, M. S. The low-density lipoprotein pathway and its relation to atherosclerosis. *Annu. Rev. Biochem.* 46 (1977) 897–930

Goldstein, J. L., Kita, T. and Brown, M. S. Defective lipoprotein receptors and atherosclerosis. Lessons from an animal counterpart of familial hypercholesterolemia. *N. Engl. J. Med.* 309 (1983) 288–296

Green, P. H. R., Tall, A. R. and Glickman, R. M. Rat intestine secretes discoidal high-density lipoprotein. *J. Clin. Invest.* 61 (1978) 528–534

Graham, D. L. and Oram, J. F. Identification and characterization of a high-density lipoprotein binding protein in cell membranes by ligand blotting. *J. Biol. Chem.* 262 (1987) 7439–7442

Gwynne, J. T. and Strauss, J. F. The role of lipoproteins in steroidogenesis and cholesterol metabolism in steroidogenic glands. *Endocrinol. Rev.* (1982) 299–329

Hamilton, R. L., Williams, M. C., Fielding, C. J. and Havel, R. J. Discoidal bilayer structure of nascent high-density lipoproteins from perfused rat liver. *J. Clin. Invest.* 58 (1976) 667–680

Havekes, L. M., Schouten, D., deWit, E. C. M., Cohen, L. H., Griffioen, M., van Hinsbergh, V. W. M. and Princen, H. M. G. Stimulation of the LDL receptor activity in the human hepatoma cell line HepG2 by high-density serum fractions. *Biochim. Biophys. Acta* 875 (1986) 236–246

Havel, R. J., Kane, J. P. and Kashyap, M. L. Interchange of apolipoproteins between chylomicrons and high-density lipoproteins during alimentary lipemia in man. *J. Clin. Invest.* 52 (1973) 32–38

Havel, R. J. Role of the liver in atherosclerosis. *Arteriosclerosis* 5 (1985) 569–580

Heimberg, M., Olubadewo, J. O. and Wilcox, H. G. Plasma lipoproteins and regulation of hepatic metabolism of fatty acids in altered thyroid states. *Endocrine Rev.* (1985) 590–607

Hoeg, J. M., Edge, S. B., Demonsky, S. J., Starzl, T. E., Triche, T., Gregg, R. E. and Brewer, H. B. Metabolism of low-density lipoproteins by cultured hepatocytes from normal and homozygous familial hypercholesterolemic subjects. *Biochim. Biophys Acta* 876 (1986) 646–657

Hui, D. Y., Innerarity, T. L. and Mahley, R. W. Lipoprotein binding to canine hepatic membranes. Metabolically distinct apoE and apoB, E receptors. *J. Biol. Chem.* 256 (1981) 5646-5655

Jackson, R. L., Pattus, F. and DeHaas, G. Mechanism of action of milk lipoprotein lipase at substrate interfaces: Effects of apolipoproteins. *Biochemistry* 19 (1980) 373–378

Kita, T., Brown, M. S., Bilheimer, D. W. and Goldstein, J. L. Delayed clearance of very low density and intermediate density lipoproteins with enhanced conversion to low density lipoprotein in WHHL rabbits. *Proc. Natl. Acad. Sci. USA* 79 (1982) 5693–5697

Laker, M. E. and Mayes, P. A. Investigations into the direct effects of insulin on hepatic ketogenesis, lipoprotein secretion and pyruvate dehydrogenase activity. *Biochim. Biophys. Acta* 795 (1984) 427–430

Mahley, R. W. and Innerarity, T. L. Lipoprotein receptors and cholesterol homeostasis. *Biochim. Biophys. Acta* 737 (1983) 197–222

Mangiapane, E. H. and Brindley, D. N. Effects of dexamethasone and insulin on the synthesis of triacylglycerols and phosphatidylcholine by monolayer cultures of rat hepatocytes. *Biochem. J.* 233 (1986) 151–160

Miller, G. J. and Miller, N. E. Plasma high-density lipoprotein concentration and development of ischaemic heart disease. *Lancet* 1 (1975) 16–19

Nestruck, A., Christine, A. N. D. and Rubinstein, D. The synthesis of apoproteins of very low density lipoproteins isolated from the Golgi apparatus of rat liver. *Can. J. Biochem.* 54 (1976) 617–628

Oram, J. F., Albers, J. J., Cheung, M. C. and Bierman, E. L. The effects of subfractions of high-density lipoprotein on cholesterol efflux from cultured fibroblasts. *J. Biol. Chem.* 256 (1981) 8348–8356

Oram, J. F. Effects of high-density lipoprotein subfractions on cholesterol homeostasis in human fibroblasts and arterial smooth muscle cells. *Arteriosclerosis* 3 (1983) 420–432

Ose, L., Ose, T., Norum, K. R. and Berg, T. Uptake and degradation of ^{125}I-labelled high-density lipoproteins in rat liver cells *in vivo and in vitro*. *Biochim. Biophys. Acta* 574 (1979) 521–536

Patsch, J. R., Gotto, A. M. Jr, Olivecrona, T. and Eisenberg, S. Formation of high-density lipoprotein 2-like particles during lypolysis of very low density lipoproteins *in vitro*. *Proc. Natl. Acad. Sci. USA* 75 (1978) 4519–4528

Patsch, W., Franz, S. and Schonfeld, G. Role of insulin in lipoprotein secretion by cultured rat hepatocytes. *J. Clin. Invest.* 71 (1983) 1161–1174

Pullinger, C. R. and Gibbons, G. F. Effects of hormones and pyruvate on the rates of secretion of very-low-density lipoprotein triacylglycerol and cholesterol by rat hepatocytes. *Biochim. Biophys. Acta* 833 (1985) 44–51

Redgrave, T. G. and Small, D. M. Quantitation of the transfer of surface phospholipid of chylomicrons to the high-density lipoprotein fraction during the catabolism of chylomicrons in the rat. *J. Clin. Invest.* 64 (1979) 162–171

Rifici, V. A. and Eder, H. A. A hepatocyte receptor for high-density lipoproteins specific for apolipoprotein A-1. *J. Biol. Chem.* 259 (1984) 13814–13818

Salter, A. M., Saxton, J. and Brindley, D. N. Characterization of the binding of human low-density lipoprotein to primary monolayer cultures of rat hepatocytes. *Biochem. J.* 240 (1986) 549–557

Salter, A. M., Bugaut, M., Saxton, J., Fisher, S. C. and Brindley, D. N. Effects of preincubation of primary monolayer cultures of rat hepatocytes with low- and high-density lipoproteins on the subsequent binding and metabolism of human low-density lipoprotein. *Biochem. J.* 247 (1987a) 79–84

Salter, A. M., Fong, B. S., Jimenez, J., Rotstein, L. and Angel, A. Regional variation in high density lipoprotein binding to human adipocyte plasma membranes of massively obese subjects. *Eur. J. Clin. Invest.* 17 (1987b) 16–22

Salter, A. M., Fisher, S. C. and Brindley, D. N. Binding of low-density lipoprotein to monolayer cultures of rat hepatocytes is increased by insulin and decreased by dexamethasone. *FEBS Lett.* 220 (1987c) 159–162

Salter, A. M., Fisher, S. C. and Brindley, D. N. Interactions of triiodothyronine, insulin and dexamethasone on the binding of human LDL to rat hepatocytes in monolayer culture. *Atherosclerosis* 71 (1988) 77–80

Salter, A. M., Saxton, J. and Brindley, D. N. Characterization of the binding of human low-density lipoprotein to cultured rat hepatocytes. *Biochem. Soc. Trans.* 15 (1987d) 253–254

Schmitz, G., Niemann, R., Brennhausen, B., Krausse, R. and Assman, G. Regulation of high-density lipoprotein receptors in cultured macrophages: role of acyl-CoA: cholesterol acyltransferase. *EMBO J.* 4 (1985) 2773–2779

Schonfeld, G., Bell, E. and Alpers, D. H. Intestinal apoproteins during fat absorption. *J. Clin. Invest.* 61 (1978) 1539–1550

Scott, J., Pease, R. J., Powell, L. M., Wallis, S. C., McCarthy, B. J., Mahley, R. W., Levy-Wilson, B. and Knott, T. J. Human apolipoprotein-B: complete cDNA sequence and identification of structural domains of the protein. *Biochem. Soc. Trans.* 15 (1987) 195–199

Sherrill, B. C., Innerarity, T. L. and Mahley, R. W. Rapid hepatic clearance of the canine lipoproteins containing only the E apoprotein by a high affinity receptor. *J. Biol. Chem.* 255 (1980) 1804–1807

Soltys, P. A., Portman, O. W. and O'Malley, J. L. Binding properties of high-density lipoprotein subfractions and low-density lipoproteins of rabbit hepatocytes. *Biochem. Biophys. Acta* 713 (1982) 300–314

Spady, D. K., Bilheimer, D. W. and Dietschy, J. M. Rates of receptor-dependent and -independent low-density lipoprotein uptake in the hamster. *Proc. Natl. Acad. Sci. USA* 80 (1983) 3499–3503

Spady, D. K., Turley, S. D. and Dietschy, J. M. Receptor-independent low-density lipoprotein transport in rat *in vivo*. Quantitation, characterization, and metabolic consequences. *J. Clin. Invest.* 76 (1985) 1113–1122

Sparks, C. F., Sparks, J. D., Bolognino, M., Salhanick, A., Strumph, P. S. and Amatruda, J. M. Insulin effects on apolipoprotein synthesis by primary cultures of rat hepatocytes. *Metabolism* 35 (1986) 1128–1136

Stalenhoef, A. F. H., Malloy, M. J., Kane, J. P. and Havel, R. J. Metabolism of apolipoproteins B-48 and B-100 of triglyceride-rich lipoproteins in patients with familial dysbetalipoproteinemia. *J. Clin. Invest.* 78 (1986) 722–728

Swift, L. L., Manowitz, N. R., Dunn, G. D. and LeQuire, V. S. Isolation and characterization of hepatic Golgi lipoproteins from hypercholesterolemic rats. *J. Clin. Invest.* 66 (1980) 415–425

Topping, D. L. and Mayes, P. A. The immediate effects of insulin and fructose on the metabolism of the perfused liver. Changes in lipoprotein secretion, fatty acid oxidation and esterification, lipogenesis and carbohydrate metabolism. *Biochem. J.* 126 (1972) 295–311

Underwood, A. H., Emmett, J. C., Ellis, D., Flynn, S. B., Leeson, P. D., Benson, G. M., Novelli, R., Pearce, N. J. and Shah, V. P. A thyromimetic that decreases plasma cholesterol levels without increasing cardiac activity. *Nature* 324 (1986) 425–429

Walton, K. W., Scott, P. J., Dykes, P. W. and Davies, J. W. L. The significance of alteration in serum lipids in thyroid dysfunction. Part 2 (Alteration in metabolism and turnover of ^{131}I-low density lipoproteins in hypothyroidism and thyrotoxicosis). *Clin. Sci.* 29 (1965) 217–238

Weisgraber, K. H. and Mahley, R. W. Subfractionation of human high-density lipoproteins by heparin–Sepharose affinity chromatography. *J. Lipid Res.* 21 (1980) 316–325

Whitton, P. D. and Hems, D. A. Glycogen synthesis in perfused liver of adrenalectomized rats. *Biochem. J.* 156 (1976) 585–592

Woodside, W. F. and Heimberg, M. Effects of anti-insulin serum, insulin and glucose on output of triglycerides and on ketogenesis by perfused rat liver. *J. Biol. Chem.* 251 (1976) 13–23

Zilversmit, D. B. Assembly of chylomicrons in the intestine cell. In Dietschy, J. M. Gotto, A. M. and Ontko, J. A. (eds.) *Disturbances of Lipid and Lipoprotein Metabolism*, American Physiological Society, Bethesda, (1978) 69–81

J. Inher. Metab. Dis. 11 Suppl. 1 (1988) 18–28

Clinical Consequences of Hyperlipidaemia

G. R. THOMPSON

MRC Lipoprotein Team, Hammersmith Hospital, Ducane Road, London W12 0HS, UK

Summary: Hyperlipidaemia is a common and important clinical entity which frequently has a genetic basis. The chief features of severe hypertriglyceridaemia are eruptive xanthomata and acute pancreatitis, whereas most forms of hypercholesterolaemia are associated with premature coronary heart disease. This applies especially to familial hypercholesterolaemia, which is also characterized by tendon xanthomata. Better recognition of hyperlipidaemia allied to recent improvements in treatment should help reduce the frequency of its disabling and sometimes fatal consequences.

An ever increasing body of epidemiological and experimental evidence points to the causal role of lipids in the aetiology of atherosclerotic coronary heart disease. Furthermore several recent studies have shown that effective treatment of hyperlipidaemia can both prevent the development of coronary heart disease and slow its rate of progression (Thompson, 1987). Yet its existence is all too often overlooked or ignored by clinicians. In part this stems from lack of appropriate training, reflecting the relatively recent emergence of lipidology on the clinical scene. This review attempts to redress the imbalance by providing a brief description of the clinical features of the major forms of hyperlipidaemia, especially those which are genetically determined.

LIPIDS AND LIPOPROTEINS IN PLASMA

The major lipids in plasma (cholesterol, triglyceride and phospholipid) are transported around the bloodstream in lipoprotein particles, in which they are rendered water-soluble by interaction with specialised proteins called apoproteins. The relative proportions of the major lipids and apoproteins vary between the different lipoproteins. Because low-density lipoprotein (LDL) is the main cholesterol-rich lipoprotein in plasma, hypercholesterolaemia commonly indicates an increased number of LDL particles, although, of course, it can also result from an increase in high-density lipoprotein (HDL). Since an increase in LDL has different implications from an increase in HDL, it is important to quantitate their respective contributions to total cholesterol. This is usually done in the laboratory by precipitating LDL and very-low-density lipoprotein (VLDL) from serum and then measuring the amount of HDL cholesterol in the supernatant. If the values of total and HDL

Journal of Inherited Metabolic Disease. ISSN 0141–8955. Copyright © SSIEM and MTP Press Limited, Queen Square, Lancaster, UK.

cholesterol and of triglyceride are known, it is then relatively simple to calculate LDL cholesterol, using the appropriate formula (Friedewald *et al.*, 1972). HDL and LDL can also be quantitated by immunoassay of their major apoproteins, apoAI and apoB, respectively. Fasting hypertriglyceridaemia usually indicates an increase in VLDL but, if very severe, can also be due to chylomicronaemia.

DEFINITION OF HYPERLIPIDAEMIA

The pragmatic approach is to adopt arbitrary criteria based on the relative risk of coronary heart disease. Most population studies show positive correlations between coronary heart disease and total or LDL cholesterol and between coronary heart disease and triglyceride, but a negative correlation between coronary heart disease and HDL cholesterol. Cholesterol and triglyceride levels rise with age but the accompanying increase in the risk of coronary heart disease relative to normolipidaemic contemporaries is less marked. Also at any given level of cholesterol the risk for men is greater than that for women.

The relative risk of coronary disease rises steeply after the serum cholesterol exceeds 6.5 mmol L^{-1} (Martin *et al.*, 1986) and this value has been recommended as the level above which dietary intervention should be instituted in Britain, with 7.8 mmol L^{-1} as the level above which drug therapy might be required (British Cardiac Society Working Group on Coronary Prevention, 1987). The need to treat is emphasised if the LDL cholesterol exceeds 5 mmol L^{-1}; conversely, hyperalphalipoproteinaemia (i.e. an HDL cholesterol of > 2 mmol L^{-1}) is harmless. Whether hypertriglyceridaemia plays a causal role in coronary disease is debatable and indications for its treatment are less well-defined than for hypercholesterolaemia. However, fasting levels persistently above the 95th percentile, i.e. 2.5–3 mmol L^{-1} may require treatment, especially if accompanied by a family history of coronary heart disease and a decrease in HDL cholesterol.

CLASSIFICATION OF HYPERLIPIDAEMIA

Fredrickson and his colleagues originally described five lipoprotein phenotypes (Fredrickson *et al.*, 1967), but their classification was later amended by the World Health Organization (WHO) to include a sixth phenotype (Beaumont *et al.*, 1970). Despite certain limitations the WHO classification provides a useful means of indicating which class of lipoproteins is present in excess in individual patients. One drawback is that it takes no account of HDL cholesterol.

As discussed elsewhere (Thompson, 1985), the type I phenotype indicates hypertriglyceridaemia due to chylomicronaemia; in type IIa the hypercholesterolaemia reflects an increase in LDL cholesterol which in type IIb is accompanied by mild to moderate hypertriglyceridaemia, due to an increase in VLDL; in type III serum cholesterol and triglyceride are both raised, due to accumulation of chylomicron and VLDL remnants; in type IV fasting hypertriglyceridaemia and mild to moderate hypercholesterolaemia are due to an increase in VLDL but the LDL-cholesterol

is normal; and in type V marked hypertriglyceridaemia is due both to chylomicronaemia and an increase in VLDL.

Each of these phenotypes can represent either primary or secondary hyperlipidaemia. Frequently, primary hyperlipidaemia is polygenically-determined and therefore rather poorly defined but several monogenic or dominantly-inherited disorders
have been described, as well as some which are recessively inherited. Apart from
xanthomata the chief clinical consequence of those phenotypes characterised by
severe hypertriglyceridaemia (types I and V) is acute pancreatitis whereas those
characterized by hypercholesterolaemia alone (type IIa) or combined with hypertriglyceridaemia (types IIb, III and IV) tend to be associated with premature coronary
heart disease.

FAMILIAL TYPE I HYPERLIPOPROTEINAEMIA

This rare disorder (McKusick 23860) is characterized by marked hypertriglyceridaemia and chylomicronaemia and usually first presents in childhood. It is due to
the inheritance of a recessive gene which results in deficiency of the enzyme
extrahepatic or adipose tissue lipoprotein lipase, the rate-limiting step in chylomicron clearance. The main clinical features are recurrent episodes of abdominal pain,
eruptive xanthomata, lipaemia retinalis and hepatosplenomegaly.

The gross chylomicronaemia results in an increase in fasting serum triglycerides
of up to 50–100 mmol L^{-1}. The diagnosis depends upon demonstrating that plasma
extrahepatic lipoprotein lipase levels following an intravenous dose of heparin of
10 iu (kg b.wt)$^{-1}$ are less than 10% of normal (Nikkila, 1983). The main line of
treatment is to minimize chylomicron formation by decreasing the intake of long-
chain triglyceride to less than 50 g per day. There appears to be no increased
susceptibility to atherosclerosis in this condition, the chief complication being acute
pancreatitis. The risk of this occurring is minimized if plasma triglyceride levels
can be kept below 20 mmol L^{-1}.

In some instances a type I phenotype is due to familial apoCII deficiency
(McKusick 20775). This is also a recessively-inherited disorder which results in the
absence from plasma of apoCII, the activator of lipoprotein lipase (Breckenridge
et al., 1978). This enzyme is present in normal amounts but cannot hydrolyse
chylomicrons or VLDL in the absence of apoCII. Addition of the latter *in vitro*
restores lipolysis to normal and its infusion *in vivo* dramatically reduces the
hypertriglyceridaemia, albeit only temporarily. Acute pancreatitis is common in
untreated individuals.

FAMILIAL HYPERCHOLESTEROLAEMIA

Familial hypercholesterolaemia (McKusick 14389), also known as familial type II
hyperlipoproteinaemia, affects approximately 0.2% of the population. Commonly
this is due to inheritance of one mutant gene encoding the LDL receptor which
causes heterozygous familial hypercholesterolaemia. Very occasionally inheritance

of two mutant genes occurs, giving rise to homozygous familial hypercholesterol-aemia.

At least ten different mutations of the LDL receptor gene have been identified (Goldstein *et al.*, 1985), some of which result in a failure to produce detectable LDL receptors whereas the remainder result in receptors which are quantitatively deficient or defective in their LDL binding properties. The LDL receptor normally plays a major role in the catabolism of LDL and deficiency of LDL receptors results in accumulation of LDL, causing hypercholesterolaemia from birth. Serum cholesterol levels range between 8 and $15 \, \text{mmol L}^{-1}$ in adult heterozygotes and between 16 and $26 \, \text{mmol L}^{-1}$ in homozygotes, as shown in Table 1. Triglyceride levels are usually normal in affected children, most of whom exhibit a type IIa phenotype, but a type IIb phenotype is quite common in adults. HDL cholesterol is normal or reduced.

Homozygous familial hypercholesterolaemia

Homozygous familial hypercholesterolaemia is characterized by extreme hyper-cholesterolaemia and the early onset of cutaneous planar or tuberose xanthomata, tendon xanthomata and corneal arcus. Levels of cholesterol in plasma correlate inversely with the severity of the LDL receptor deficit, which varies according to the nature of the underlying gene defect (Sprecher *et al.*, 1985). The deficit is more marked when there is an inability to produce receptors (receptor-negative) than when mature but abnormal receptors are formed (receptor-defective).

Atheromatous involvement of the aortic root is always evident by puberty as manifested by an aortic systolic murmur, a gradient across the aortic valve (Table 1) and angiographic narrowing of the aortic root (Allen *et al.*, 1980). Coronary ostial stenosis is commonly seen on angiography or two-dimensional echocardiography. Sudden death from acute coronary insufficiency during the late teens or early twenties used to be the rule but the introduction of plasma exchange has markedly improved survival (Thompson *et al.*, 1985). At postmortem the aortic valve, sinuses of Valsalva and ascending arch of the aorta are grossly infiltrated with atheroma. Less severe changes are found in the abdominal aorta, pulmonary artery, carotid arteries and circle of Willis.

The chief determinant of the age of onset of coronary heart disease and the likelihood of premature death appears to be LDL receptor status (Goldstein and Brown, 1983). Pooled data show that 60% of receptor-negative homozygotes exhibited coronary heart disease before the age of 10 years whereas this was never observed in receptor-defective patients until after that age. Furthermore, 26% of receptor-negative subjects had died from coronary heart disease before the age of 25 years compared with only 4% of receptor-defective homozygotes. Gender plays little part in determining the age of onset of the cardiovascular complications of homozygous familial hypercholesterolaemia, in contrast to its prominent role in heterozygotes (Gagné *et al.*, 1979) and the population at large. This may be a reflection of the lack in homozygotes of the usual sex difference in HDL cholesterol (Seftel *et al.*, 1980).

Table 1 Clinical characteristics of seven familial hypercholesterolaemic homozygotes studied at Hammersmith Hospital (Allen et al., 1980)

Patient	Sex	Age (years)	Cholesterol (mmol L⁻¹)	Triglyceride (mmol L⁻¹)	Onset of Xs* (years)	Aortic gradient (mmHg)	Receptor status
P.A-S.	F	23†	26	0.5	0.5	75	Negative
N.E.	M	19†	25	—	1.5	80	Defective
R.W.	M	31†	21	0.9	5	40	Defective
D.L.	M	24	21	1.1	9	20	Defective
M.M.	M	23	19	0.9	5	0	Defective
Y.M.	M	23†	17	—	5	30	Defective
A.R.	F	36	16	2.2	12	34	Defective

*xanthomata; †died

Heterozygous familial hypercholesterolaemia

This condition can be detected early by screening the children or siblings of an affected subject but often remains undiagnosed until the onset of cardiovascular symptoms in adult life. In addition to hypercholesterolaemia there are often visible signs of cholesterol deposition, such as corneal arcus, xanthelasma and tendon xanthomata. Characteristic sites for the latter are the extensor tendons on the back of the hands and elbows, the Achilles tendons and the patellar tendon insertion into the pretibial tuberosity.

Tendon xanthomata, the clinical hallmark of familial hypercholesterolaemia, are an age-related phenomenon. Thus only 7% of heterozygotes below the age of 19 years exhibited tendon xanthomata in one series whereas these lesions were present in 75% of their parents (Kwiterovich *et al.*, 1974). Heterozygotes with tendon xanthomata have higher LDL cholesterol levels than those of the same age without tendon xanthomata (Heiberg, 1976). The frequency of tendon xanthomata is equal in both sexes despite the fact that HDL cholesterol levels are higher in female patients (Gagné *et al.*, 1979). Therefore age and LDL cholesterol are the major determinants of the presence of tendon xanthomata in untreated heterozygotes.

The high frequency and premature onset of coronary heart disease in heterozygous familial hypercholesterolaemia has been well documented as has its much lower incidence in females as compared with males, in whom the onset of symptoms occurs 9–10 years earlier (Slack, 1969; Gagné *et al.*, 1979). If left untreated up to 50% of males but only 12% of females with heterozygous familial hypercholesterolaemia develop coronary heart disease by the age of 50. On angiography, over 70% of male heterozygotes have triple vessel disease, including 32% with disease of the left main stem (Sugrue *et al.*, 1981). It has been estimated that coronary heart disease occurs about 20 years earlier in familial hypercholesterolaemia than in the remainder of the population.

In one large series (Gagné *et al.*, 1979) no cases of coronary heart disease were observed in patients without tendon xanthomata. However, in subjects with tendon xanthomata, neither total nor LDL cholesterol differed between those with and without coronary heart disease. An inverse correlation between HDL cholesterol and coronary heart disease has been observed in patients of both sexes (Streja *et al.*, 1978). HDL cholesterol values are lower in male heterozygotes with a type IIb phenotype than in those with a type IIa pattern, reflecting an inverse correlation between HDL cholesterol and plasma triglyceride; also their frequency of myocardial infarction is significantly higher (Moorjani *et al.*, 1986).

The influence of age, sex and lipid levels on the presence of tendon xanthomata and coronary heart disease is summarized in Table 2. An LDL cholesterol above $7.8 \, \text{mmol} \, L^{-1}$ seems to be a prerequisite for the development of both tendon xanthomata and coronary heart disease before the age of 50 years. Coronary heart disease rarely occurs in the absence of tendon xanthomata but their presence is not necessarily a marker for this complication. There is a strong association between being male and developing coronary heart disease, and it seems that this is in some way related to the lower HDL cholesterol or higher triglyceride levels found in

Table 2   Factors associated with development of tendom xanthomata and coronary heart
disease in heterozygotes by age of 50 years

	Tendon xanthomata	Coronary heart disease
Increasing age	++	++
Male sex	−	++
Increased LDL-cholesterol*	+	+
Decreased HDL-cholesterol	−	+
Increased triglyceride	−	±
Tendon xanthomata		±

*>7.8 mmol L^{-1}

men with familial hypercholesterolaemia. The apparent absence of any relationship
between HDL cholesterol and tendon xanthomata is noteworthy.

Treatment of familial hypercholesterolaemia

The management of homozygous familial hypercholesterolaemia presents a major
therapeutic challenge. Dietary and drug regimens have little impact on the hypercholesterolaemia and the only reliable means of reducing cholesterol levels is to
undertake plasma exchange at 2-weekly intervals (Thompson *et al.*, 1975). Liver
transplantation remedies the hepatic deficiency of LDL receptors and results in
near normal cholesterol levels (Bilheimer *et al.*, 1984) but is hazardous and
necessitates long-term immunosuppression. Although plasma exchange slows down
the rate of progression of aorto-coronary atherosclerosis it may be necessary to
undertake coronary artery bypass grafting for coronary ostial stenosis and replace
the aortic valve if this becomes significantly fibrosed.

The treatment of heterozygous familial hypercholesterolaemia usually involves
drug therapy with an anion-exchange resin such as cholestyramine. In patients with
a type IIb phenotype this may need supplementing with nicotinic acid or one of the
fibric acid derivatives. Partial ileal bypass can be useful in patients who are
intolerant of resins. The effectiveness of all these approaches is markedly enhanced
by concomitant administration of HMG CoA reductase inhibitor, such as lovastatin
(mevinolin) (Thompson *et al.*, 1986). The recent advent of LDL apheresis, which
selectively removes LDL but not HDL (Yokoyama *et al.*, 1985), holds great
promise, especially when used in conjunction with a cholesterol synthesis inhibitor.
Despite these measures it is sometimes necessary to resort to coronary artery bypass
grafting, especially in those with left main stem disease. Post-operative control of
hypercholesterolaemia is vital if graft atherosclerosis is to be avoided (Thompson
and Sapsford, 1985).

FAMILIAL TYPE III HYPERLIPOPROTEINAEMIA

This disorder (McKusick 14450) is characterized by the accumulation in plasma of
chylomicron and VLDL remnants which fail to get cleared by the liver because
they contain apoE$_2$ instead of the normal isoforms, apoE$_3$ or apoE$_4$. The defect in

this disorder resides in the ligand not the receptor, unlike familial hypercholester-olaemia. Most type III patients are homozygous for $apoE_2$ but some are hetero-zygous, exhibiting an $apoE_2/E_3$ or $apoE_2/E_4$ phenotype. Inheritance of the apoprotein abnormality alone is not sufficient to cause the full clinical picture of the disorder, however, and other factors must also be present, such as obesity, hypothyroidism, diabetes or familial combined hyperlipidaemia (Brown *et al.*, 1983).

Type III hyperlipoproteinaemia seldom presents in males before puberty and in females it is rare before the menopause. Clinical features include corneal arcus, xanthelasma, tubero-eruptive xanthomata and, characteristically, palmar striae. Serum cholesterol and triglyceride are both elevated, usually to about $10 \, \text{mmol L}^{-1}$, and lipoprotein electrophoresis shows the 'broad β' band characteristic of remnant particles. The diagnosis should be confirmed by apoE phenotyping whenever possible. Vascular disease occurs in over 50% of patients, involving not only the coronary tree but also peripheral and cerebral vessels.

Management of type III hyperlipoproteinaemia involves remedying any obvious precipitating factor, such as hypothyroidism, diabetes or obesity. In addition, most patients will require therapy with a fibric acid derivative such as bezafibrate or gemfibrozil. Providing body weight can be controlled by diet, administration of one of these drugs often results in virtual normalization of serum lipids and rapid regression of cutaneous xanthomata.

FAMILIAL COMBINED HYPERLIPIDAEMIA

Goldstein and colleagues (1973) were the first to describe familial combined hyperlipidaemia (McKusick 14425) which they considered was the commonest cause of hyperlipidaemia in patients with coronary heart disease below the age of 60 years. Family studies suggested that it was monogenically-inherited although this has been disputed (Nikkila and Aro, 1973). Overall roughly 50% of the relatives of affected subjects are hyperlipidaemic, of whom one third have hypercholesterol-aemia (type IIa), one third have hypertriglyceridaemia (type IV or V), and one third have both abnormalities (type IIb). Affected children present with type IV or IIb phenotypes but not with a type IIa phenotype, unlike familial hypercholesterolaemia.

The precise cause of familial combined hyperlipidaemia is unknown but it is characterized by increased synthesis of both VLDL and LDL. The underlying genetic defect seems to be over-production of the apoprotein common to both VLDL and LDL, namely apoB. Until a genetic marker becomes available familial combined hyperlipidaemia will continue to be diagnosed mainly by exclusion. In general adults with a type IIb phenotype who do not have familial hypercholester-olaemia or any secondary cause can be regarded as having this disorder. The response of such individuals to diet should always be tried but if inadequate then fibric acid derivatives such as gemfibrozil or bezafibrate are often effective.

FORESTALLING THE VASCULAR CONSEQUENCES OF HYPERLIPIDAEMIA

The beneficial effects of lowering serum cholesterol by diet or drugs have been demonstrated in several trials of primary prevention of coronary heart disease and of secondary intervention (Thompson, 1987). These showed that reducing serum cholesterol by as little as $1\,mmol\,L^{-1}$ resulted in significant decreases in the incidence of coronary heart disease and the rate of progression of coronary lesions. Ratios of HDL to total or LDL cholesterol were the best predictors of angiographic change in the intervention trials, these ratios being negatively correlated with progression of coronary disease. In view of the potential benefit to patients with coronary heart disease of measures which increase HDL cholesterol and decrease LDL cholesterol it seems probable that the advent of newer and more potent drugs such as the HMG CoA reductase inhibitors will eventually enable hyperlipidaemia to be treated as widely and effectively as is hypertension (Martin *et al.*, 1986).

This raises the question as to how best to detect hyperlipidaemia in the population. According to the US Consensus Statement on Lowering Blood Cholesterol to Prevent Heart Disease (Consensus Conference, 1985) every adult who happens to consult their family doctor for any reason should have their serum lipids measured. Alternatively, selective screening of those with a high probability of having hyper-lipidaemia should be undertaken. This includes all individuals below the age of 60 years with coronary heart disease, or with a family history of premature coronary disease or hyperlipidaemia, or with xanthelasma, premature corneal arcus, diabetes or hypertension. Screening of family members is especially important where a parent is known to have familial hypercholesterolaemia and the putative spouse of a known heterozygote should be screened before marriage to avoid the risk of producing homozygous children.

Having detected hyperlipidaemia in an individual, what action should be taken? Secondary hyperlipidaemia should be treated by correction of the underlying cause but all other forms of hyperlipidaemia should be treated by dietary means initially. Drug therapy should be contemplated only if there is a failure to respond to diet, which often indicates that the disorder has a genetic basis and that drug therapy will be necessary to correct the metabolic defect.

REFERENCES

Allen, J. M., Thompson, G. R., Myant, N. B., Steiner, R. and Oakley, C. M. Cardiovascular complications of homozygous familial hypercholesterolaemia. *Br. Heart J.* 44 (1980) 361–368

Beaumont, J. L., Carlson, L. A., Cooper, G. R., Fejfar, Z., Fredrickson, D. S. and Strasser, T. Classification of hyperlipidaemias and hyperlipoproteinaemias. *Bull. WHO* 43 (1970) 891–908

Bilheimer, D. W., Goldstein, J. L., Grundy, S. M., Starzl, T. E. and Brown, M. S. Liver transplantation to provide low-density-lipoprotein receptors and lower plasma cholesterol in a child with homozygous familial hypercholesterolemia. *N. Engl. J. Med.* 311 (1984) 1658–1664

Breckenridge, W. C., Little, A., Steiner, G., Chow, A. and Poapst, M. Hypertriglyceridemia associated with deficiency of apolipoprotein CII. *N. Engl. J. Med.* 298 (1978) 1265–1273

British Cardiac Society Working Group on Coronary Prevention. Conclusions and recommendations. *Br. Heart J.* 57 (1987) 188–189

Brown, M. S., Goldstein, J. L. and Fredrickson, D. S. Familial type 3 hyperlipoproteinemia (dysbetalipoproteinemia). In Stanbury, J. B., Wyngaarden, J. B., Fredrickson, D. S., Goldstein, J. L. and Brown, M. S. (eds.), *The Metabolic Basis of Inherited Disease* 5th edition, McGraw-Hill, New York, 1983, pp. 655–671

Consensus Conference. Lowering blood cholesterol to prevent heart disease. *J. Am. Med. Assoc.* 253 (1985) 2080–2086

Fredrickson, D. S., Levy, R. I. and Lees, R. I. Fat transport in lipoproteins – an integrated approach to mechanisms and disorders. *N. Engl. J. Med.* 276 (1967) 148–156

Friedewald, W. T., Levy, R. I. and Fredrickson, D. S. Estimation of the concentration of low-density lipoprotein cholesterol in plasma, without use of the preparative ultracentrifuge. *Clin. Chem.* 18 (1972) 499–502

Gagné, C., Moorjani, S., Brun, D., Toussaint, M., Lupien, P-J. Heterozygous familial hypercholesterolemia. Relationship between plasma lipids, lipoproteins, clinical manifestations and ischaemic heart disease in men and women. *Atherosclerosis* 34 (1979) 13–24

Goldstein, J. L. and Brown, M. S. Familial hypercholesterolemia. In Stanbury, J. B., Wyngaarden, J. B., Fredrickson, D. S., Goldstein, J. L. and Brown, M. S. (eds.), *The Metabolic Basis of Inherited Disease* 5th edition, McGraw-Hill, New York, 1983, pp. 672–712

Goldstein, J. L., Brown, M. S., Anderson, R. G. W., Russell, D. W., Schneider, W. J. Receptor-mediated endocytosis: concepts emerging from the LDL receptor system. *Annu. Rev. Cell Biol.* 1 (1985) 1–39

Goldstein, J. L., Schrott, H. G., Hazzard, W. R., Bierman, E. L. and Motulsky, A. G. Hyperlipidemia in coronary heart disease. II. Genetic analysis of lipid levels in 176 families and delineation of a new inherited disorder, combined hyperlipidemia. *J. Clin. Invest.* 52 (1973) 1544–1568

Heiberg, A. Inheritance of xanthomatosis and hyper-betalipoproteinaemia. A study of seven large kindreds. *Clin. Genet.* 9 (1976) 92–111

Kwiterovich, P. O., Fredrickson, D. S. and Levy, R. I. Familial hypercholesterolaemia (one form of familial type II hyperlipoproteinaemia). *J. Clin. Invest.* 53 (1974) 1237–1249

Martin, M. J., Hulley, S. B., Browner, W. S., Kuller, L. H. and Wentworth, D. Serum cholesterol, blood pressure and mortality: implications from a cohort of 361 662 men. *Lancet* 2 (1986) 933–936

Moorjani, S., Gagné, C., Lupien, P-J and Brun, D. Plasma triglycerides related decrease in high density lipoprotein cholesterol and its association with myocardial infarction in heterozygous familial hypercholesterolemia. *Metabolism* 35 (1986) 311–316

Nikkila, E. A. Familial lipoprotein lipase deficiency and related disorders of chylomicron metabolism. In Stanbury, J. B., Wyngaarden, J. B., Fredrickson, D. S., Goldstein, J. L. and Brown, M. S. (eds.), *The Metabolic Basis of Inherited Disease*, 5th edition, McGraw-Hill, New York, 1983, pp. 622-642

Nikkila, E. A. and Aro, A. Family study of serum lipids and lipoproteins in coronary heart disease. *Lancet* 1 (1973) 954-959

Seftel, H. C., Baker, S. G., Sandler, M. P., Forman, M. B., Joffe, B. I., Mendelsohn, D., Jenkins, T. and Mieny, C. J. A host of hypercholesterolaemic homozygotes in South Africa. *Br. Med. J.* 281 (1980) 633–636

Slack, J. Risk of ischaemic heart disease in familial hyperlipoproteinaemic states. *Lancet* 2 (1969) 1380–1382

Sprecher, D. L., Hoeg, J. M., Schaefer, E. J., Zech, L. A., Gregg, R. E., Lakatos, E. and Brewer, H. B. Jr. The association of LDL receptor activity, LDL cholesterol level, and clinical course in homozygous familial hypercholesterolaemia. *Metabolism* 34 (1985) 294–299

Streja, D., Steiner, G. and Kwiterovich, P. O. Plasma high density lipoproteins and ischemic heart disease: studies in a large kindred with familial hypercholesterolemia. *Ann. Intern. Med.* 89 (1978) 871–880

Sugrue, D. D., Thompson, G. R., Oakley, C. M., Trayner, I. M. and Steiner, R. E. Contrasting patterns of coronary atherosclerosis in normocholesterolaemic smokers and patients with familial hypercholesterolaemia. *Br. Med. J.* 283 (1981) 1358–1360

Thompson, G. R. The hyperlipidaemias. In Lloyd, J. K. and Scriver, C. R. (eds.), *Genetic and Metabolic Disease in Pediatrics*, Butterworths, London, 1985, pp. 211–233

Thompson, G. R. Evidence that lowering serum lipids favourably influences coronary heart disease. *Q. J. Med. (New Series)* 62 (1987) 87–95

Thompson, G. R., Ford, J., Jenkinson, M. and Trayner, I. Efficacy of mevinolin as adjuvant therapy for refractory familial hypercholesterolaemia. *Q. J. Med.* 60 (1986) 801–809

Thompson, G. R., Lowenthal, R. and Myant, N. B. Plasma exchange in the management of homozygous familial hypercholesterolaemia. *Lancet* 1 (1975) 1208–1211

Thompson, G. R., Miller, J. P. and Breslow, J. L. Improved survival of patients with homozygous familial hypercholesterolaemia treated by plasma exchange. *Br. Med. J.* 291 (1985) 1671–1673

Thompson, G. R. and Sapsford, R. Coronary artery bypass grafting and hyperlipidaemia. *Br. Heart J.* 53 (1985) 237–239

Yokoyama, S., Hayashi, R., Satani, M. and Yamamoto, A. Selective removal of low density lipoprotein by plasmapheresis in familial hypercholesterolemia. *Arteriosclerosis* 5 (1985) 613–622

J. Inher. Metab. Dis. 11 Suppl. 1 (1988) 29–32

Lipase Deficiencies

H. GRETEN and F. U. BEIL
Medizinische Klinik, Universitats-Krankenhaus Eppendorf. Martinistrasse 52, 2000 Hamburg 20-07, FRG.

Summary: Two enzymes, lipoprotein lipase and hepatic triglyceride lipase, are involved in the hydrolysis of triglycerides from chylomicrons and very low density lipoprotein (VLDL). Lipoprotein lipase has an absolute requirement for apolipoprotein CII for activity. Three inborn errors of metabolism which give rise to hypertriglyceridaemia have been described. The biochemical and clinical aspects of these disorders, lipoprotein lipase deficiency (familial type I hyperlipoproteinaemia), hepatic triglyceride lipase deficiency and apo-CII deficiency are discussed.

The hydrolysis of triglycerides in chylomicrons and very low density lipoproteins (VLDL) is catalysed by at least two different lipases, lipoprotein lipase and hepatic triglyceride lipase. Both enzymes are glycoproteins and linked to the endothelian cells of the capillary system. Lipoprotein lipase (E.C. 3.1.1.34) has been purified from plasma after heparin injection from milk and adipose tissue. It has an apparent monomeric molecular weight of about 60 000 daltons on SDS gel electrophoresis and 48 300 daltons by sedimentation–equilibrium ultracentrifugation (Augustin and Greten, 1979). It preferentially attacks the 1.3-ester bonds of triglycerides. Lipoprotein lipase is found in many tissues of the body other than liver including heart, skeletal muscle, lung, mammary gland, and of course most particularly adipose tissue. Lipoprotein lipase has to be differentiated from hepatic triglyceride lipase (EC 3.1.1.3) with similar molecular weight. The main difference between these two enzymes is the requirement of lipoprotein lipase for apolipoprotein CII (apoCII) to give full activity (LaRosa *et al.*, 1970). ApoCII serves as a co-factor for lipoprotein lipase whereas hepatic triglyceride lipase activity does not depend on apoCII. Lipoprotein lipase and hepatic lipase are separable by chromatography on heparin–Sepharose. Antisera specific for one or the other have been prepared and employed to measure specific and separate activities (Greten *et al.*, 1976). The hepatic enzyme is resistant to protamine and sodium chloride. Hepatic triglyceride lipase is probably identical with phospholipase also found in plasma following heparin injection. Lipoprotein lipase has to be distinguished from other triglyceride hydrolases such as acid hydrolase of lysosomal origin, pancreatic lipase and 'hormone-sensitive' lipase. The latter is activated by cyclic mononucleotides.

Chylomicron and very low density lipoprotein (VLDL) metabolism occurs through multiple interactions of various enzymes, hydrolysis of triglycerides, transfer of fatty acids and apoproteins with concomitant decrease of the core volume.

Journal of Inherited Metabolic Disease. ISSN 0141–8955. Copyright © SSIEM and MTP Press Limited, Queen Square, Lancaster, UK.

During this process the core of these particles changes from a more triglyceride-rich to a predominantly cholesterol ester-rich one, low density lipoprotein (LDL). These LDL are finally eliminated in peripheral tissues by receptor-mediated endocytosis. ApoC molecules are progressively removed from VLDL to be associated with HDL until they recirculate to newly formed VLDL or chylomicrons which enter the circulation. This transfer only occurs when VLDL-triglycerides are catabolized. HDL serves as a reservoir for physiologically important apolipoproteins which regulate the activities of various enzymes: apoAI activates lecithin:cholesterol acyl transferase (EC 2.3.1.43) and apoCII activates lipoprotein lipase. The three plasma enzymes lipoprotein lipase, hepatic lipase and lecithin:cholesterol acyl transferase regulate a cascade of reactions including hydrolysis of phospholipids at the surface of VLDL, breakdown of triglycerides in the core, transfer of free cholesterol and formation of cholesterol esters from spherical HDL. The exact sequence of events is not known and many of these assumptions are still hypothetical. Importantly, the exact role of lipoprotein lipase and hepatic lipase during delipidation of VLDL and chylomicrons remains to be established. Thus, lipolytic enzymes have a key role in energy distribution (fatty acids) and distribution of cholesterol to LDL and HDL.

There exist at least three inborn errors of metabolism described so far leading to hypertriglyceridaemia which are due to either a deficiency of one of these enzymes or to a deficiency of apoCII, the co-factor for lipoprotein lipase (Nikkilä, 1983).

I. FAMILIAL TYPE I HYPERLIPOPROTEINAEMIA (FAMILIAL LIPOPROTEIN LIPASE DEFICIENCY (McKusick 23860))

This is a rare genetic syndrome caused by decreased activity of lipoprotein lipase. The removal of lipoprotein particles containing dietary fat is severely retarded in the affected subjects. They all develop massive chylomicronaemia with low levels of other lipoproteins and most subjects have severe abdominal attacks. These attacks are often associated with pancreatitis. As chylomicrons may also accumulate in macrophages, liver and spleen usually are enlarged. Most of these patients have eruptive xanthomas, which are small, yellowish nodules. They are not painful and are seen on the skin at one time or another. In addition to the skin, foam cells appear in tissues rich in reticulo-endothelial cells such as bone marrow, spleen and liver. Severe hypertriglyceridaemia is often associated with abdominal pain and these attacks are usually preceded by excessive fat intake. Patients with type I hyperlipoproteinaemia are easily identified by a characteristic appearance of their plasma after storage overnight at 4°C, when a white layer of chylomicrons develops at the top.

Familial lipoprotein lipase deficiency is best treated by restriction of dietary fat. Usually most patients will be free of symptoms if they restrict their fat intake to between 40 and 60 g per day. There is no drug treatment available at the present time.

II. FAMILIAL HEPATIC TRIGLYCERIDE LIPASE DEFICIENCY (McKusick 24665)

Primary hepatic triglyceride lipase deficiency has recently been described in a few adult patients (Breckenridge *et al.*, 1982). The clinical symptoms are similar to those found in Type I hyperlipoproteinaemia. Patients present with severe abdominal pain and pancreatitis. Specific measurements of lipoprotein lipase and hepatic triglyceride lipase make it possible to distinguish between the two disorders. Hepatic triglyceride lipase deficiency secondary to liver dysfunction has been described before (Klose *et al.*, 1977). In most cases with liver disease hepatic triglyceride lipase activity is usually low leading to mild to moderate hypertriglyceridaemia.

III. ApoCII DEFICIENCY (McKusick 20775)

ApoCII deficiency as a clinical syndrome was originally described by Breckenridge and colleagues in 1978 and since then in a few more kindreds.

As outlined above this protein serves as a co-factor for lipoprotein lipase activity. Several mutations of apoCII molecule have been identified (Connelly *et al.*, 1987). The clinical symptoms of this rare disease are similar to lipoprotein lipase and hepatic lipase deficiency. Usually patients present with severe abdominal pain mimicking an acute abdomen. Severe hypertriglyceridaemia is due to the presence of chylomicrons. The addition of normal plasma to post-heparin plasma of the patients yields normal enzymatic activity. Both lipoprotein lipase and hepatic lipase activity are found to be normal in most cases. Treatment of severe hypertriglycerid-aemia can be accomplished by intravenous administration of fresh frozen plasma containing sufficient amounts of apoCII for activation of lipoprotein lipase. Upon normalization of elevated triglyceride levels a diet is prescribed which is low in dietary fat.

The pathogenesis of lipid disorders accompanied by hypertriglyceridaemia has not been completely elucidated. Especially, the exact role of lipoprotein lipase and hepatic lipase during delipidation of VLDL in chylomicrons remains to be established. Physicians, especially paediatricians, should be aware of these rare disorders as patients with these syndromes often present with acute abdominal attacks being misinterpreted and leading to unnecessary surgical procedures.

REFERENCES

Augustin, J. and Greten, H. In Paoletti, R. and Golto, A. M. (eds.) *Atherosclerosis Reviews*, Vol. 5, Raven Press, New York (1979).

Breckenridge, W. C., Little, J. A., Steiner, G., Chow, A. and Poapst, M. Hypertriglycerid-aemia associated with deficiency of apolipoprotein C-II. *N. Engl. J. Med.* 298 (1978) 1265–1273

Breckenridge, W. C., Little, J. A., Alaupovic, P., Wang, C. S., Kuksis, A., Kakis, G., Lindgrene, F. and Gardner, G. Lipoprotein abnormalities associated with a familial deficiency of hepatic lipase. *Atherosclerosis* 45 (1982) 161–179

Connelly, P. W., Maguire, G. F. and Little, J. A. Apolipoprotein C-II St Michael. Familial apolipoprotein C-II deficiency associated with premature vascular disease. *J. Clin. Invest.* 80 (1987) 1597–1606

Greten, H., de Gretta, R., Klose, G., Roscher, W., deGennes, J. L. and Gjone, E. Measurement of two plasma triglyceride lipases by an immunochemical method: studies in patients with hypertriglycerdaemia. *J. Lipid Res.* 17 (1976) 203–210

Klose, G., Windelband, J., Weizel, A. and Greten, H. Secondary hypertriglyceridaemia in patients with parenchymal liver disease. *Eur. J. Clin. Invest.* 7 (1977) 557–562

LaRosa, J. C., Levy, R. I., Herbert, P., Lux, S. E. and Frederickson, D. S. A specific apolipoprotein activator for lipoprotein lipase. *Biochem. Biophys. Res. Commun.* 41 (1970) 57–62

Nikkilä, F. A. Familial lipoprotein lipase deficiency and related disorders of chylomicron metabolism. In Stanbury, J. B., Wyngaarden, J. B., Frederickson, D. S., Goldstein, J. L. and Brown, M. S. (eds.) *The Metabolic Basis of Inherited Disease.* 5th Edn., McGraw-Hill Book Company, New York (1983) pp. 622–642

J. Inher. Metab. Dis. 11 Suppl. 1 (1988) 33–44

The Use of Recombinant DNA Techniques for the Diagnosis of Familial Hypercholesterolaemia

S. Humphries[1], R. Taylor[1], M. Jeenah[1] and M. Seed[2]
[1]*Charing Cross Sunley Research Centre, Lurgan Avenue, Hammersmith, London, W6 8LW and [2]Charing Cross Hospital Medical School, Hammersmith, London W6 8RP, UK*

Summary: In the UK, about 5% of patients with familial hypercholesterol-aemia have a detectable deletion or rearrangement of part of the LDL-receptor gene. This results in the detection of shorter or abnormal sized fragments of the LDL-receptor gene in a Southern blot hybridization. This can be used to follow the inheritance of the defective gene, and for diagnosis in the families of these individuals. In the families of the rest of the patients, diagnosis may be possible using linked restriction fragment length polymor-phisms (RFLPs) detected with the LDL-receptor probe. There are now ten common RFLPs of the LDL-receptor gene, with variable sites in the 3′ half of the gene. Over 80% of patients are heterozygous for at least one of these RFLPs, and therefore potentially informative for DNA diagnosis. For a foetus at risk of homozygous familial hypercholesterolaemia, antenatal diagnosis may also be possible using these methods. However, family studies require samples to be available from affected or unaffected relatives of the patient, and this limits the applicability of the tests. For some mutations, the base pair change causing the defect in the LDL-receptor itself creates or destroys a site for a restriction enzyme. Such 'mutation-specific' RFLPs could be used for population screening, but so far have only been reported for the familial hypercholesterolaemia mutation that is common in Lebanon. In the future it may be possible to develop mutation-specific oligonucleotide probes for the diagnosis of familial hypercholesterolaemia. These would be appropriate for population screening or screening patients with hyperlipidaemia. This information may be useful if different mutations require different therapeutic strategies.

INTRODUCTION

Familial hypercholesterolaemia (McKusick 14389) contributes significantly to the number of individuals suffering from premature coronary artery disease. If individuals with familial hypercholesterolaemia could be identified before they develop symptomatic disease, they could be treated prophylactically to reduce their future

Journal of Inherited Metabolic Disease. ISSN 0141–8955. Copyright © SSIEM and MTP Press Limited, Queen Square, Lancaster, UK.

risk of myocardial infarction. To date, however, the disease cannot always be diagnosed unequivocally in early childhood.

The identification of potentially affected individuals in families has, in the past, relied upon the measurement of serum and LDL-cholesterol concentrations determined from peripheral blood and cord blood (Glueck *et al.*, 1971; Kwiterovich *et al.*, 1983; Leonard *et al.*, 1977). However, serum cholesterol concentrations do not always allow unequivocal diagnosis of familial hypercholesterolaemia, especially when determined from cord blood. The values obtained in this way sometimes lie within the normal reference ranges and in some children may not rise to levels at which unequivocal diagnosis of familial hypercholesterolaemia would be made until later in life (A. Kessling and M. Seed, unpublished observation). This problem could be circumvented by carrying out serial blood cholesterol determinations but this would be inconvenient and expensive. The use of a direct genetic test for familial hypercholesterolaemia could provide an unequivocal diagnosis. Diagnosis has also been attempted by growing fibroblasts from a skin biopsy for a direct measurement of the binding of labelled LDL to the cell LDL-receptors. This method is technically difficult and, again, overlapping values can be obtained from normal individuals and individuals with heterozygous familial hypercholesterolaemia. There exists, therefore, the possibility that familial hyper-cholesterolaemic children would not be diagnosed by these methods until coronary artery disease was already initiated as a result of rising cholesterol concentrations.

The cloning of the human LDL-receptor cDNA (Yamamoto *et al.*, 1984) has now made it possible to analyse the defects at the DNA level. Here we describe four patients who have a deletion in the LDL-receptor gene, and a number of common restriction fragment length polymorphisms (RFLP) of the gene. These polymorphisms can be used for presymptomatic and antenatal diagnosis in families with familial hypercholesterolaemia.

DNA REARRANGEMENTS

In order to identify gross alterations in the LDL-receptor gene, we have digested DNA samples from 55 unrelated heterozygous and five homozygous familial hypercholesterolaemic patients (= 65 defective genes) with the restriction enzymes *Xba*I or *Bgl*II and probed with LDL-receptor cDNA. Two different probes were used, one for the 3' part of the gene (*Xba*I digest), and one for the 5' part of the gene (*Bgl*II digest) (Figure 1). In four of the patients, the probes detect gene fragments both of the expected size and an additional smaller fragment. In each patient these smaller fragments are the result of a partial deletion of one of the alleles of the LDL-receptor gene. We have carried out a detailed analysis of the defect in the patient T.D. The deletion removes about 5 kb of DNA, including exon 13 and 14 (Horsthemke *et al.*, 1985). The defective gene has been isolated and the sequence of DNA in the region determined (Horsthemke *et al.*, 1987a). Analysis indicates that the deletion has occurred as a result of a recombination event between two repeated 'Alu' sequences in the introns of the gene. From the sequence we can also predict that the defective gene should produce a truncated

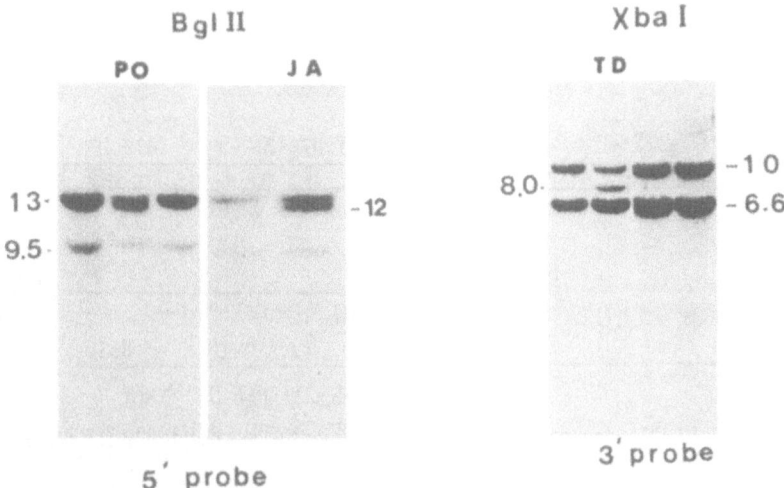

Figure 1 Genomic Southern blot analysis of the LDL-receptor gene in familial hypercholesterolaemia patients. 5 μg of DNA from each patient was digested with *Xba*I or *Bgl*II size-fractionated on an agarose gel, transferred to filter membranes and hybridized with the radioactively labelled LDL-receptor probe as described (Humphries *et al.*, 1985). The 3' probe consists of a 1.9 kb *Bam*HI fragment (base pairs 1573–3486) of the 3' part of the LDL-receptor cDNA. The 5' probe consists of a 1.8 kb for *Hind*III/*Bgl*II fragment from plasmid pLDLR-3, and covers nucleotides 1–1700

protein that lacks the carboxy-terminal 230 amino acids (Horsthemke *et al.*, 1987a). This region of the protein is involved in anchoring the protein in the cell membrane, and we would predict that the protein, if transported to the membrane, would not be retained in the coated pit. However using Western blotting we have not been able to detect a truncated LDL-receptor protein on the surface of the patient's fibroblasts or in cytoplasmic extracts of the cells (Horsthemke *et al.*, 1987a). It thus appears that the truncated protein may be rapidly turned over within the cell.

We do not yet have a detailed map of the deletions detected with the 5' probe. The deletions appear to be 1–2 kb in length (Horsthemke *et al.*, 1987b), and may remove regions of the gene coding for part of the LDL-binding domain or part of the epidermal growth factor 'stem' region (Figure 2). Several other deletions or insertions in the LDL-receptor gene of familial hypercholesterolaemia patients have been reported (e.g. Lehrman *et al.*, 1985a, 1987a). These deletions have probably also been caused by a recombination event involving two 'Alu' sequences. Presymptomatic diagnosis based on DNA analysis will thus be possible in all these families, as demonstrated in the family of T.D. (Horsthemke *et al.*, 1985).

COMMON DNA POLYMORPHISMS

Although our screening method would not identify small deletions or insertions, the results suggest that the majority of defects in the LDL-receptor gene are due to point mutations rather than gross alterations in the gene. Many of these have

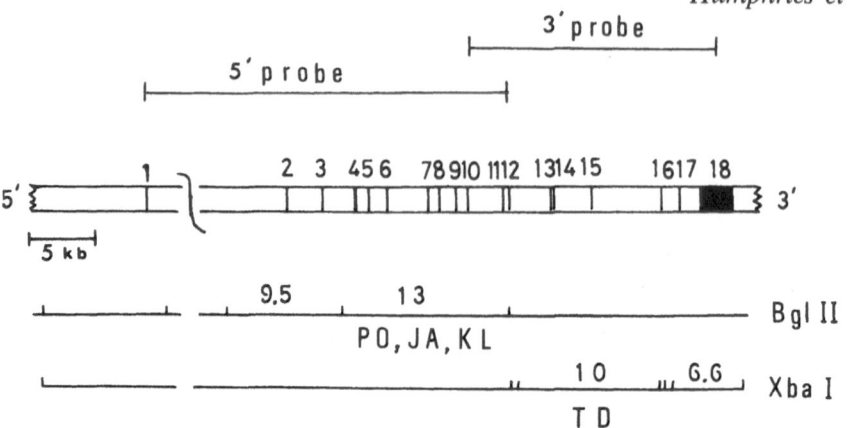

Figure 2 Map of LDL-receptor gene showing regions that are deleted in the patients T.D. (3′ probe), and P.O., J.A., and K.L. (5′ probe)

now been described in detail (e.g. Lehrman *et al.*, 1985b). Presymptomatic diagnosis in these cases could be based on genetic linkage analysis using RFLPs.

To date common RFLPs of the LDL-receptor gene have been detected using

Table 1 Common RFLPs of the LDL-receptor gene

Enzyme	Rare allele frequency	Location of varying site	Comments	Reference
*Apa*L1	0.42/0.39	intron 15 + 3′ flanking	Two RFLPs Three alleles	Leitersdorf *et al.*, 1987
*Ava*II	0.44	exon 13	Requires double digest to detect easily	Hobbs *et al.*, 1987a
*Bst*EII	0.25	3′ half	Large fragments	Steyn *et al.*, 1987
*Msp*I	0.23	3′ half	Two RFLPs	Geisel *et al.*, 1987
*Nco*I	0.36	exon 18	Frequency altered in UK-FH	Kotze *et al.*, 1987
*Pst*I	0.40	3′ flanking	Use oligo to detect	Funke *et al.*, 1986a
*Pvu*II	0.23	intron 15	Frequency altered in Afrikaaner-FH	Humphries *et al.*, 1985
*Stu*I	0.07	exon 8	Most 5′ RFLP reported	Kotze *et al.*, 1986

FH = familial hypercholesterolaemia

eight different restriction enzymes (Table 1). For some enzymes there are several different alleles, and taken together these RFLPs should be informative in more than 80% of families. The relative usefulness of these different RFLPs is increased both if frequency of the minor allele in the population is high and if there is linkage

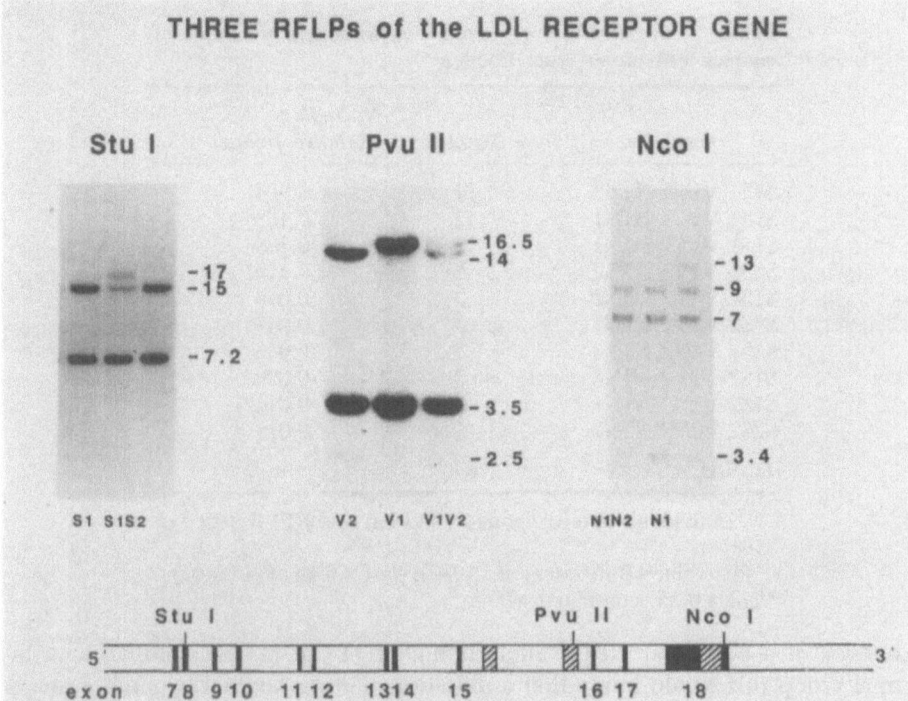

Figure 3 Map of the LDL-receptor gene showing the approximate location of the variable sites for *Pvu*II, *Nco*II and *Stu*I, and an example of a Southern blot hybridization of the DNA fragments observed

equilibrium (random association) of the alleles of the two RFLPs with each other. We have examined this for the RFLPs detected with *Pvu*II, *Stu*I and *Nco*I (Figure 3) in a sample of 69 normal individuals from London. In this sample we detected significant linkage disequilibrium between the *Pvu*II and *Nco*I RFLPs and the *Pvu*II and *Stu*I and *Nco*I polymorphisms. Over such small genetic distances the degree of linkage disequilibrium is not related primarily to the physical distance between the varying genetic markers but rather due to the evolutionary history of the three polymorphisms. However, the data are consistent with our preliminary physical map of the location of the polymorphisms, with the *Stu*I varying site being in the 5′ region of the gene, and the *Nco*I and *Pvu*II sites being closer together in the 3′ region of the gene (Figure 3).

It is possible to get an estimate of the relative usefulness of these polymorphisms taken together, by asking how many individuals are heterozygous for at least one RFLP. In this sample 52 of the 69 individuals (75%) are heterozygous for one or more of the polymorphisms (Table 2). This suggests that using these three RFLPs alone roughly three-quarters of families will be informative for genetic studies and for early detection of familial hypercholesterolaemia.

It is, however, possible that the frequency of these RFLPs may differ in groups of patients with familial hypercholesterolaemia and the normal population. If the

Table 2   Three-RFLP genotype distribution in a sample of
normal individuals from London

Genotype	Number	Normals Relative frequency
S1S1 V1V1 N1N2	21	0.304*
S1S1 V1V1 N1N1	13	0.188
S1S1 V1V2 N1N1	13	0.188
S1S1 V1V2 N1N2	9	0.130*
S1S1 V1V1 N2N2	3	0.043
S1S2 V1V2 N1N1	3	0.043*
S1S2 V2V2 N1N1	3	0.043*
S1S2 V1V2 N1N2	2	0.029*
S1S2 V1V1 N1N1	1	0.014*
S2S2 V2V2 N1N1	1	0.014
TOTAL	69	

* Individuals heterozygous for at least one RFLP (52/69 or
75%)

PIC value (Botstein *et al.*, 1980): *Stu*I = 0.10, *Pvu*I = 0.30,
*Nco*I = 0.35, combined = 0.65

frequency of a particular RFLP allele is higher in the patient compared to the
normal group this would imply that a mutation causing familial hypercholesterolaemia occurred in an LDL-receptor gene with this RFLP allele and that the
association has been maintained over recent generations (i.e. linkage disequilibrium). The observation of an altered frequency would also imply that a particular
mutation, associated with this RFLP, must be making up a significant proportion
of all familial hypercholesterolaemia mutations in the population of patients. If
many different mutations have occurred in the population, they would by chance
have occurred on chromosomes containing either alleles of the polymorphism. This
would cancel out the effect of the associations, and no difference in frequency
between patients and normals would be observed.

There are a number of problems with the interpretation of such results. For
example, it is possible that the frequency of these RFLPs may be different in
different ethnic groups. Significant differences in allele frequency have been reported for polymorphism of the apoprotein AI/CIII/AIV genes on chromosome
11, even within different parts of the UK (Morris *et al.*, 1985). Thus if the ethnic
origins of the patients and control group were not the same, it might be possible
to observe a difference in RFLP allele frequency for this reason alone.

The frequency of the *Stu*I, *Pvu*II and *Nco*I RFLPs in a sample of normal
individuals and familial hypercholesterolaemia patients from London is presented
in Table 3. The frequency of the *Nco*I polymorphism (but not the *Pvu*II or *Stu*I
polymorphisms), is indeed altered in the London familial hypercholesterolaemia
group compared to the London controls. The observed difference in frequency
might be explained by a founder effect such as has been proposed to have occurred
in South Africa (Brink *et al.*, 1986), but this needs to be confirmed by studies with
larger samples.

Table 3 Relative allele frequency of *Puv*II, *Stu*I and *Nco*I RFLPs in controls and familial hypercholesterolaemia patients from London

The alleles are designated:
  *Pvu*II *V1* = 16.5+3.5 kb; *V2* = 14.0+2.5+3.5 kb;
  *Stu*I *S1* = 15.0+7.2 kb; *S2* = 17.0+7.2 kb;
  *Nco*I *N1* = 3.4+9.0+7.0 kb; *N2* = 13.0+ 9.0+7.0 kb

	*Stu*I		*Pvu*II		*Nco*I	
	Number	*S1/S2*	*Number*	*V1V2*	Number	*N1/N2*
UK controls	69	0.92/0.08	150	0.75/0.25	117	0.74/0.26
						*
UK patients	85	0.96/0.04	72	0.75/0.25	109	0.62/0.38

* by gene counting $\chi^2 = 9.06$, $p < 0.05$
R. Taylor and M. Jeenah, unpublished observations

USE OF THE RFLPS FOR DIAGNOSIS

There have been several reports of linkage studies in families with familial hypercholesterolaemia using a common polymorphism of the LDL-receptor gene detected with the enzyme *Pvu*II (Humphries *et al.*, 1985; Berg *et al.*, 1985; Leppert *et al.*, 1986; Armston *et al.*, 1987). The varying site occurs in an intron of the gene, and does not itself alter the amino acid sequence of the protein (Hobbs *et al.*, 1985). In the UK, defective LDL-receptor genes can be found on both alleles of the *Pvu*II polymorphism (Humphries *et al.*, 1985; Armston *et al.*, 1987) and the frequency of the polymorphism is similar in a sample of normal individuals, and a group of patients with familial hypercholesterolaemia (Humphries *et al.*, 1985, and Table 2). This indicates that there is no strong linkage disequilibrium between the polymorphism and mutations causing familial hypercholesterolaemia. In the UK population roughly 35% of patients are heterozygous and therefore potentially informative for diagnosis using family studies. However, family studies require samples to be available from affected or unaffected relatives, and also require that the spouse of the patient is not also heterozygous for the polymorphism. Thus in practice the *Pvu*II polymorphism may be informative in only 20–25% of families (Armston *et al.*, 1987). For such families other RFLPs of the gene will need to be used, and information from several RFLPs can be combined to distinguish the defective and normal alleles and follow the inheritance of the genes in the family.

As well as being useful in some situations for diagnosis in children, these RFLPs may also be applicable for prenatal diagnosis for a foetus at risk of homozygous familial hypercholesterolaemia. Figure 4 shows a pedigree of a family who have already had one child with homozygous familial hypercholesterolaemia. The parents, who are of Asian origin, are first cousins and part of a multiple consanguineous family. Both parents are homozygous for all RFLPs tested except for *Nco*I. The genotype of the father is *N1N2* and that of the mother and daughter is *N1N1*. From this we can infer that the defective allele of the LDL-receptor in both the father and the mother is on the chromosome carrying the *N1* allele of the polymorphism. It is most likely that the parents both carry the same defective

PRENATAL DIAGNOSIS of FH USING THE Nco I RFLP

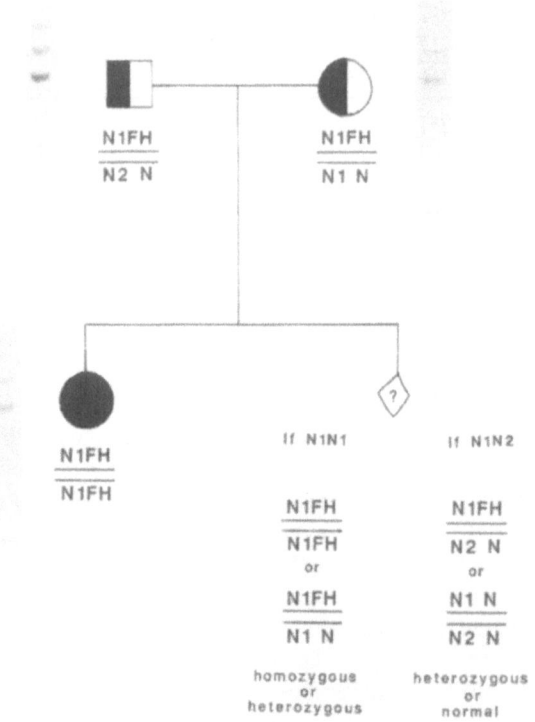

Figure 4 Use of RFLPs for diagnosis of homozygous familial hypercholesterolaemia. RFLP in a model familial hypercholesterolaemia family: N, normal LDL-receptor gene; FH, defective LDL-receptor gene. (■), (◑): male and female individuals with heterozygous familial hypercholesterolaemia; (●): individual with homozygous familial hypercholester-olaemia

LDL-receptor gene (inherited from a common ancestor) and that their daughter is homozygous at the DNA level for the same defect. This information does not allow us to carry out diagnosis unequivocally. If a foetus has the genotype *N1N2*, the child must have inherited the normal LDL-receptor gene from the father, and either the normal or defective gene from the mother. If however the genotype is *N1N1* the foetus may be either a familial hypercholesterolaemia heterozygote or a homozygote like the first child. Fetal samples for testing are now routinely obtained by biopsy of the developing placenta at 8–10 weeks of pregnancy (Old *et al.*, 1982). It may therefore be possible to use classical techniques of labelled LDL binding to fetal cells to distinguish these possibilities, and offer unequivocal diagnosis using the combined methods.

DIAGNOSIS IN THE GENERAL POPULATION

It would be extremely useful to develop a rapid DNA test for familial hypercholesterolaemia, so that any individual detected with hyperlipidaemia could be screened for defects in the LDL-receptor gene without resorting to family studies. At the present time this is not technically possible and current evidence indicates that this is unlikely to be feasible in the near future.

There are several methods that have been developed for detecting single base changes in a gene. The easiest approach is if the mutation creates or destroys the cleavage site for a particular restriction enzyme. This will result in a 'mutation-specific' DNA polymorphism of the LDL-receptor gene that can be detected using Southern blotting. For example if the mutation creates a new cutting site for the enzyme the normal gene will show only one band and the defective gene will be cleaved and show two bands. To date the only case where this approach has been successful is the mutation that commonly causes familial hypercholesterolaemia in Lebanon, which can be detected with a cDNA probe and the enzyme *Hinf*I (Lehrman *et al.*, 1987b).

The second method to detect a single base pair change uses a pair of oligonucleotides, one specific for the normal gene sequence and the other specific for the mutated sequence. This 'oligonucleotide melting' technique allows the identification of this same mutation in other patients. This approach has been successfully applied to the detection of thalassaemias (Conner *et al.*, 1983), in α-1-antitrypsin deficiency and even variants of the apolipoprotein apoE (Funke *et al.*, 1986b). The problem is that there are many mutations reported that cause familial hypercholesterolaemia, at least 20 have been documented and there are probably many more. This means that to detect all known mutations a battery of at least 20 or more specific oligonucleotides would have to be used. At best this approach would only be able to exclude all known mutations as being involved in causing hyperlipidaemia in the individual being tested. This problem would be simplified in situations where one or two mutations make up a significant proportion of all familial hypercholesterolaemia defects in a particular population and oligonucleotide tests could be developed for these mutations.

Another diagnostic approach, which has proved fruitful in the study of the thalassaemia (Antonarakis *et al.*, 1985) has been to look for an association between a specific mutation and an individual's genotype as defined by several RFLPs in conjunction (the RFLP haplotype). This would happen because a specific mutation causing familial hypercholesterolaemia must have occurred on a chromosome with a particular haplotype and by a founder effect, all descendants from this individual with familial hypercholesterolaemia will have this same haplotype. This haplotype will also exist in individuals without familial hypercholesterolaemia and this approach may not be particularly useful for population screening, but will help in patients to distinguish different mutations causing the disorder. This information may be useful in developing therapeutic strategies if, for example, patients with a particular defect respond best to a certain drug. This haplotype approach may only be applicable within defined geographical areas (Antonarakis *et al.*, 1985), and we are determining the haplotypes of patients with familial hypercholesterolaemia in London, to see if this approach may be useful for familial hypercholesterolaemia.

There are three examples known where the majority of familial hypercholesterolaemia patients in an area have the same mutation in the LDL-receptor gene, presumably as a result of geographical or cultural isolation. The first example is in the Lebanese and Syrian population, where it appears that the majority of patients with familial hypercholesterolaemia have a mutation that creates a site for the restriction enzyme *Hin*fI (Lehrman *et al.*, 1987b). This can be easily detected using Southern blotting techniques. The second example is of French-Canadian patients with familial hypercholesterolaemia who share a common deletion of part of the LDL-receptor gene, which again can be easily detected (Hobbs *et al.*, 1987b). The third reported example of this is in the Afrikaaner population in South Africa, where in one study 19/20 familial hypercholesterolaemia genes were of the *Stu*I/*Pvu*II haplotype *S1V1* (Brink *et al.*, 1987). Once the specific mutation causing familial hypercholesterolaemia in these patients is known a test can be developed using the oligonucleotide melting technique and this might detect more than 90% of patients with the disorder in the Afrikaaner population. Thus, in certain geographical areas or countries it may be possible to develop tests that will identify the majority of but not all patients with familial hypercholesterolaemia, and this, coupled with family studies for the remainder of the patients, may be a useful approach.

CONCLUSIONS

There are now enough common DNA polymorphisms of the LDL-receptor gene to be able to carry out family studies, where appropriate, for unequivocal diagnosis of familial hypercholesterolaemia in childhood. The challenge for the future is to develop methods of detecting defects causing familial hypercholesterolaemia in the population without having to resort to family studies. At the present time this appears technically difficult. However, once identified, affected individuals could be treated prophylactically and encouraged from an early age to take a low-fat diet, take physical exercise and avoid smoking. The recent results of the Primary Coronary Prevention Study offer the hope that early therapeutic intervention will effectively lower the risk of developing premature atherosclerosis (Lipid Research Clinics Program, 1984).

ACKNOWLEDGEMENTS

This work was supported by the Charing Cross Sunley Research Trust and grants from the British Heart Foundation (RG5) and the Family Heart Association. M.J. is supported by a grant from the South African MRC. The authors would like to thank Drs D. W. Russell, M. Brown and J. Goldstein for supplying the LDL-receptor probe, and Dr G. Thompson, Professor B. Lewis and Professor V. Wynn for access to patient samples, Naila Loqueman for skilled technical assistance, the British Medical Association, London, for permission to reproduce material from published papers and P. Wells for help in preparing the manuscript.

REFERENCES

Antonarakis, S. E., Kazazian, H. H. and Orkin, S. H. DNA polymorphism and molecular pathology of the human globin gene clusters. *Hum. Genet.* 69 (1985) 1–14

Armston, A. E., Iverson, S. A., Burke, J. F. Diagnosis of familial hypercholesterolaemia using DNA probes for the low-density lipoprotein (LDL) receptor gene. *Ann. Clin. Biochem.* 25 (1988) 142–149

Berg, K., Pedersen, J. C., Borresen, A. I., Heiberg, A. and Solaas, M. H. Close linkage between a common DNA polymorphism of the low-density lipoprotein (LDL) receptor gene and its use in diagnosis. *Cytogenet. Cell. Genet.* 40 (1985) 581–582

Botstein, D., White, R. L., Skolnick, M. and Davis, R. W. Construction of a genetic linkage map using restriction fragment length polymorphisms. *Ar·. J. Hum. Genet.* 32 (1980) 314–331

Brink, P. A., Steyn, L. T., Bester, A. J. and Steyn, K. Linkage disequilibrium between a marker on the low-density lipoprotein receptor and high cholesterol levels. *South Afr. Med. J.* 70 (1986) 80–82

Brink, P. A., Steyn, L. T., Coetzee, G. A. and Van der Westhuyzen, D. R. Familial Hypercholesterolaemia in South African Afrikaaners: *Pvu*II and *Stu*I DNA polymorphisms in the LDL-receptor gene consistent with a predominating founder gene effect. *Hum. Genet.* 77 (1987) 32–35

Conner, B. J., Reyes, A. A., Morin, C., Itakura, K., Teplitz, R. L. and Wallace, R. B. Detection of sickle cell βS globin allele by hybridisation with synthetic oligonucleotides. *Proc. Natl. Acad. Sci. USA* 80 (1983) 278–282

Funke, H., Klug, J., Frossard, P., Coleman, R. and Assmann, G. *Pst*I RFLP close to the LDL receptor gene. *Nucl. Acid Res.* 14 (1986a) 7820

Funke, H., Rust, S. and Assmann, G. Detection of apolipoprotein E variants by an oligonucleotide "melting" procedure. *Clin. Chem.* 32 (1986b) 1285–1289

Geisel, J., Weisshaar, B., Oette, K., Mechtel, M. and Doerfler, W. Double *Msp*I RFLP in the human LDL-receptor gene. *Nucl. Acid. Res.* 15 (1987) 3943

Gluek, C. J., Heckmann, F., Schoenfeld, M., *et al.* Neonatal familial type II hyperlipoproteinaemia: cord blood cholesterol in 1800 births. *Metabolism* 20 (1971) 597–608

Hobbs, H. H., Lehrman, M. A., Yamamoto, T. and Russell, D. W. Polymorphism and evolution of Alu sequences in the human low density lipoprotein receptor gene. *Proc. Natl. Acad. Sci. USA* 82 (1985) 7651–7655

Hobbs, H. H., Esser, V. and Russell, D. W. *Ava*II polymorphism in the human LDL receptor gene. *Nucl. Acid Res.* 15 (1987a) 379

Hobbs, H. H., Brown, M. S., Russell, P. W., Davigon, J. and Goldstein, J. L. Deletion in LDL receptor gene occurs in majority of French Canadians with FH. *N. Engl. J. Med.* 317 (1987b) 734–737

Horsthemke, B., Kessling, A. M., Seed, M., Wynn, V., Williamson, R. and Humphries, S. E. Identification of a deletion in the low density lipoprotein (LDL) receptor gene in a patient with familial hypercholesterolaemia. *Hum. Genet.* 71 (1985) 75–78

Horsthemke, B., Beisiegel, U., Dunning, A., Williamson, R. and Humphries, S. Non-homologous crossing-over between two alu-repetitive DNA sequences in the LDL-receptor gene: A possible mechanism for a novel mutation in a patient with familial hypercholesterolaemia. *Eur. J. Biochem.* 164 (1987a) 77–81

Horsthemke, B., Dunning, A. and Humphries, S. Identification of deletions in the human low-density lipoprotein (LDL) receptor. *J. Med. Genet.* 24 (1987b) 144–147

Humphries, S. E., Kessling, A. M., Horsthemke, B., Donald, J. A., Seed, M., Jowett, N., Holm, M., Galton, D. J., Wynn, V. and Williamson, R. A common DNA polymorphism of the low density lipoprotein (LDL) receptor gene and its use in diagnosis. *Lancet* 1 (1985) 1003–1005

Kotze, M. J., Langenhoven, E., Dietzsch, E., and Retief, A. E. An RFLP associated with the low-density lipoprotein receptor gene (LDLR). *Nucl. Acid Res.* 15 (1987) 37

Kotze, M. J., Retief, A. E., Brink, P. A. and Weich, H. F. H. A DNA polymorphism in the human low-density lipoprotein receptor gene. *S. Afr. J. Med.* 70 (1986) 77–79

Kwiterovich, P. O., Levy, R. I. and Frederickson, D. S. Diagnosis of familial type-II hyperlipoproteinaemia. *Lancet* 1 (1973) 118–121

Lehrman, M. A., Schneider, W. J., Sudhof, T. C., Brown, M. S., Goldstein, J. L. and Russell, D. W. Mutations in LDL-receptor: Alu-Alu recombination deletes exons encoding transmembrane and cytoplasmic domains. *Science* 227 (1985a) 140–146

Lehrman, M. A., Goldstein, J. L., Brown, M. S., Russell, D. W. and Schneider, W. J. Internalization-defective LDL-receptors produced by genes with nonsense and frameshift mutations that truncate the cytoplasmic domain. *Cell* 47 (1985b) 735–743

Lehrmann, M. A., Goldstein, J. L., Russell, D. W. and Brown, M. S. Duplication of seven exons in LDL receptor gene caused by Alu-Alu recombination in a subject with familial hypercholesterolaemia. *Cell* 48 (1987a) 827–835

Lehrman, M. A., Schneider, W. J., Brown, M. S., Davis, C. G., Elhammer, A., Russell, D. W. and Goldstein, J. L. The Lebanese allele at the low-density lipoprotein receptor locus. Nonsense mutation produces truncated receptor that is retained in endoplasmic reticulum. *J. Biol. Chem.* 262 (1987b) 401–410

Leitersdorf, E. and Hobbs, H. H. Human LDL receptor gene: two *Apa*LI RFLPs. *Nucl. Acid Res.* 15 (1987) 2782

Leonard, J. V., Whitelaw, A. G. L., Wolff, O. H., Lloyd, J. K. and Slack, J. Diagnosing familial hypercholesterolaemia in children by measuring serum cholesterol. *Br. Med. J.* (1977) 1566–1568

Leppert, M. F., Hasstedt, S. J., Holm, T., O'Connell, P., Wu, L., Ash, O., Williams, R. R. and White, R. A DNA probe for the LDL receptor gene is tightly linked to hypercholesterolaemia in a pedigree with early coronary disease. *Am. J. Hum. Genet.* 39 (1986) 300–306

Lipid Research Clinics Program. The Lipid Research Clinics Coronary Primary Prevention Trial Results. II. The relationship of reduction in incidence of coronary heart disease to cholesterol lowering. *J. Am. Med. Assoc.* 251 (1984) 356–374

Morris, S. W. and Price, W. H. DNA sequence polymorphisms in the apolipoprotein A-I/C-III gene cluster. *Lancet* 2 (1985) 1127–1128

Old, J. M., Ward, R. H. T., Karagozlu, F., Petrou, M., Modell, B. and Weatherall, D. J. First-trimester fetal diagnosis for haemoglobinopathies: three cases. *Lancet* 2 (1982) 1413–1416

Steyn, L. T., Pretorius, A., Brink, P. A. and Bester, A. J. RFLP for the human LDL receptor gene (LDLR): *Bst*EII. *Nucl. Acid. Res.* 15 (1987) 4702

Yamamoto, T., Davis, L. G., Brown, M. S., Schneider, W. J., Casey, M. L., Goldstein, J. L. and Russell, D. W. The human LDL-receptor: a cysteine-rich protein with multiple Alu sequences in its mRNA. *Cell* 39 (1984) 27–38

J. Inher. Metab. Dis. 11 Suppl. 1 (1988) 45–56

Familial LCAT Deficiency and Fish-Eye Disease

N. McIntyre
Academic Department of Medicine, Royal Free Hospital and Medical School, Rowland Hill Street, London, NW3 2QG, UK

Summary: Familial LCAT deficiency is due to deficiency of plasma lecithin–cholesterol acyltransferase. The plasma is rich in free cholesterol and lecithin while cholesterol ester and lysolecithin levels are reduced. Analysis of the abnormal lipoproteins has helped our understanding of plasma lipid and lipoprotein metabolism in normals and in patients with liver disease.

Proteinuria and anaemia are common and there is marked corneal lipid deposition. Eventually renal function deteriorates and dialysis and/or renal transplantation may be necessary.

The human LCAT gene has been sequenced and been shown to be present on chromosomal segment 16q22 – the region predicted on the basis of recombination studies as the site of the LCAT deficiency gene. The gene defect has been identified in some cases, but the mechanism remains unclear as the mutations were not in the region presumed to be the enzyme's active site.

Only three cases of fish-eye disease have been described; all were elderly and had obvious corneal opacities. They had fasting hypertriglyceridaemia and increased VLDL. IDL and LDL were increased and were triglyceride rich. HDL, reduced by 90%, was mainly HDL_3 – with a high free and low ester cholesterol.

LCAT activity in fish-eye plasma was normal but when measured in an exogenous substrate it was only 10–15% of normal. Fish-eye HDL is a substrate for purified LCAT, but fish-eye LCAT does not esterify free cholesterol of HDL (normal or fish-eye), although it esterifies free cholesterol of VLDL and LDL. It has been suggested that one type of LCAT activity acts on HDL (α-LCAT) and another on VLDL and LDL (β-LCAT) – and that fish-eye disease is due to α-LCAT deficiency, and classical familial LCAT deficiency due to lack of both components.

FAMILIAL LCAT DEFICIENCY

Lecithin–cholesterol acyltransferase (LCAT: E.C. 2.3.1.43) is a plasma enzyme catalysing the transfer of a fatty acid from the 2-position of lecithin to the 3-β-hydroxy group of free cholesterol to produce cholesteryl ester and lysolecithin (Glomset *et al.*, 1983). Sperry (1935) first recognized that when human serum or

Journal of Inherited Metabolic Disease. ISSN 0141–8955. Copyright © SSIEM and MTP Press Limited, Queen Square, Lancaster, UK.

blood was incubated the free cholesterol concentration fell markedly without change in total cholesterol. He attributed this to enzymatic esterification of free cholesterol because the effect was abolished by heating serum to 55–60°C. He also suggested that the fatty acid must come from hydrolysis of another fatty acid-containing compound (as he was not aware then of the presence of free fatty acids in blood).

Despite great interest in the relationship between plasma cholesterol and coronary artery disease little was done to characterize the cholesterol esterifying activity of plasma. The seminal papers which led to our current understanding of LCAT and familial LCAT deficiency (McKusick 24590) did not appear until the 1960s, and they were written by three Norwegians, Glomset, Gjone and Norum.

Glomset, though American born, is of Norwegian origin and studied medicine in Uppsala in Sweden. In the early 1960s when working in Seattle he published a now classic series of papers. First he found that the fall in free cholesterol with incubation of plasma was indeed due to its esterification with long-chain fatty acid derived from phospholipid or triglyceride (Glomset *et al.*, 1962); then that there was an almost equimolar decrease in lecithin, and that the fatty acid composition of the newly synthesized cholesteryl ester reflected that of the fatty acid at the 2-position of the lecithin (Glomset, 1962). The high initial rate of the reaction suggested a major role for it in the turnover of plasma cholesterol and lecithin. Further studies demonstrated the importance of high-density lipoproteins (HDL) in the reaction (Glomset, 1963), and the importance of LCAT, as it was now called, in the metabolism of HDL (Glomset *et al.*, 1966). In 1964 Glomset and Wright introduced a radioassay for plasma LCAT activity using an excess of heat inactivated plasma containing [^{3}H]cholesterol as substrate. Despite its obvious importance this work, like that of Sperry, received relatively little initial recognition.

The situation changed dramatically in 1967. Norum and Gjone (from Oslo) described unusual biochemical findings in three sisters. Their plasma contained very little cholesteryl ester and a large amount of free cholesterol; the plasma lecithin was high while lysolecithin was low. All three had an elevated plasma triglyceride and the two older patients had lipaemic plasma. However, on paper electrophoresis there was no pre-β nor α band. These patients were found to lack plasma LCAT activity and this was felt to be the underlying abnormality. Norum and Gjone suggested that plasma cholesterol esterification must normally be a major source of plasma cholesteryl ester and showed, by feeding labelled cholesterol, that the small amount of cholesteryl ester present in the plasma of their patients was mainly of intestinal origin.

The clinical features in these patients, which typify subsequent findings in patients with familial LCAT deficiency, were presented in a separate paper (Gjone and Norum, 1968). The first patient had had moderate proteinuria from the age of 19 years. Fourteen years later she presented with general asthenia, ankle oedema and severe anaemia (Hb 7.6 g dL^{-1}) and was referred to the Rikshospitalet in Oslo where the only additional physical sign was 'marked lipoid arcus present in both corneae'.

Subsequently another 26 families have been found, in many parts of the world, with 50 members homozygous for LCAT deficiency (Gjone, 1987, personal communication). The four families identified in Norway all come from the same area (More) near Molde and Alesund in the north-west, where the frequency of heterozygous carriers is about 4%. It has been calculated that this results from a single mutational event occurring before the year 1700. It is a curious coincidence that the village of Glomset, the ancestral home of the Glomset family, lies in the middle of this area!

Eye changes

It was soon realized that the corneal abnormality was rather unusual (Gjone and Bergaust, 1969). The whole cornea appeared cloudy but with a more pronounced annular opacity near the limbus. In the parenchyma of the cornea there were innumerable minute grayish dots. Over the pupil they were evenly distributed in all layers but in the surrounding area they were concentrated at the back of the cornea. Near the limbus a much denser accumulation caused the 'arcus', but its outer border was less sharply defined than classical arcus lipoides senilis. Closer to the limbus there was a narrow zone of clear cornea without stromal dots. Despite these changes the patients had normal visual acuity.

Ultrastructural examination of sections removed by superficial keratectomy revealed numerous vacuoles, many containing electron dense 'membranous' deposits, in Bowman's layer and in the stroma (Bron *et al.*, 1975; Bethell *et al.*, 1975). The exact nature of the deposits is still not known but the tissue contains an excess of free cholesterol and phospholipid.

We now know that diffuse corneal opacity is a relatively common finding in genetic disorders with some form of HDL deficiency, e.g. in familial apoAI and CIII deficiency, Tangier disease, HDL deficiency with planar xanthomas and in fish-eye disease (see later) but it occurs only in homozygotes with these conditions (Schaefer, 1984). The mechanism of this lipid deposition is unknown.

Fundal changes have also been described in two patients with familial LCAT deficiency and include retinal haemorrhage, aneurysmal dilatation of the retinal veins, disc protrusion and rupture of Bruch's membrane resembling that seen in angioid streaks (Horven *et al.*, 1974). It is not clear whether these are really part of the syndrome.

Renal abnormalities

Most patients with familial LCAT deficiency have proteinuria of about 0.5–1.0 g L^{-1}; it is detectable early in life and persists. There is little reduction in serum albumin initially and indices of renal function remain normal for many years, as does the blood pressure. Eventually renal function deteriorates, often rapidly, and hypertension develops; proteinuria increases and there may then be hypoalbuminaemia. Because of renal failure dialysis and renal transplantation may be necessary.

Macroscopically the kidneys are slightly enlarged and rather pale. On light microscopy Bowman's capsule is thickened and the mesangium is prominent. There are foam cells in the glomerular tufts. The renal arteries and arterioles show subendothelial lipid deposits and a thickened basement membrane. Electron microscopy reveals capillary lumina partly filled with a meshwork of membranes and of particles with an amorphous mottled structure. The capillary walls are abnormal; endothelial cells may be missing or may be fenestrated, the basal lamina is of irregular thickness, foot processes are fused, and there are membrane surrounded particles (approximately 1000 Å in diameter) in both the subendothelial and subepithelial regions.

Glomeruli have been isolated from the kidneys of patients and have been found to contain a marked excess (5 × normal) of free cholesterol and of lecithin.

Haematological abnormalities

Most patients have a normochromic anaemia with a haemoglobin around $10 \, \mathrm{g \, dL^{-1}}$; target cells are prominent in the peripheral blood film. Red cell free cholesterol and lecithin are markedly increased, but total phospholipid is normal (so the cholesterol:phospholipid ratio is increased). Red cell half-life was reduced to about 16 or 17 days in two patients (normal 23–35); the spleen was not particularly active in red cell destruction (Glomset *et al.*, 1983).

Surprisingly platelet lipids were grossly normal in three patients (Nordoy and Gjone, 1971). In liver disease, in which the lipoprotein and red cell lipid changes are similar, the platelet cholesterol:phospholipid ratio is increased though not to the same extent as that of the red cells (Owen *et al.*, 1981).

The bone marrow suggests a combination of moderate haemolysis with a reduction in compensatory erythropoiesis. Giemsa stain reveals the presence in bone marrow and spleen of 'sea-blue histiocytes' in which the granules are composed of a lamellar arrangement of membranes which are presumably made up of cholesterol and lecithin (Jacobsen *et al.*, 1972). A small number of foam cells are also seen in the marrow and spleen as they are in other tissues.

Atherosclerosis

Early atherosclerosis has been found in the aorta and large arteries of many patients with familial LCAT deficiency. Sections show lipid accumulation, like that found in other organs, present in all layers of the vessel wall; foam cells are also present. Only 35% of the cholesterol is cholesteryl ester compared with about 75% in normal atheroma. Furthermore the fatty acid pattern of the cholesteryl ester is unusual; the oleic:linoleic ratio is markedly increased, reflecting the abnormal pattern seen in plasma cholesteryl ester (and that normally found in cells as a result of the action of acyl CoA:cholesterol acyltransferase).

Plasma lipoprotein changes (Glomset *et al.*, 1983)

There are striking abnormalities in the different lipoprotein fractions in familial LCAT deficiency. Very-low-density lipoprotein (VLDL) may be high and contain large amounts of free cholesterol but with a low total protein and particularly low amounts of apoCII and III. As the latter are the most electronegative components of normal VLDL this finding probably accounts for the slow electrophoretic mobility of VLDL in familial LCAT deficiency. Intermediate-density lipoprotein (IDL or LDL_1, d = 1.006–1.019) contains both normal-sized IDL and unusually large particles, whose composition is intermediate between that of VLDL of corresponding size and of LDL_2.

LDL_2 (d = 1.019–1.063) usually contains three subfractions on 2% agarose gel filtration (only one is present normally). The void volume contains particles with a multilamellar structure and a very high free cholesterol:phospholipid ratio of 2:1. Their concentration appears to be related to that of chylomicrons and both decrease when the patients consume fat free diets.

A second subfraction contains particles of 30–80 nm diameter: these have a disc-shaped structure on electron microscopy, tend to form stacks and are made up mainly of free cholesterol, phospholipid and apoC (mainly apoCI); they are similar to the Lp-X of biliary obstruction. It also contains spherical particles with triglyceride, cholesteryl ester and apoB and apoE, which resemble normal remnants of chylomicrons and VLDL.

In the third subfraction of LDL_2 there are spherical particles of the same size as normal LDL (20–22 nm). However they contain large amounts of triglyceride and their cholesteryl ester content is correspondingly reduced; their protein is apoB as in normal LDL.

High-density lipoprotein (HDL) particles are also of unusual shape and distribution. Some are disc-shaped and contain free cholesterol, lecithin and apoE or apoAI; they are similar to nascent HDL isolated from rat liver perfusates. The spherical HDL are only about 6 nm in diameter and contain free cholesterol, lecithin, a small amount of core lipid and apoAI (two molecules per 94 600 dalton particle).

The analysis of the various types of lipoprotein abnormality found in familial LCAT deficiency has proved of great value for our understanding of normal plasma lipid and lipoprotein metabolism (but space does not permit consideration of this topic in this paper). It has also helped to interpret the lipid and lipoprotein changes seen in liver disease. LCAT is produced in the liver and when the liver is damaged plasma LCAT activity falls, as it has a plasma half-life of only a few days. In patients with obstructive jaundice the lipoprotein changes observed are very similar to those seen in familial LCAT deficiency but only when plasma LCAT activity is low (Agorastos *et al.*, 1978). Some of the changes are seen in the patients with parenchymal liver disease (Day *et al.*, 1979) and again their presence appears related to low plasma LCAT activity. We believe that the lipoprotein changes of liver disease may have important pathophysiological consequences, because of their secondary effects on cell membranes (Owen *et al.*, 1984); if this is true then correction of the low LCAT activity may be beneficial.

Genetics and molecular biology

Familial LCAT deficiency clearly has an autosomal recessive mode of inheritance. Linkage of the LCAT deficiency gene was sought, in the first three Norwegian families, against many genetic markers; matching against α-haptoglobin (Hp) subtypes it was found that the gene was closely linked to the serum Hp locus, known to be located between the middle and end of the long arm of chromosome 16. By matching with the Hp[15] subtype a combined lod (log odds) score of 3.41 was obtained with a recombinant fraction of 0.00 (Teisberg and Gjone, 1974, 1981).

Working with Tata and Humphries at St. Mary's Hospital, London, and with Waterfield at ICRF, we purified human LCAT, sequenced some fragments, and then made an oligonucleotide probe with which we isolated some incomplete cDNA clones from an adult human cDNA liver library. Their DNA sequence allowed prediction of the entire amino acid sequence of mature LCAT (Tata *et al.*, 1987); this confirmed that of McLean and colleagues (1986a). The mature protein has 416 amino acids with several stretches of hydrophobic residues and four potential sites for glycosylation (plasma LCAT is a heavily glycosylated protein with carbohydrate making up about 25% of its mass).

McLean *et al.* (1986b) subsequently sequenced the whole LCAT gene. It has six exons spanning about 4200 base pairs, and exon 5 codes for amino acids homologous to the interfacial active site of several lipases (this is assumed to be the active site of the enzyme), and also for an amphipathic alpha helix resembling the carboxy terminus of apoE. Blot hybridization data suggested that only one LCAT gene is present in humans. The 1550 base LCAT mRNA was detected in liver and in HepG2 cells but not in small intestine, spleen, pancreas, placenta or adrenal tissue.

With groups in Paris and Munster (Azoulay *et al.*, 1987) we used plasmid pLCAT14B and Southern blotting techniques to locate the human LCAT gene using rodent × human somatic cell hybrids. The results were compatible with its presence on chromosome 16. We also did *in situ* hybridization studies adding the LCAT probe (labelled with ^{3}H) to human metaphase chromosome spreads from normal peripheral blood lymphocytes. Autoradiography revealed about 16% of the total number of grains associated with about 0.8% of the genome; they clustered around chromosomal segment 16q22, the region predicted on the basis of recombination studies to be the site of the LCAT deficiency gene. It seems likely, therefore, that the genetic abnormality in familial LCAT deficiency lies in the *structural* gene for LCAT, at least in those families in whom close linkage with Hp locus has been demonstrated.

There is, however, considerable genetic heterogeneity in homozygous patients with familial LCAT deficiency, in terms of LCAT mass and LCAT activity (Glomset *et al.*, 1983). The plasma of the Norwegian patients, which has no LCAT activity, appears to contain an immunologically identifiable LCAT protein, present at a concentration of about 10–20% of the LCAT protein present in normal subjects. The Sardinian patients have about 5–10% of the normal mass while in one Canadian family there appears to be no detectable LCAT mass in plasma.

We have studied four unrelated individuals with familial LCAT deficiency. Rocket electrophoresis, using a polyclonal antibody to LCAT, revealed circulating LCAT protein at a concentration about 20–40% lower than in normal subjects but it had essentially undetectable activity. In two patients the molecular weight of the plasma LCAT appeared normal at about 68 000. Restriction enzyme analysis of the LCAT gene in these patients, using many enzymes, showed no fragments distinguishable from those found in normal subjects. It did not appear therefore that the abnormality was caused by a large deletion or a major rearrangement of the LCAT gene sequences (Humphries *et al.*, 1988). The group in Oslo have also failed to find restriction enzyme polymorphisms in the Norwegian families with LCAT deficiency (Rogne *et al.*, 1987).

These results accord with more detailed studies of the gene sequence done on homozygotes from other families (McLean, personal communication). In a Japanese family with very low enzyme activity and LCAT mass there is an extra nucleotide close to the 5′ end of the first exon, at the position coding for proline 9 of LCAT. The resulting frame shift causes termination after amino acid 16. The remainder of the LCAT molecule either may not be secreted effectively from the hepatocyte or may be removed more rapidly from the circulation. In another Japanese family (LCAT activity about 8%, mass about 50% of normal) there is a point mutation at the beginning of exon 6 causing a methionine to isoleucine conversion. In a North American Indian family (12% activity and 27% mass) there is a point mutation in the same region causing an arginine to glycine conversion. The mechanism underlying the loss of activity is not clear as these mutations are not in the region presumed to contain the active site of the enzyme (i.e. exon 5). In another family in which there appears to be no LCAT mass in plasma there is no evidence of an mRNA transcript and the defect may therefore be in the promoter region. The mass measurements in all the above studies must, however, be interpreted with caution as they are determined immunologically, and an amino acid substitution may affect the affinity of the antibody for the LCAT molecule.

Advances in the molecular biology of LCAT give hope that the enzyme may eventually become available for the treatment of patients with familial LCAT deficiency, or for patients with other conditions in which there is secondary deficiency of the enzyme.

FISH-EYE DISEASE

Fish-eye disease is also a Scandinavian disease. In 1979 Carlson and Philipson reported their clinical and biochemical findings in two sisters from a village in northern Sweden. Both had obvious corneal opacities, as did their late father and sister. Fellow villagers said they had 'fish-eye disease' because their eyes resembled those of boiled fish. The index patient, at the age of 71 years, was referred to Carlson with hypertriglyceridaemia (which was also found in her elder sister).

Poor vision had developed late in life and was particularly troublesome in dark winter months. As in familial LCAT deficiency physical examination was otherwise unremarkable. The index patient had ST depression on an exercise ECG and later developed a left bundle branch block; the second patient and the father both suffered myocardial infarctions at the age of 76 years.

Subsequently a third, unrelated patient, also elderly, was identified as having fish-eye disease when she attended the ophthalmology department at the Karolinska Institute with failing vision.

The eye in fish-eye disease

In fish-eye disease both corneae are opaque and the irises visible only as indistinct shadows. With a biomicroscope small white–yellow dots, forming a mosaic pattern, are seen in all layers of the cornea. The peripheral cornea is most opaque; there is no distinct arcus but a thin yellow ring-shaped opacity is present superficially about 1 mm from the limbus.

Lipid and lipoprotein abnormalities

All three patients had fasting hypertriglyceridaemia, varying from 2.7 to 7.3 mmol L^{-1} (Carlson, 1982). The serum cholesterol was normal in the two sisters; in the third patient it was initially elevated but fell to normal after treatment of hypothyroidism with thyroxine. There was a moderate (four to five fold) increase in serum VLDL, but its composition was normal. It was later found (Forte and Carlson, 1984) that the mean VLDL particle diameter was greater than normal, mainly because the largest particles (up to 102 nm) were bigger than those seen in controls (up to 74 nm).

There was increased lipoprotein in the fraction d = 1.006–1.063. There was about three times as much IDL compared with normal subjects; it was triglyceride-rich, the triglyceride:cholesteryl ester molar ratio (c.1.1) being about twice normal but the particles appeared normal on electron microscopy. The concentration of LDL_2 was also increased and its triglyceride:cholesteryl ester molar ratio (0.61 and 0.73) was much greater than in controls (0.07–0.08). These LDL particles were slightly smaller on electron microscopy; some larger translucent structures (50–85 nm diameter) were also seen in this fraction (Forte and Carlson, 1984).

Striking abnormalities were found in HDL (d = 1.063–1.21). HDL cholesterol was reduced by 90% and HDL apoproteins by 80–90%, with greater reduction of apoAII than of apoAI (Carlson, 1982). With high resolution two-dimensional electrophoresis a normal amount of proapoAI was found but with markedly reduced amounts of the other apoAI isoforms; this suggested normal cellular apoAI synthesis in fish-eye disease but with impaired metabolic conversion of lipoproteins in plasma (Marshall *et al.*, 1985). Restriction enzyme analysis of the apoAI gene in two patients with fish-eye disease showed no evidence of major deletions or insertions (Rees *et al.*, 1984).

HDL contained almost exclusively HDL_3, but this migrated further on gradient gel electrophoresis than normal HDL_3. The apparent molecular weight of these particles was about 115000 (normal HDL_3 is about 180000). Their non-polar lipid content (triglyceride+cholesteryl ester) was only about 9% of their total volume due to a marked reduction in cholesteryl ester content. Cholesteryl ester was only about 20–26% of total cholesterol in this HDL, compared to about 76% in normal HDL, so there was an unusually high content of free cholesterol (Carlson and Holmquist, 1983).

The metabolism of VLDL, IDL and LDL in fish-eye disease was studied using lipoproteins in which apoB was labelled with radioactive iodine. There was a marked reduction in the rate of fractional conversion of VLDL-B to IDL-B and of

IDL-B to LDL-B, suggesting that the dyslipoproteinaemia resulted partly from accumulation in plasma of products of VLDL catabolism. *In vitro* studies revealed no defect in the enzyme, activator or substrate components of the lipoprotein lipase or hepatic lipase systems (Turner *et al.*, 1984). Furthermore there was no evidence of a deficiency of the plasma cholesteryl ester:triglyceride transfer protein (Calvert and Carlson, 1983). The total production of bile acids and the net steroid balance (reflecting total body cholesterol synthesis) were both within the normal range, and the biliary lipid composition and cholesterol saturation of bile were normal (Angelin and Carlson, 1986). However the ratio between the synthesis of cholic and chenodeoxycholic acids was 0.54 compared to a mean of 1.81 in controls; the significance of this observation is not clear.

Recently Mackness and colleagues (1987) assayed the plasma activity of the A-esterases which hydrolyse toxic organophosphate anticholinesterases. This activity is normally associated with HDL and in fish-eye disease the reduction in plasma paraoxonase activity (about 90%) was found to be similar to the reduction in HDL apoproteins. Of the paraoxonase activity present in fish-eye disease only about 15% was found in the HDL fraction compared with about 80% in controls. There is no known natural substrate for A-esterase activity and it is interesting to speculate whether it may be involved in some way in reactions involving the synthesis or hydrolysis of cholesteryl ester.

LCAT and fish-eye disease

Perhaps the most interesting findings in fish-eye disease are those relating to plasma LCAT activity. Plasma cholesterol esterification, measured *in vitro* by the method of Stokke and Norum and by the reduction in plasma free cholesterol over a period of 24 h, was found to be normal (Carlson, 1982; Carlson and Holmquist, 1985a), and the fatty acid composition of the cholesteryl ester of VLDL, LDL and HDL was almost identical in fish-eye disease to that of normal subjects; in familial LCAT deficiency there is a marked reduction in the proportion of linoleic acid. But when plasma LCAT activity was measured using an exogenous substrate, by the methods of Glomset and Wright and of Piran and Morin, LCAT activity was only about 10–15% of that found in normal plasma.

Isolated HDL from patients with fish-eye disease was found to be an excellent substrate for normal LCAT present in lipoprotein depleted plasma from control subjects; nearly all the free cholesterol was esterified and the small HDL particles were converted into larger particles similar in size to those of normal HDL. However, lipoprotein free plasma of patients with fish-eye disease was not able to esterify the free cholesterol of normal HDL or of fish-eye HDL (Carlson and Holmquist, 1985b). Clearly this made it difficult to explain the apparently normal cholesterol esterification occurring with incubation of plasma. In a more recent study it has been found that addition of a 'subnormal' amount of highly purified normal LCAT to fish-eye plasma leads to esterification of HDL cholesterol and normalization of the HDL, suggesting that the presence or absence of activators or inhibitors of LCAT is relatively unimportant in fish-eye disease (Holmquist and Carlson, in press).

J. Inher. Metab. Dis. 11 (1988)

In a further series of experiments Carlson and Holmquist (1985c) removed HDL from plasmas of normal subjects and patients with fish-eye disease by sequential ultracentrifugation and reconstitution, adding the fraction containing VLDL and LDL (d<1.063) to the lipoprotein free fraction (d>1.21). When these reconstituted fractions were incubated at 37°C there was esterification of the remaining free cholesterol of the combined VLDL and LDL fractions and the shapes of the cholesterol esterification rate curves were similar whether the reconstituted fractions were from controls or from patients with fish-eye disease. In cross incubation studies the lipoprotein free fraction from patients with fish-eye disease esterified the free cholesterol in VLDL and LDL of normal subjects: it also esterified free cholesterol in an isolated total lipoprotein fraction from a patient with familial LCAT deficiency; however, it did not appear to affect the LCAT-deficient HDL particles although it did increase the amount of LDL in the incubate.

On the basis of these findings Carlson and Holmquist suggested that two types of LCAT activity must be present in normal plasma – one with an action only on HDL (α-LCAT), and the other with an action only on VLDL and LDL (β-LCAT). They suggested that fish-eye disease was due to a deficiency of α-LCAT, and that classical familial LCAT deficiency was due to lack of both components. These results are difficult to interpret. In view of our current knowledge of the molecular biology of LCAT one would expect familial LCAT deficiency to be the consequence of an abnormality of only one polypeptide chain. If this polypeptide chain has two different actions, which can be differentially affected in two genetic conditions, one must ask whether it has two active sites or two binding sites (one for HDL and one for LDL/VLDL). Alternatively the two actions may result from differing post-translational modification of the enzyme, perhaps related to the degree of glycosylation. Fish-eye disease might then be due to failure of the modification resulting in α-LCAT activity. Not surprisingly, analysis of the LCAT gene from patients with fish-eye disease has revealed no evidence of restriction enzyme polymorphisms (Rogne *et al.*, 1987).

The study of fish-eye disease, like that of familial LCAT deficiency, clearly has important implications for our further understanding of plasma lipoprotein metabolism. It is certainly posing many questions about the action of LCAT, and the relationship between cholesterol esterification in HDL and the transfer of cholesteryl ester between HDL and VLDL and LDL. We must hope that in the near future some of these questions will be answered.

ACKNOWLEDGEMENTS

Work described in this paper has been supported by the Medical Research Council, the Wellcome Trust, the British Heart Foundation and by INSERM. I am grateful to Egil Gjone, Lars Carlson and John McLean for their help in the preparation of this paper.

REFERENCES

Agorastos, J., Fox, C., Harry, D. S. and McIntyre, N. Lecithin–cholesterol acyltransferase and the lipoprotein abnormalities of obstructive jaundice. *Clin. Sci. Mol. Med.* 54 (1978) 369–379

Angelin, B. and Carlson, L. A. Bile acids and plasma high-density lipoproteins: bilary lipid metabolism in fish-eye disease. *Eur. J. Clin. Invest.* 16 (1986) 157–162

Azoulay, M., Henry, I., Tata, F., Weil, D., Grezchik, K. H., Chaves, M. E., McIntyre, N., Williamson, R., Humphries, S. E. and Junien, C. The structural gene for lecithin: cholesterol acyltransferase (LCAT) maps to 16q22'. *Ann. Hum. Genet.* 51 (1987) 129– 136

Bethell, W., McCulloch, C. and Ghosh, M. Lecithin-cholesterol acyltransferase deficiency: light and electron microscopic findings from two corneas. *Can. J. Ophthalmol.* 10 (1975) 494–501

Bron, A. J., Lloyd, J. K., Fosbrooke, A. S., Winder, A. F. and Tripathi, R. C. Primary LCAT deficiency disease. *Lancet* 1 (1975) 928–929

Calvert, G. D. and Carlson, L. A. Plasma lipid transfer in fish-eye disease. *Acta Med. Scand.* 213 (1983) 253–254

Carlson, L. A. and Philipson, B. Fish-eye disease: a new familial condition with massive corneal opacities and dyslipoproteinaemia. *Lancet* 2 (1979) 921–924

Carlson, L. A. Fish-eye disease: a new familial condition with massive corneal opacities and dyslipoproteinaemia. Clinical and laboratory studies in two afflicted families. *Eur. J. Clin. Invest.* 12 (1982) 41–53

Carlson, L. A. and Holmquist, L. Studies on high density lipoproteins in fish-eye disease. *Acta Med. Scand.* 213 (1983) 177–182

Carlson, L. A. and Holmquist, L. Paradoxical elevation of plasma cholesterol in fish-eye disease. *Acta Med. Scand.* 217 (1985a) 491–499

Carlson, L. A. and Holmquist, L. Evidence for deficiency of high density lipoprotein lecithin:cholesterol acyltransferase activity (α-LCAT) in fish-eye disease. *Acta Med. Scand.* 218 (1985b) 189–196

Carlson, L. A. and Holmquist, L. Evidence for the presence in human plasma of lecithin: cholesterol acyltransferase activity (β-LCAT) specifically esterifying free cholesterol of combined preβ and β-lipoproteins. Studies of fish-eye disease patients and control subjects. *Acta Med. Scand.* 218 (1985c) 197–205

Day, R. C., Harry, D. S., Owen, J. S. and McIntyre, N. Lecithin–cholesterol acyltransferase and the lipoprotein abnormalities of parenchymal liver disease. *Clin. Sci.* 56 (1979) 575– 583

Forte, T. M. and Carlson, L. A. Electron microscopic structure of serum lipoproteins from patients with fish-eye disease. *Arteriosclerosis* 4 (1984) 130–137

Gjone, E. and Bergaust, B. Corneal opacity in familial plasma cholesterol ester deficiency. *Acta Ophthalmol.* 47 (1969) 222–227

Gjone, E. and Norum, K. R. Familial serum cholesterol ester deficiency: clinical study of a patient with a new syndrome. *Acta Med. Scand.* 183 (1968) 107–112

Glomset, J. A. The mechanism of the plasma cholesterol esterification reaction: plasma fatty acid transferase. *Biochim. Biophys. Acta* 65 (1962) 128–135

Glomset, J. A. Further studies on the mechanism of the plasma cholesterol esterification reaction. *Biochim. Biophys. Acta* 70 (1963) 389–395

Glomset, J. A., Parker, F., Tjaden, M. and Williams, R. H. The esterification *in vitro* of free cholesterol in human and rat plasma. *Biochim. Biophys. Acta* 58 (1962) 398–406

Glomset, J. A. and Wright, J. L. Some properties of a cholesterol esterifying enzyme in human plasma. *Biochim. Biophys. Acta* 89 (1964) 266–276

Glomset, J. A., Janssen, E. T., Kennedy, R., and Dobbins, J. Role of plasma lecithin– cholesterol acyltransferase in the metabolism of high density lipoproteins. *J. Lipid. Res.* 7 (1966) 639–648

Glomset, J. A., Norum, K. R. and Gjone, E. Familial lecithin–cholesterol acyltransferase deficiency. In Stanbury, J. B., Wyngaarden, J. B., Fredrickson, D. S., Goldstein, J. L., and Brown, M. S., (eds.) *The Metabolic Basis of Inherited Disease*, 5th edition, McGraw-Hill, New York, 1983, pp. 643–654

Holmquist, L. and Carlson, L. A. *In vitro* normalization of cholesteryl ester content and size of high density lipoprotein particles in fish-eye disease plasma by purified normal human lecithin:cholesterol acyltransferase. *Lipids* 22 (1987) 305–311

Horven, I., Egge, K. and Gjone, E. Corneal and fundus changes in familial LCAT deficiency. *Acta Ophthalmol.* 52 (1974) 201–210

Humphries, S. E., Chaves, M. E., Tata, F., Lima, V. L. M., Owen, J. S., Borysiewicz, L. K., Catapano, A., Vergani, C., Gjone, E., Clemens, M. R., Williamson, R. and McIntyre, N. A study of the structure of the gene for lecithin:cholesterol acyltransferase (LCAT) in four unrelated individuals with familial LCAT deficiency. *Clin. Sci.* (in press)

Jacobsen, C. D., Gjone, E. and Hovig, T. Sea-blue histiocytes in familial lecithin–cholesterol acyltransferase deficiency. *Scand. J. Haematol.* 9 (1972) 106–113

Mackness, M. I., Walker, C. H. and Carlson, L. A. Low A-esterase activity in serum of patients with fish-eye disease. *Clin. Chem.* 33 (1987) 587–588

McLean, J., Fielding, C., Drayna, D., Dieplinger, H., Baer, B., Kohr, W., Henzel, W. and Lawn, R. Cloning and expression of the human lecithin–cholesterol acyltransferase cDNA. *Proc. Natl. Acad. Sci. USA* 83 (1986a) 2335–2339

McLean, J., Wion, K., Drayna, D., Fielding, C. and Lawn, R. Human lecithin–cholesterol acyltransferase gene: complete gene sequence and sites of expression. *Nucl. Acid Res.* 14 (1986b) 9387–9406

Marshall, T., Williams, K. M., Holmquist, L., Carlson, L. A. and Vesterberg, O. Plasma apolipoprotein pattern in fish-eye disease examined by high-resolution two-dimensional electrophoresis. *Clin. Chem.* 31 (1985) 2032–2035

Nordoy, A. and Gjone, E. Familial plasma lecithin–cholesterol acyltransferase deficiency: a study of the platelets. *Scand. J. Clin. Lab. Invest.* 27 (1971) 263–268

Norum, K. R. and Gjone, E. Familial plasma lecithin–cholesterol acyltransferase deficiency. *Scand. J. Clin. Lab. Invest.* 20 (1967) 231–243

Owen, J. S., Hutton, R. A., Day, R. A., Bruckdorfer, K. R. and McIntyre, N. Platelet lipid composition and platelet aggregation in human liver disease. *J. Lipid Res.* 22 (1981) 423–430

Owen, J. S., McIntyre, N. and Gillett, M. P. T. Lipoproteins, cell membranes and cellular functions. *Trends Biochem. Sci.* 9 (1984) 238–242

Piran, U. and Morin, R. J. A rapid radioassay procedure for plasma lecithin–cholesterol acyltransferase. *J. Lipid Res.* 20 (1979) 1040–1043

Rees, A., Stocks, J., Schoulders, C., Carlson, L. A., Baralle, F. E. and Galton, D. J. Restriction enzyme analysis of the apolipoprotein A-1 gene in fish-eye disease and Tangier disease. *Acta Med. Scand.* 215 (1984) 235–237

Rogne, S., Skretting, G., Larsen, L., Myklebost, O., Mevag, B., Carlson, L. A., Holmquist, L., Gjone, E. and Prydz, H. The isolation and characterization of a cDNA clone for human lecithin–cholesterol acyl transferase and its use to analyse the genes in patients with LCAT deficiency and fish-eye disease. *Biochem. Biophys. Res. Commun.* 148 (1987) 161–169

Schaefer, E. J. Clinical, biochemical, and genetic features in familial disorders of high density lipoprotein deficiency. *Arteriosclerosis* 4 (1984) 303–322

Sperry, W. M. Cholesterol esterase in blood. *J. Biol. Chem.* 111 (1935) 467–478

Stokke, K. T. and Norum, K. R. Determination of lecithin–cholesterol acyltransfer in human blood plasma. *Scand. J. Lab. Clin. Invest.* 27 (1971) 21–27

Tata, F., Chavez, E., Markham, A. F., Scrace, G. D., Waterfield, M. D., McIntyre, N., Williamson, R. and Humphries, S. E. The isolation and characterisation of cDNA and genomic clones for human lecithin:cholesterol acyltransferase. *Biochim. Biophys. Acta* (in press)

Teisberg, P. and Gjone, E. Probable linkage of LCAT locus in man to the α-haptoglobin locus on chromosome 16. *Nature* 249 (1974) 550–551

Teisberg, P. and Gjone, E. Genetic heterogeneity in familial lecithin cholesterol (LCAT) deficiency. *Acta Med. Scand.* 210 (1981) 1–2

Turner, P. R., Carlson, L. A., Cortese, C., Rao, S., Marenah, C. B., Miller, N. E. and Lewis, B. Studies of lipoprotein metabolism in a patient with fish-eye disease. *Eur. J. Clin. Invest.* 14 (1984) 273–277

J. Inher. Metab. Dis. 11 Suppl. 1 (1988) 57–73

Biochemical, Clinical, Genetic and Metabolic Studies of Hyperapo-β-lipoproteinaemia

P. O. Kwiterovich Jr.
Lipid Research-Atherosclerosis Unit, Departments of Pediatrics and Medicine, The Johns Hopkins University School of Medicine, 600 North Wolfe Street, Baltimore, Maryland 21205, USA

Summary: Hyperapo-β-lipoproteinaemia is a common lipoprotein disorder characterized by an elevated plasma level of the major apolipoprotein, B (apoB) of low-density β lipoproteins (LDL), combined with a low ratio of LDL cholesterol to LDL apoB. Hyperapo-β-lipoproteinaemia is due to the overproduction of LDL apoB that results from an enhanced synthesis of very low-density (pre-β) lipoprotein (VLDL) in liver. The plasma levels of high-density (α) lipoprotein (HDL) and its major apolipoprotein, A-I, are often low in hyperapo-β-lipoproteinaemia. Hyperapo-β-lipoproteinaemia is often familial and aggregates in children and adults from families with premature coronary artery disease. The precise defect(s) that cause hyperapo-β-lipoproteinaemia are not known. In a family with premature coronary artery disease and hyperapo-β-lipoproteinaemia, a mutation in codon 4046 in exon 29 of the apolipoprotein B gene, a CGG to TGG transition produced a change from arginine, a positively charged amino acid, to tryptophan, a hydrophobic amino acid, at position 4,019 of the mature apolipoprotein B protein. Decreased incorporation of free fatty acids into triglycerides of adipocytes has been described *in vitro*, and *in vivo* studies suggested a defect in clearance of postprandial lipoproteins associated with decreased uptake of plasma free fatty acids.

INTRODUCTION

An elevated concentration of plasma total cholesterol is strongly associated with increased risk of coronary artery disease (Kannel *et al.*, 1979). The major carrier of cholesterol in blood is low-density (β) lipoprotein (LDL). LDL is a large spherical lipoprotein in which the principal lipid, cholesteryl ester, is contained in its core surrounded by a surface layer made up of unesterified cholesterol, phospholipid and the major protein of LDL, apolipoprotein B. LDL contains only small amounts of triglyceride in its core.

Until recently, the plasma level of LDL was usually determined by measuring its cholesterol content and expressing the concentration of LDL as LDL cholesterol (LDL C). As is now well recognized, LDL vary in size, hydrated density, molecular

Journal of Inherited Metabolic Disease. ISSN 0141–8955. Copyright © SSIEM and MTP Press Limited, Queen Square, Lancaster, UK.

weight and chemical composition. LDL composition may be influenced by a number of genetic, metabolic and therapeutic factors (Kwiterovich *et al.*, 1987). While the content of cholesteryl ester within the core of LDL may increase or decrease, each LDL particle contains only one molecule of apolipoprotein B, a large polypeptide of molecular weight 550 000 Da. The measurement of apolipoprotein B in LDL, expressed as LDL apoB, thus provides a more accurate assessment of the number of LDL particles than measurement of LDL C alone. LDL apoB is usually measured by immunochemical methods, and, in the studies to be described, LDL apoB was determined using a radial immunodiffusion assay (Sniderman *et al.*, 1980; Kwiterovich *et al.*, 1987). Finally, measurement of both LDL C and LDL apoB provides an assessment of LDL composition. Larger, cholesterol-enriched LDL particles have a high LDL C/LDL apoB ratio (>1.6); smaller denser LDL particles have a low LDL C to LDL apoB ratio (<1.3); in normal humans, LDL C to LDL apoB ratio usually ranges between 1.3 and 1.6.

Previously, we have found that many patients with coronary artery disease have elevated plasma concentrations of LDL apoB, with normal or near normal LDL C levels (Sniderman *et al.*, 1980, 1982). A high plasma LDL apoB concentration and a low LDL C to LDL apoB ratio reflect the presence of increased numbers of smaller denser LDL particles which contain less core cholesteryl ester than normal and which are relatively enriched in LDL apoB. We named this lipoprotein phenotype hyperapo-β-lipoproteinaemia. This paper reviews certain biochemical, clinical, genetic and metabolic studies that have been performed in patients with hyperapo-β-lipoproteinaemia.

CLINICAL STUDIES

Patients with coronary atherosclerosis

Hyperapo-β-lipoproteinaemia was initially described in 100 patients undergoing elective diagnostic coronary arteriography (Sniderman *et al.*, 1980). Fifty-nine patients had coronary artery disease (>50% stenosis in at least one major coronary artery). The mean plasma LDL apoB concentration in this group of 118 mg/dl was significantly higher than that of 82 mg/dl in the 41 patients without coronary artery disease (Table 1). The mean LDL apoB concentration in the patients without coronary artery disease was very similar to that in three unrelated control groups (Table 1). The mean LDL cholesterol concentration in those with coronary artery disease of 134 mg/dl* was significantly higher than that of 112 mg/dl in those without coronary artery disease. The mean LDL apoB concentrations of those with and without coronary artery disease differed by two standard deviations, while the mean LDL C concentrations differed by one standard deviation. Thus, the LDL apoB plasma level was a greater discriminator for coronary artery disease than the LDL C level.

Familial hypercholesterolaemia (McKusick 14389) is a Mendelian dominant disorder characterized by markedly elevated plasma total and LDL cholesterol

---

*To convert cholesterol in mg/dl to mmol/L, multiply the result by 0.026.

Table 1 **Plasma LDL cholesterol and LDL apoB protein levels in patients undergoing coronary arteriography, unrelated controls, and in subjects with heterozygous familial hypercholesterolaemia**

	Plasma levels (mg/dl)			
	LDL cholesterol		LDL apoB	
Group (number of subjects)	Mean	SD	Mean	SD
Patients undergoing coronary arteriography				
Without coronary artery disease (41)	112	30	82	15
With coronary artery disease (59)	134	27	118	22
Unrelated controls				
Normal coronary arteries (30)	114	25	84	13
Male physicians (35)	118	15	83	11
Medical students (90)	102	28	72	17
Heterozygotes for familial hypercholesterol-aemia (40)	250	55	138	20

Data are from Sniderman *et al.* (1980)

concentrations due to a defect in the LDL receptor (Goldstein and Brown, 1982). We also studied 40 familial hypercholesterolaemia heterozygous patients and found that their mean LDL apoB concentration of 138 mg/dl was similar to that in the patients with coronary artery disease (Table 1). However, the mean LDL cholesterol of 250 mg/dl in those with familial hypercholesterolaemia was considerably higher than the 134 mg/dl found in those with coronary artery disease (Table 1).

About 50% of those with coronary artery disease had a LDL apoB concentration above 120 mg/dl, the upper limit of normal. Thus, the LDL apoB concentration in many patients with coronary artery disease was as high as that in the patients with familial hypercholesterolaemia. In contrast, the LDL cholesterol levels in those with coronary artery disease were usually between 130 mg/dl and 160 mg/dl, within the normal range. Those with familial hypercholesterolaemia had markedly elevated levels of LDL cholesterol, between 200 and 320 mg/dl. Thus, two phenotypes of elevated LDL levels were apparent: in one phenotype, hyperapo-β-lipoprotein-aemia, there were increased numbers of smaller denser LDL particles (elevated LDL apoB level, low LDL C to LDL apoB ratio); in the other phenotype, type II hyperlipoproteinemia, found in the familial hypercholesterolaemia heterozygotes, there were increased numbers of larger less-dense LDL particles (elevated LDL apoB, high LDL C to LDL apoB ratio).

Survivors of myocardial infarction

One hundred patients who had previously suffered a myocardial infarction were next studied (Sniderman *et al.*, 1982). Their mean age was 55 years, 78% were male and they were studied at least three months postinfarction. The mean plasma total cholesterol concentration in this group was 227 mg/dl, not markedly elevated

Table 2   Plasma lipid and LDL apoB protein levels in survivors of myocardial infarction

	Plasma levels (mg/dl)					
	Total cholesterol		LDL apoB		Total triglyceride	
Group (number of subjects)	Mean	SD	Mean	SD	Mean	SD
All patients (100)	227	41	151	34	255	159
Hypertriglyceridaemics (53)	236	41	155	36	350	158
Normotriglyceridaemics (47)	212	37	145	32	143	37

Data from Sniderman et al. (1982)

(Table 2). In contrast, the mean LDL apoB concentration was 151 mg/dl and 75% of these patients had hyperapo-β-lipoproteinaemia. The mean total triglyceride concentration was 255 mg/dl*. When the group was divided into those with hypertriglyceridaemia (defined as a plasma triglyceride concentration above 200 mg/dl) and those without hypertriglyceridaemia, the mean LDL apoB concentration in both subgroups was significantly elevated; 43 of 53 hypertriglyceridaemics, and 32 of 47 normotriglyceridaemics had hyperapo-β-lipoproteinaemia. This study indicated that the hyperapo-β-lipoproteinaemia phenotype was very common in postinfarction patients, and that patients with hyperapo-β-lipoproteinaemia may be either hypertriglyceridaemic or normotriglyceridaemic.

Hypertriglyceridaemia and hyperapo-β-lipoproteinaemia

In the fasted state, most of the plasma triglyceride is carried on very low-density (pre-β) lipoproteins (VLDL). The finding that hypertriglyceridaemia was present in some patients with the hyperapo-β-lipoproteinaemia phenotype indicated that an elevated VLDL level often accompanies the increased number of small dense LDL particles. To study this further, we determined the LDL apoB concentrations in 162 patients with type IV lipoprotein patterns (elevated VLDL cholesterol, normal LDL C level) (Sniderman et al., 1982). When the LDL apoB data were analysed by cluster analysis (a statistical technique of random sampling design which can divide a group into two such that the least overlap occurs), the means providing the greatest separation were plasma LDL apoB concentrations of 84 mg/dl and 129 mg/dl. On the basis of these two means, two subgroups, normal LDL apoB and hyperapo-β-lipoproteinaemia, were selected (Table 3). The group with normal LDL apoB concentrations made up approximately the lower quartile, and those with elevated LDL apoB concentrations the upper quartile of LDL apoB distribution. The clinical manifestation of atherosclerotic disease was five times more prevalent in those with hyperapo-β-lipoproteinaemia than in the normal LDL apoB subgroup (10 versus 2 cases) ($p < 0.01$ after age correction). Thus, within this group of patients selected because they had type IV lipoprotein patterns, the LDL apoB concentration was a potent predictor of coronary artery disease.

*To convert triglyceride in mg/dl to mmol/L, multiply the result by 0.0113.

Table 3 Plasma lipid, LDL cholesterol and LDL apoB protein levels in patients with endogenous hypertriglyceridaemia (type IV lipoprotein pattern)

	Plasma levels (mg/dl)							
	Total cholesterol		LDL cholesterol		LDL apoB		Total triglyceride	
Groups (number of subjects)	Mean	SD	Mean	SD	Mean	SD	Mean	SD
All patients (162)	226	40	129	35	108	28	289	160
Normal LDL apoB (36)	197	35	94	26	69	26	370	256
Hyperapo-β-lipo-proteinaemia (38)	254	35	157	25	144	12	266	115

Data from Sniderman *et al.* (1982)

All but one of the hyperapo-β-lipoproteinaemia patients in this study had a plasma triglyceride concentration less than 500 mg/dl, while 12 of the 36 in the normal LDL apoB group had triglyceride concentrations between 500 mg/dl and 1000 mg/dl (Kwiterovich and Sniderman, 1983). Thus, the patients with moderate hypertriglyceridaemia had hyperapo-β-lipoproteinaemia and coronary artery disease, while those with more severe hypertriglyceridaemia did not. This may explain why the plasma triglyceride level is not usually found to be an independent predictor of coronary artery disease in epidemiologic studies. Hypertriglyceridaemia may be a marker for underlying risk of coronary artery disease when it is accompanied by hyperapo-β-lipoproteinaemia.

PHYSICOCHEMICAL STUDIES

Chemical composition

The chemical basis for the disproportionate levels of LDL C and LDL apoB in plasma, characteristic of hyperapo-β-lipoproteinaemia was examined (Teng *et al.*, 1983). LDL subfractions were isolated by density gradient ultracentrifugation from 10 normal subjects, 20 patients with hyperapo-β-lipoproteinaemia (10 normotriglyceridaemic and 10 hypertriglyceridaemic) and 7 patients with familial hypercholesterolaemia. LDL was separated into two major subfractions, called fraction 1 (light LDL) and fraction 2 (heavy LDL). Patients with hyperapo-β-lipoproteinaemia had a greater percentage of their LDL in the heavy LDL fraction (Table 4). The ratio of LDL C to LDL apoB in those with hyperapo-β-lipoproteinaemia was low: the ratio was the lowest (0.88) in those with hypertriglyceridaemic hyperapo-β-lipoproteinaemia (Table 4). The patients with familial hypercholesterolaemia contained a greater percentage of their LDL in the light LDL fraction than the normal and hyperapo-β-lipoproteinaemia groups. The ratio of LDL C to LDL apoB in the light LDL particles was 1.8 in familial hypercholesterolaemia. Those with hyperapo-β-lipoproteinaemia, therefore, had a greater number of denser,

Table 4 Distribution and chemical composition of two ultracentrifugal subfractions of LDL

	Fraction 1 (light)		Fraction 2 (heavy)	
	Percent of LDL apoB	*Ratio of LDL C/ LDL apoB*	*Percent of LDL apoB*	*Ratio of LDL C/ LDL apoB*
Group (number of subjects)				
Normal (10)	34	1.53	54	1.33
Hyperapo-β-lipoproteinaemia				
Normotriglyceridaemic (10)	30	1.52	62	1.16
Hypertriglyceridaemic (10)	30	1.44	59	0.88
Familial hypercholesterolaemic heterozygote (6)	45	1.81	46	1.23

Data are from Teng *et al.* (1983)

cholesterol-depleted LDL particles, a phenotype that was more pronounced in those with hypertriglyceridaemic hyperapo-β-lipoproteinaemia than in those with normotriglyceridaemic hyperapo-β-lipoproteinaemia. In familial hypercholesterolaemia there was a shift toward a cholesterol-enriched LDL particle, reflecting an increased number of larger more buoyant molecules. The major changes in the chemical composition of these different LDL subfractions were due to changes in the content of the core cholesteryl ester, accompanied by relative changes in the apolipoprotein B content (data not shown) (Teng *et al.*, 1983).

Physicochemical characteristics

The physicochemical characteristics of LDL from patients with the hyperapo-β-lipoproteinaemia phenotype have been studied (Kwiterovich *et al.*, 1987). Using analytical ultracentrifugation, the average molecular weight of hyperapo-β-lipoproteinaemia LDL was found to be lower (about 1.9–2.0×10^6 Da) than normal (about 2.5×10^6 Da). The fact that hyperapo-β-lipoproteinaemia LDL is smaller than normal LDL has also been confirmed by electron microscopy and gradient gel electrophoresis (data not shown). The physicochemical properties of hyperapo-β-lipoproteinaemia LDL have been correlated using equilibrium density gradient ultracentrifugation, analytical ultracentrifugal and chemical analysis (Kwiterovich *et al.*, 1987). The cholesteryl ester content of LDL is directly related to molecular weight and rate of flotation, while the apolipoprotein B content is inversely correlated with these two characteristics. To summarize, cholesteryl ester-depleted, relatively apolipoprotein B-enriched LDL particles have a lower average molecular weight and size, and slower ultracentrifugal flotation rates.

METABOLIC STUDIES

LDL is derived primarily from the metabolism of its triglyceride-rich precursor, VLDL (Figure 1). VLDL are synthesized in the liver where its major lipid constituent, triglyceride, is packaged along with a smaller amount of cholesteryl ester, into

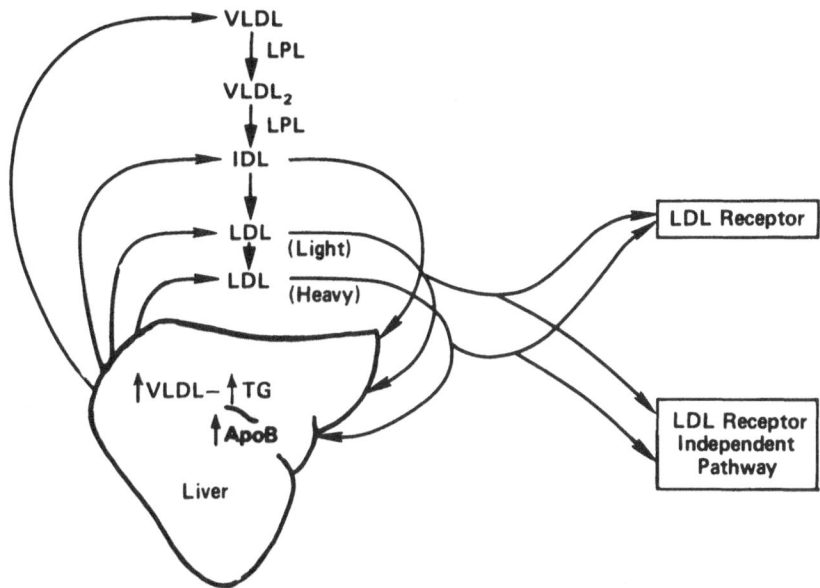

Figure 1 Metabolic scheme for the metabolism of VLDL and the production of LDL subfractions, LDL light and LDL heavy

the core of VLDL, which is surrounded by the two major apolipoproteins of VLDL, apolipoprotein B and apolipoprotein E. Following secretion from the liver into the vascular compartment, the core triglyceride of VLDL is hydrolysed by lipoprotein lipase, first producing VLDL remnants ($VLDL_2$) and then IDL (intermediate-density lipoproteins) (Figure 1). Some IDL particles are removed directly by the liver through the interaction of apolipoprotein E with the LDL (B,E) receptor. Other particles do not enter the liver but are converted into LDL. LDL are then removed through the interaction of apolipoprotein B with the LDL (B,E) receptor on liver or peripheral cells (Figure 1). Some LDL particles are removed through the LDL receptor-independent pathway (Figure 1).

In familial hypercholesterolaemia the increase in plasma of cholesterol-enriched LDL particles results primarily from a defect in the LDL receptor (removal defect). The metabolic basis underlying the increase in plasma of cholesterol-depleted LDL particles in hyperapo-β-lipoproteinaemia might result from a defect in overproduction of LDL apoB (overproduction defect). It was also necessary to establish whether the heavy LDL subfraction in hyperapo-β-lipoproteinaemia was derived from the light LDL subfraction or whether the heavy LDL was secreted directly by the liver (Figure 1).

Teng and co-workers (1986) studied the turnover of plasma apolipoprotein B in VLDL, IDL, LDL, and the light and heavy subfractions of LDL in 7 patients with hyperapo-β-lipoproteinaemia, 6 normolipidaemic controls and 5 patients with heterozygous familial hypercholesterolaemia. After receiving an injection of [125I]VLDL, hyperapo-β-lipoproteinaemic patients were found to have a higher

Table 5   Kinetics of apoB turnover in VLDL

Group (number)	Rate of synthesis mg/kg/day		Fractional cata- bolic rate (FCR)		Pool size (mg)	
	Mean	SD	Mean	SD	Mean	SD
Hyperapo-β- lipoproteinaemia (5)	40.1	9.2*	0.230	0.088†	598	254†
Normals (6)	21.5	3.3	0.366	0.017	191	25

Data from Teng *et al.* (1986)
*$p<0.05$, †$p<0.01$

Table 6   Kinetics of apoB turnover in LDL

Group (number)	Rate of synthesis mg/kg/day		Fractional cata- bolic rate (FCR)		Pool size (mg)	
	Mean	SD	Mean	SD	Mean	SD
Hyperapo-β- lipoproteinaemia (5)	23.1	5.1†	0.386	0.076	4304	767*
Normal (6)	13.0	1.7	0.366	0.030	2732	515
FH heterozygotes (5)	22.7	6.0†	0.192	0.039†	8000	1262†

Data from Teng *et al.* (1986)
*$p<0.01$, †$p<0.001$

rate of synthesis of VLDL apoB than controls but a reduced fractional catabolic rate (Table 5). After receiving an injection of [^{131}I]LDL, hyperapo-β-lipoproteinaemic patients had higher rates of LDL apoB synthesis than controls (Table 6). The familial hypercholesterolaemic patients also had higher rates of LDL apoB synthesis than normal. However, the fractional catabolic rate of LDL was similar in the hyperapo-β-lipoproteinaemic patients and controls, but the fractional catabolic rate was markedly decreased in the familial hypercholesterolaemic patients (Table 6).

Most subjects exhibited precursor–product relationships between VLDL and IDL, and all did between IDL and light LDL (data not shown). A precursor–product relationship between light and heavy LDL was evident in most hyperapo-β-lipoproteinaemic patients and controls. Such a precursor–product relation was not found in the familial hypercholesterolaemic patients. Patients with hyperapo-β-lipoproteinaemia had increased synthesis of light with normal conversion into heavy LDL; in contrast, in those with familial hypercholesterolaemia, the conversion of light LDL into heavy LDL was reduced and there was an independent synthesis of heavy LDL (Teng *et al.*, 1986).

In both normals and hyperapo-β-lipoproteinaemic patients, the heavy LDL was removed at a slower rate than light LDL. Thus, while there is no apparent defect in the LDL receptor in hyperapo-β-lipoproteinaemia, the shift to the denser LDL particles is not advantageous since dense LDL has a longer residence time in plasma than light LDL. Such depletion of cholesterol in LDL in hyperapo-β-lipoproteinaemia is felt to be secondary to enhanced transfer of cholesteryl esters from LDL to VLDL, secondary to enhanced VLDL secretion (Sniderman and Kwiterovich, 1987). It is not known if dense LDL has less affinity for the LDL receptor than light LDL, or if dense LDL are more readily taken by the LDL receptor-independent pathway (Figure 1).

In summary, the increased concentration of LDL apoB in hyperapo-β-lipoproteinaemia is due to increased LDL synthesis, which is secondary to increased VLDL synthesis. In contrast to familial hypercholesterolaemia LDL is removed from plasma at a normal rate in hyperapo-β-lipoproteinaemia.

GENETIC STUDIES

Human genetic studies

We studied LDL apoB concentrations in 240 members of an Amish pedigree, ascertained through a 12-year-old boy who died suddenly from coronary artery disease (Beaty *et al.*, 1986). This family was found to have two traits: sitosterolaemia (McKusick 21025), which was inherited as a Mendelian recessive, and hyperapo-β-lipoproteinaemia, which was inherited as a Mendelian dominant. The results of one of the complex segregation ratio analyses performed in this family, namely,

Table 7   Segregation analysis of the ratio of LDL sterol to LDL apoB protein in 230 relatives in an Amish pedigree

Model	Estimated parameters				
	Mean	Standard deviation	Frequency	Heritability	Relative log likelihood
Sporadic	1.15	0.189	—	—	0.0
Polygenic	1.11	0.128	—	0.559	5.26
Mendelian	1.08	0.151	0.867	—	5.46
	1.23	0.151			
	1.54	0.151			
Mixed	1.08	0.114	0.930	0.528	7.99
	1.27	0.114			
	1.89	0.114			

Data from Beaty *et al.* (1986)

the ratio of LDL C to LDL apoB are summarized in Table 7. A combined Mendelian dominant–polygenic model best fits the data, at a high level of statistical significance. Similar observations were made for the mean level of LDL apoB (data not shown). In this family, both polygenic and single gene factors influenced the concentration of plasma LDL apoB and the ratio of LDL C to LDL apoB.

The familial aggregation of hyperapo-β-lipoproteinaemia and its expression in children and young adults was shown in a Montreal study of 66 offspring of 24 families where the index parents had both hyperapo-β-lipoproteinaemia and premature coronary artery disease (before the age of 55 years) (Sniderman *et al.*, 1986). Mean plasma total and LDL cholesterol, LDL apoB and triglyceride concentrations were significantly higher in the offspring than in 207 unrelated

Table 8 Plasma lipids, lipoproteins and LDL apoB levels (mg/dl) in offspring of parents with premature CAD and hyperapo-β-lipoproteinaemia

	Offspring (n = 66)		Controls (n = 208)	
	Mean	SD	Mean	SD
Age (years)	17	7	20	8
Total cholesterol	174	34†	164	27
Total triglyceride	103	58‡	81	32
LDL C	103	33*	93	27
HDL C	51	13	52	11
LDL apoB	110	36‡	82	18

Data from Sniderman *et al.* (1985)
$*p<0.05$, $†p<0.01$, $‡p<0.001$

controls (Table 8). Plasma LDL apoB best differentiated the offspring from the normal controls. Eighteen of the 66 children of the affected hyperapo-β-lipoproteinaemic parents also had hyperapo-β-lipoproteinaemia. Most of the hyperapo-β-lipoproteinaemic offspring were beyond puberty, suggesting some delayed expression of hyperapo-β-lipoproteinaemia

In a separate study in Baltimore, 20 paediatric probands (<20 years of age) were ascertained who had a hyperapo-β-lipoproteinaemic phenotype (Kwiterovich *et al.*,

Table 9 Plasma lipid, lipoprotein cholesterol and LDL apoB levels in paediatric hyperapo-β-lipoproteinaemic probands

Probands	Total cholesterol	Total triglycerides	HDL C	LDL C	LDL apoB	LDL C/ LDL apoB
Males	219±24*	120±94	52±11	144±25	145±22	1.01±0.17
(n = 9)	(187–267)	(54–371)	(39–69)	(104–199)	(119–180)	(0.74–1.20)
Females	220±34	131±93	47±11	153±36	146±18	1.04±0.16
(n = 11)	(162–277)	(40–404)	(26–67)	(88–217)	(126–179)	(0.70–1.24)

*mean±1 SD (range) in mg/dl
Data from Kwiterovich *et al.* (1988)

1988) (Table 9). About half of the parents of these children had hyperapo-β-lipoproteinaemia (Table 10). However, further segregation ratio analysis showed that only 8 of 20 parental matings were of the type: hyperapo-β-lipoproteinaemia×normal (Table 11). In 6 of the 20 matings, both parents were normal. Thus, the basis for the hyperapo-β-lipoproteinaemic phenotype is undoubtedly heterogeneous and is under the influence of a number of genetic and environmental (such as diet and obesity) influences.

Table 10   Lipoprotein patterns in parents of paediatric hyperapo-β-lipoproteinaemic probands

	Parents			
Probands	Normal	Hyperapo-β-lipoprotein-aemia	Type IIa	Type IV
Males (n = 9)	8	8	1	1
Females (n = 11)	13	8	0	1
Total	21	16	1	2

Data from Kwiterovich *et al.* (1988)

Table 11   Parental matings of paediatric hyperapo-β-lipoproteinaemic probands

	Hyperapo-β-lipo-proteinaemic probands	
Parental matings	Males	Females
Hyperapo-β-lipoproteinaemia×N	3	5
Hyperapo-β-lipoproteinaemia×hyperapo-β-lipoproteinaemia	2	1
Hyperapo-β-lipoproteinaemia×type IIa	1	0
Hyperapo-β-lipoproteinaemia×type IV	0	1
Type IV×N	1	0
N×N	2	4
Total	9	11

Data from Kwiterovich *et al.* (1988)
N = normal

Relationship between hyperapo-β-lipoproteinaemia and familial combined hyperlipidaemia

Familial combined hyperlipidaemia (McKusick 14425) is a Mendelian dominant disorder, originally defined as marked hypercholesterolaemia (>99th percentile), hypertriglyceridaemia (>99th percentile), or both hypercholesterolaemia and hypertriglyceridaemia in families of patients with premature coronary artery disease (Goldstein *et al.*, 1973). More recently, an elevated plasma total apolipoprotein B level, and a low LDL C to LDL apoB ratio have been found in hypertriglyceridaemic patients with familial combined hyperlipidaemia (Brunzell *et al.*, 1983). Increased hepatic synthesis of apolipoprotein B occurs in familial combined hyperlipidaemia (Chait *et al.*, 1980; Janus *et al.*, 1980), as well as in hyperapo-β-lipoproteinaemia (see also above). We and others (Brunzell *et al.*, 1984) have postulated that, at least in some hypertriglyceridaemic patients, the phenotype of hyperapo-β-lipoproteinaemia may indicate the presence of familial combined hyperlipidaemia. The presence of hyperapo-β-lipoproteinaemia in familial combined hyperlipidaemia has been studied in detail in one kindred (Kwiterovich *et al.*, 1987).

Molecular biology of apolipoprotein B gene

The apolipoprotein B gene spans about 45 kb of DNA on the short arm of chromosome 2 and consists of 29 exons. A number of investigators have contributed to our knowledge of the exon–intron organization of this large gene (Knott *et al.*, 1986; Yang *et al.*, 1986).

The Johns Hopkins Coronary Artery Disease Study was designed to study the role of the apolipoprotein B gene in premature coronary artery disease and in hyperapo-β-lipoproteinaemia. The population consists of 100 males, 50 years of age or less, and 100 females, 60 years of age or less, undergoing elective coronary arteriography and their spouse controls. Using the cDNA probe, AB6, which spans exons 26 to 29, an abnormal *Msp*I site was identified in a 43-year-old woman with triple vessel coronary artery disease and a hyperapo-β-lipoproteinaemia phenotype (Ladias *et al.*, 1987). Restriction mapping analysis revealed that the absent *Msp*I site was located round the codon 4046 in exon 29 of the apolipoprotein B gene. To elucidate the molecular defect in this *Msp*I site, the DNA region surrounding it was amplified and sequenced, as recently described by Dr Erlich and co-workers in the Cetus corporation. The normal sequence at this *Msp*I site is CCGG, and, in the index case, the sequence was CTGG. The mutation, therefore, was a C to T transition. It has been shown previously that CG dinucleotides are hot spots for mutation because of spontaneous deamination of methyl-cytosine to thymine.

This point mutation changes codon 4046 in the exon 29 of the apoB gene from CGG to TGG and produces a change from arginine, a polar positively-charged amino acid, to tryptophan, a non-polar hydrophobic amino acid (Figure 2). This amino acid is at position 4019 of the mature apolipoprotein B protein. Further study of six other family members with this mutation is underway to determine its relationship to the hyperapo-β-lipoproteinaemic phenotype and its effect on LDL metabolism.

Possible fundamental defects in hyperapo-β-lipoproteinaemia

The increased synthesis of VLDL apoB in the liver might be due to a defect in the apolipoprotein B gene, for example, an abnormality in the promoter region resulting in faulty regulation of synthesis of apoB, or a defect in the structural region producing an apoB protein with decreased affinity for lipid, or an apoB protein that is abnormally recognized by the LDL receptor, leading to faulty intracellular regulation of lipoprotein synthesis in liver (Figure 1). As mentioned earlier, studies are currently in progress to address the possible role of the apolipoprotein B gene in hyperapo-β-lipoproteinaemia.

Mutation in Codon 4046
of the Apolipoprotein B Gene

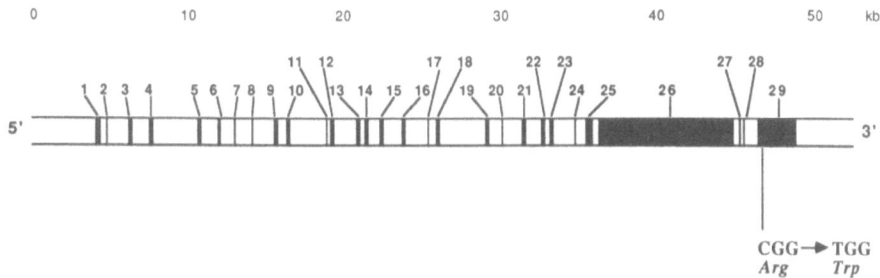

CGG → TGG
Arg   Trp

Figure 2 The exon (dark bars) and intron (white bars) organization of the apolipoprotein B gene. A mutation (CGG → TGG) in codon 4,046 in exon 29, producing the substitution of arginine with tryptophan at amino acid residue 4,019 of the mature apolipoprotein B, was found in a 43-year-old woman with coronary artery disease and hyperapo-β-lipoproteinaemia

Another defect resulting in overproduction of VLDL apoB in hyperapo-β-lipoproteinaemia may reside in faulty metabolism of free fatty acids and triglyceride.

In preliminary studies, Dr Allan Sniderman and colleagues (Sniderman *et al.*, 1986) described decreased incorporation of radiolabelled fatty acids into triglycerides in both adipocytes and cultured fibroblasts from patients with hyperapo-β-

Hypothesis: Pathophysiology of Hyperapo B

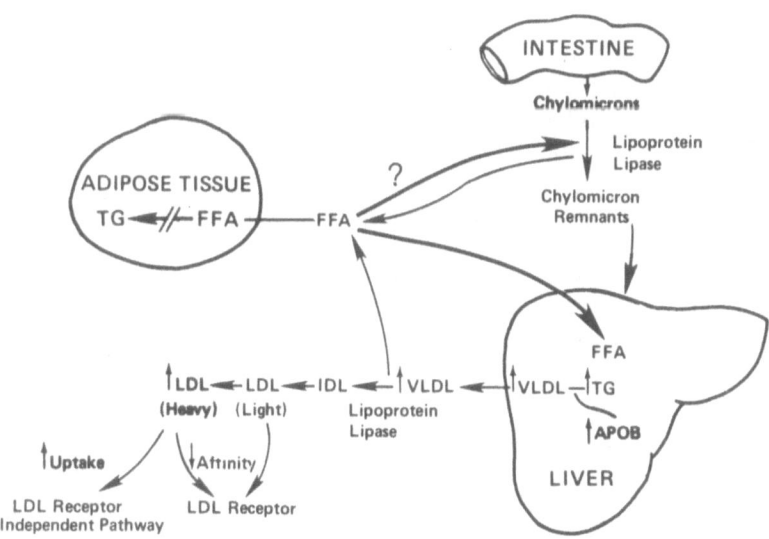

Figure 3 A hypothetical scheme for the pathophysiology of hyperapo-β-lipoproteinaemia

lipoproteinaemia (Figure 3). The mechanism underlying these observations is not known, but it is of interest that these *in vitro* observations are compatible with *in vivo* studies in which the response to an oral fat load was studied in 6 normotriglycer-idaemic hyperapo-β-lipoproteinaemic patients and 6 normolipidaemic controls (Genest *et al.*, 1986). In both groups, as expected, postabsorptive peak triglyceride concentrations four hours after the meal was significantly higher than fasting concentrations. However, the peak increase was significantly greater in the hyper-apo-β-lipoproteinaemic patients than in the normals. The mean triglyceride concen-tration in the normals did not differ significantly from the fasting concentration, but, in the hyperapo-β-lipoproteinaemic group, the mean plasma triglyceride at 7 hours was still twice as high as the fasting level. The delayed clearance of postpran-dial triglycerides in the hyperapo-β-lipoproteinaemic patients was due to higher amounts of both $S_f > 400$ triglycerides (chylomicrons), and S_f 20–400 triglycerides (VLDL).

Of particular interest was the observation that, by three hours after the meal, the plasma free fatty acids had risen significantly in the hyperapo-β-lipoproteinaemic group compared with normal, and remained elevated for up to 7 hours (Sniderman and Kwiterovich, 1987). This observation suggests a decreased uptake of free fatty acids from plasma in hyperapo-β-lipoproteinaemia (Figure 3). Such a finding is compatible with the *in vitro* observations, detailed above, where a decreased incorporation of fatty acid into triglycerides in both adipocytes and fibroblasts from hyperapo-β-lipoproteinaemic patients was found.

Decreased uptake and incorporation of free fatty acids into adipose tissue, resulting in increased plasma free fatty acid concentrations, may have two effects: (1) inhibition of lipoprotein lipase, slowing the clearance of postprandial chylomic-ron particles; and (2) enhanced flux of free fatty acid to liver driving triglyceride and apolipoprotein B synthesis, leading to increased VLDL production (Figure 3). As the triglyceride in VLDL is hydrolysed by lipoprotein lipase, the released free fatty acids in this hypothetical scheme, will be taken up more slowly by adipose tissue, and flux back to the liver driving further the production of VLDL (Figure 3). Thus, a cycle is produced in which free fatty acids derived from the hydrolysis of VLDL triglyceride, stimulate further VLDL production. The overproduction of VLDL results in the increased synthesis of LDL and a shift toward the heavy LDL subfraction (Figure 3). This panlipoprotein disorder is exacerbated after the consumption of a meal high in fat.

What might be the basis for the abnormality in free fatty acid uptake and/or incorporation into triglycerides in hyperapo-β-lipoproteinaemic cells? Hyperapo-β-lipoproteinaemic patients might be deficient in a plasma protein that promotes free fatty acid uptake or incorporation, or there may be a defect in the cellular recognition of the protein. Cianflone *et al.* (1987a, 1987b) have recently isolated a partially purified plasma protein which is free of albumin and which stimulates the incorporation of oleate into triglyceride in both normal fibroblasts and adipocytes. The possible role of this protein in the pathogenesis of hyperapo-β-lipoproteinaemia is under study.

Metabolism of HDL in hyperapo-β-lipoproteinaemia

The levels of HDL and its major apolipoprotein A-I are lower in patients with hyperapo-β-lipoproteinaemia than in normals (Kwiterovich and Sniderman, 1983).

J. Inher. Metab. Dis. 11 (1988)

These findings may be secondary to the hypertriglyceridaemia and higher VLDL and IDL levels found in hyperapo-β-lipoproteinaemia. Following a postprandial fat load (see also above), HDL cholesterol and its subfractions, HDL_2 and HDL_3, did not change significantly in normals. In the hyperapo-β-lipoproteinaemic group, on the other hand, total HDL cholesterol decreased significantly, due to a very sharp decrease in HDL_2 cholesterol; HDL_3 cholesterol increased slightly (Genest *et al.*, 1986). The role of HDL and apolipoprotein A-I metabolism in the pathogenesis of hyperapo-β-lipoproteinaemia requires further study.

CONCLUSION

Hyperapo-β-lipoproteinaemia is a common metabolic disorder associated with premature coronary artery disease. In some families, hyperapo-β-lipoproteinaemia behaves as a Mendelian dominant trait. The increased numbers of small dense LDL particles results from overproduction of VLDL apoB in liver. The biochemical and genetic basis for the overproduction of VLDL apoB is under further study, and includes both investigation of the apolipoprotein B gene and studies of free fatty acid, triglyceride and HDL metabolism.

ACKNOWLEDGEMENTS

Many of the studies reviewed here have resulted from the collaboration of a number of investigators. Researchers at the Royal Victoria Hospital, McGill University, Montreal, include Allan Sniderman, Babie Teng, Katherine Cianflone and Jacques Genest; at the Hammersmith Hospital, London, Gilbert Thompson; at The Johns Hopkins University, John Ladias, Stylianos Antonarakis, Paul Bachorik, Thomas Pearson, and Hazel Smith; at UCLA, Los Angeles, Jake Lusis. I thank Carol McGeeney for the preparation of this manuscript. This review was supported in part by the following US Public Health Service grants from the National Institutes of Health: HL31497; and General Clinical Research Center Program, RR-52.

REFERENCES

Beaty, T. H., Kwiterovich, P. O., Jr., Khoury, M. J., White, S., Bachorik, P. S., Smith, H. H., Teng, B. and Sniderman, A. D. Genetic analysis of plasma sitosterol, apoprotein B and lipoproteins in a large Amish pedigree with sitosterolemia. *Am. J. Hum. Genet.* 38 (1986) 492–504

Brunzell, J. D., Albers, J. J., Chait, A., Grundy, S., Groszek, E. and McDonald, G. B. Plasma lipoproteins in familial combined hyperlipidemia and monogenic familial hypertriglyceridemia. *J. Lipid Res.* 24 (1983) 147–155

Brunzell, J. D., Sniderman, A., Albers, J. J. and Kwiterovich, P. O., Jr. Apoproteins B and A-1 and coronary artery disease in humans. *Arteriosclerosis* 4 (1984) 79–83

Chait, A., Albers, J. J. and Brunzell, J. D. Very low density lipoprotein overproduction in genetic forms of hypertriglyceridemia. *Eur. J. Clin. Invest.* 10 (1980) 17–22

Cianflone, K., Kwiterovich, P. O., Jr., Walsh, M., Forse, A., Rodriguez, M. A. and Sniderman, A. D. Stimulation of fatty acid uptake and triglyceride synthesis in human cultured skin fibroblasts and adipocytes by a serum protein. *Biochim. Biophys. Res. Commun.* 144 (1987a) 94–100

Cianflone, K. M., Rodriguez, M. A., Walsh, M. J., Vu, H. T., Kwiterovich, P. O. and Sniderman, A. D. Effect of acylation-stimulating protein on triglyceride synthesis in cultured skin fibroblasts from normals and patients with hyperapobetalipoproteinemia. *Arteriosclerosis* 7 (1987b) 496a

Genest, J., Sniderman, A. D., Cianflone, K., Teng, B., Wacholder, S., Marcel, Y. and Kwiterovich, P. O., Jr. Hyperapobetalipoproteinemia. Plasma lipoprotein responses to oral fat load. *Arteriosclerosis* 6 (1986) 297–304

Goldstein, J. L. and Brown, M. S. The LDL receptor defect in familial hypercholesterolemia. Implication for pathogenesis and therapy. *Med. Clin. N. Am.* 66 (1982) 335–362

Goldstein, J. L., Schrott, H. G., Hazzard, W. R., Bierman, E. L. and Motulsky, A. G. Hyperlipidemia in coronary heart disease. II. Genetic analysis of lipid levels in 176 families and delineation of a new inherited disorder, combined hyperlipidemia. *J. Clin. Invest.* 52 (1973) 1544–1568

Janus, E. D., Nicoll, A. M., Turner, P. R., Magill, P. and Lewis, B. Kinetic bases of the primary hyperlipidemia: studies of apolipoprotein B turnover in genetically defined subjects. *Eur. J. Clin. Invest.* 10 (1980) 161–172

Kannel, W. B., Castelli, W. P. and Gordon, T. Cholesterol in the prediction of atherosclerotic disease. New perspectives based on the Framingham Study. *Ann. Intern. Med.* 90 (1979) 85–91

Knott, T. J., Pease, R. J., Powell, L. M., Wallis, S. C., Rall, S. C., Jr., Innerarity, T. L., Blackhart, B., Taylor, W. H., Marcel, Y., Milne, R., Johnson, D., Fuller, M., Lusis, A. J., McCarthy, B. J., Mahley, R. W., Levy-Wilson, B. and Scott, J. Complete protein sequence and identification of structural domains of human apolipoprotein B100. *Nature (London)* 323 (1986) 734–738

Kwiterovich, P. O., Jr. and Sniderman, A. D. Atherosclerosis and apoproteins B and apoA-I. *Prevent. Med.* 12 (1983) 815–834

Kwiterovich, P. O., Jr., White, S., Forte, T., Bachorik, P. S., Smith, H. and Sniderman, A. Hyperapo-β-lipoproteinaemia in a kindred with familial combined hyperlipidemia and familial hypercholesterolemia. *Arteriosclerosis* 7 (1987) 211–225

Kwiterovich, P., Beaty, T., Bachorik, P., Chen, J., Franklin, F., Georgopolous, L. and Sniderman, A. Pediatric hyperlipoproteinemia: The phenotypic expression of hyperapo-β-lipoproteinaemia in young probands and their parents. In Widholm, K. and Naito, H. K. (eds.) *Recent Aspects of Diagnosis and Treatment of Lipoprotein Disorders: Impact on the Prevention of Atherosclerotic Disease*, Alan R. Liss, Inc., New York, 1988, pp. 89–105

Ladias, J., Kwiterovich, P., Lusis, A., Smith, H. and Antonarkakis, S. A missense mutation arginine$_{4019} \rightarrow$ tryptophan in apolipoprotein B in a family with hyperapoB and premature atherosclerosis. *Arteriosclerosis* 7 (1987) 492a

Sniderman, A., Shapiro, S., Marpole, D., Malcolm, I., Skinner, B. and Kwiterovich, P. O., Jr. The association of coronary atherosclerosis and hyperapo-β-lipoproteinaemia (increased protein but normal cholesterol content in human plasma low density lipoprotein). *Proc. Natl. Acad. Sci. (USA)* 97 (1980) 604–608

Sniderman, A., Wolfson, C., Teng, B., Franklin, F., Bachorik, P. and Kwiterovich, P. O., Jr. Association of hyperapo-β-lipoproteinaemia with endogenous hypertriglyceridemia and atherosclerosis. *Ann. Intern. Med.* 97 (1982) 833-839

Sniderman, A. D., Teng, B., Genest, J., Cianflone, K., Wacholder, S. and Kwiterovich, P. O., Jr., Familial aggregation and early expression of hyperapo-β-lipoproteinaemia. *Am. J. Cardiol.* 55 (1985) 291–295

Sniderman, A., Teng, B., Forse, A., Cianflone, K. and Kwiterovich, P. O., Jr. Depressed glyceride biosynthesis in adipocytes and fibroblasts in hyperapo-β-lipoproteinaemia. *Clin. Res.* 34 (1986) 668A

Sniderman, A. and Kwiterovich, P. O., Jr. Hyperapo-β-lipoproteinaemia and LDL and

HDL$_2$ heterogeneity. *Proceedings of the Workshop on Lipoprotein Heterogeneity*, U.S. Department of Health and Human Services, NIH Publication No. 87–2646, 1987, pp. 293–304

Teng, B., Thompson, G. R., Sniderman, A. D., Forte, T. M., Krauss, R. M. and Kwiterovich, P. O., Jr. Composition and distribution of low density lipoprotein fractions in hyperapo-β-lipoproteinaemia, normolipidemia and familial hypercholesterolemia. *Proc. Natl. Acad. Sci. (USA)* 80 (1983) 6662–6666

Teng, B., Sniderman, A. D., Soutar, A. K. and Thompson, G. R. Metabolic basis of hyperapo-β-lipoproteinaemia. Turnover of apolipoprotein B in low density lipoprotein and its precursors and subfractions compared with normal and familial hypercholesterolemia. *J. Clin. Invest.* 77 (1986) 663–672

Yang, C.-Y., Chen, S-H., Gianturco, S. H., Bradley, W. A., Sparrow, J. T., Tanimura, M., Li, W.-H., Sparrow, D. A., DeLoof, H., Rosseneu, M., Lee, F.-S., Gu, Z.-W., Gotto, A. M., Jr. and Chan, L. Sequence, structure, receptor-binding domains and internal repeats of human apolipoprotein B100. *Nature (London)* 323 (1986) 738–742

J. Inher. Metab. Dis. 11 Suppl. 1 (1988) 74–86

Apolipoprotein Polymorphism and Multifactorial Hyperlipidaemia

G. Utermann

Institute of Medical Biology and Genetics, University of Innsbruck, Austria

Summary: Apolipoproteins AIV, B, E, and the Lp(a) glycoprotein are genetically polymorphic in humans. Three common alleles $\varepsilon2$, $\varepsilon3$ and $\varepsilon4$ control the polymorphism of apolipoprotein E. These code for proteins which differ in functional properties, e.g. receptor binding activity and *in vivo* catabolism. This explains the significant effect of the apoE gene locus on the variability of plasma lipoprotein concentrations and moreover the implication of apoE alleles in the aetiology of multifactorial forms of hyperlipidaemia e.g. familial type III hyperlipidaemia (apoE2; arg158→cys) and polygenic hypercholesterolaemia (apoE4; cys112→arg). A further gene locus controls the concentrations in plasma of the Lp(a) lipoprotein that is composed of an LDL-like particle containing apoB-100 and the disulphide-bonded Lp(a) glycoprotein. The latter exhibits a genetic size polymorphism (MW~400 kD–700 kD) that is controlled by at least seven autosomal alleles. These alleles at the same time are involved in determining the plasma concentrations of the lipoprotein that range from <1 mg/dl to >200 mg/dl. Thus there is evidence that genetic variability in apolipoproteins relates to the variability of lipoprotein concentrations in the population and is implicated in the aetiology of multifactorial hyperlipidaemias.

INTRODUCTION

Much has been learned during recent decades of the aetiology and pathogenesis of rare monogenic inborn errors of metabolism, in contrast to the relatively poor knowledge of common metabolic disorders, e.g. diabetes mellitus and hyperlipidaemias. Hyperlipidaemias that are widespread among the populations of industrialized western countries may be categorized into three major types (Figure 1). The first encompasses monogenic disorders of lipid metabolism. The classical example is familial hypercholesterolaemia where multiple mutations in the LDL-receptor gene disrupt cellular uptake of low-density lipoprotein (LDL)(Brown and Goldstein, 1986). The resulting increase of these cholesterol-rich lipoproteins in plasma may ultimately result in premature atherosclerosis. Other examples of monogenic dyslipidaemias are familial hyperchylomicronaemia, familial deficiency of lecithin–cholesterol acyltransferase and abetalipoproteinaemia, all of which are rare. The second group of hyperlipidaemias, called secondary hyperlipidaemia, is

Journal of Inherited Metabolic Disease. ISSN 0141–8955. Copyright © SSIEM and MTP Press Limited, Queen Square, Lancaster, UK.

at the other end of the spectrum. These hyperlipidaemias occur secondary to other primary disorders, e.g. cholestatic liver disease, nephropathies, uncontrolled diabetes mellitus and hypothyroidism. In between there is the third and largest group that includes most of the common forms of hyperlipidaemia. These are believed to be multifactorial in origin resulting from the interaction of predisposing genetic factors and environmental factors, e.g. nutrition. As will be outlined below type III hyperlipidaemia is a prototype for this group. Obviously however there is no sharp borderline between these categories.

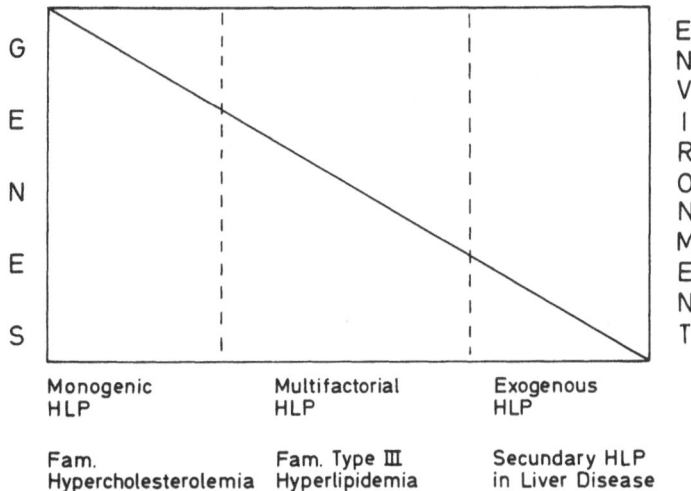

Figure 1 Classification of hyperlipidaemias (HLP). For explanation see text

Most structures involved in plasma lipoprotein transport and metabolism are known and well characterized and the respective genes have been cloned (Breslow, 1985). This detailed knowledge makes the plasma lipoprotein system an excellent model for studying the interaction of defined genes and environment in producing a phenotype. Following the working hypothesis that polymorphic gene loci that contribute to the variability of lipoprotein concentrations in the population are involved in the aetiology of multifactorial hyperlipidaemia we have started to search for genetic polymorphism of apolipoproteins.

APOLIPOPROTEIN POLYMORPHISM

Apolipoproteins are important programmers of plasma lipoprotein metabolism (Dolphin, 1985). Specific functions have been ascribed to certain members of this protein class. Some are cofactors for lipolytic enzymes, e.g. apoCII for lipoprotein lipase and apoAI and AIV for lecithin–cholesterol acyltranferase. Others are ligands for high-affinity cell surface receptors and facilitate endocytosis of lipoproteins by cells, e.g. apoB and apoE. For some apolipoproteins the precise *in vivo* function is not yet known but all have been cloned and sequenced (Breslow, 1985). Four of the known apolipoproteins are genetically polymorphic, namely apoAIV, apoB (the Ag-polymorphism), apoE and the Lp(a) glycoprotein (apo(a)) (Utermann *et al.*, 1977a, 1982, 1987; Zannis and Breslow, 1981; Menzel *et al.*, 1982). In the context of this paper I will discuss only the polymorphisms of apolipoprotein E and apo(a).

APOLIPOPROTEIN E POLYMORPHISM

Apolipoprotein E is synthesized in different tissues and cells throughout the body and is a constituent of different lipoprotein particles. It is present on triglyceride-rich lipoproteins originating from intestine (plasma chylomicrons), liver (VLDL) and their lipolytic degradation products called remnants or IDL. Moreover apoE is found in a subfraction of HDL called HDL-1 or HDLc. This fraction is thought to carry cholesterol from peripheral cells to the liver. ApoE is a ligand with high affinity for the LDL-receptor and for the chylomicron remnant receptor thus allowing the specific uptake of lipoprotein particles carrying apoE by the liver (Mahley *et al.*, 1984). Though at present several details are not well undersood it seems clear that apoE by virtue of its functional properties is involved in three important pathways of lipoprotein metabolism: (i) the uptake of dietary cholesterol from chylomicrons by the liver, (ii) the re-utilization of liver-derived cholesterol from VLDL by the liver and (iii) the transport of cholesterol carried in HDL-1 from peripheral cells to the liver (see Figure 2).

Since its discovery the genetic polymorphism of apolipoprotein E has been associated with a specific form of dyslipidaemia called primary dysbetalipoprotein-aemia (Utermann *et al.*, 1977a). This metabolic abnormality is characterized by accumulation of remnants but with low LDL concentrations in plasma. The polymorphism of apoE is controlled by three common alleles resulting in six phenotypes that can be distinguished by electrofocusing of delipidated serum followed by Western blotting (Figure 3). Except in American Indians the three alleles $\varepsilon2$, $\varepsilon3$ and $\varepsilon4$ have been found in all populations investigated and $\varepsilon3$ is the most common in each (see Table 1). Rall and coworkers (1984) have shown that the apoE isoforms differ by single aminoacid substitutions. Experiments from several laboratories have demonstrated that the three common isoforms apoE2 (arg158→cys), apoE3 and apoE4 (cys112→arg) also differ in their functional properties.

ApoE2 has less than 2% of the binding activity of apoE3 to the LDL-receptor (Schneider *et al.* 1981; Weisgraber *et al.*, 1982). ApoE4, though indistinguishable in *in vitro* binding activity from apoE3, exhibits an enhanced *in vivo* catabolism (Gregg *et al.*, 1986). These functional differences are clearly reflected in differences of lipoprotein metabolism in the various apoE phenotypic classes (Utermann *et al.*, 1977a, 1979a; Robertson and Cumming, 1985; Sing and Davignon, 1985; Boerwinkle and Utermann, 1988). ApoE alleles strongly effect total cholesterol, LDL-cholesterol, apoB and apoE concentrations in the population. In the German population the average effect of the $\varepsilon2$ allele is to *raise* apoE concentrations by 9.5 mg/L and to *lower* apoB by 9.46 mg/dl and total cholesterol by 14.2 mg/dl. The average effect of the $\varepsilon4$ allele is to *decrease* apoE concentrations by 1.9 mg/L and

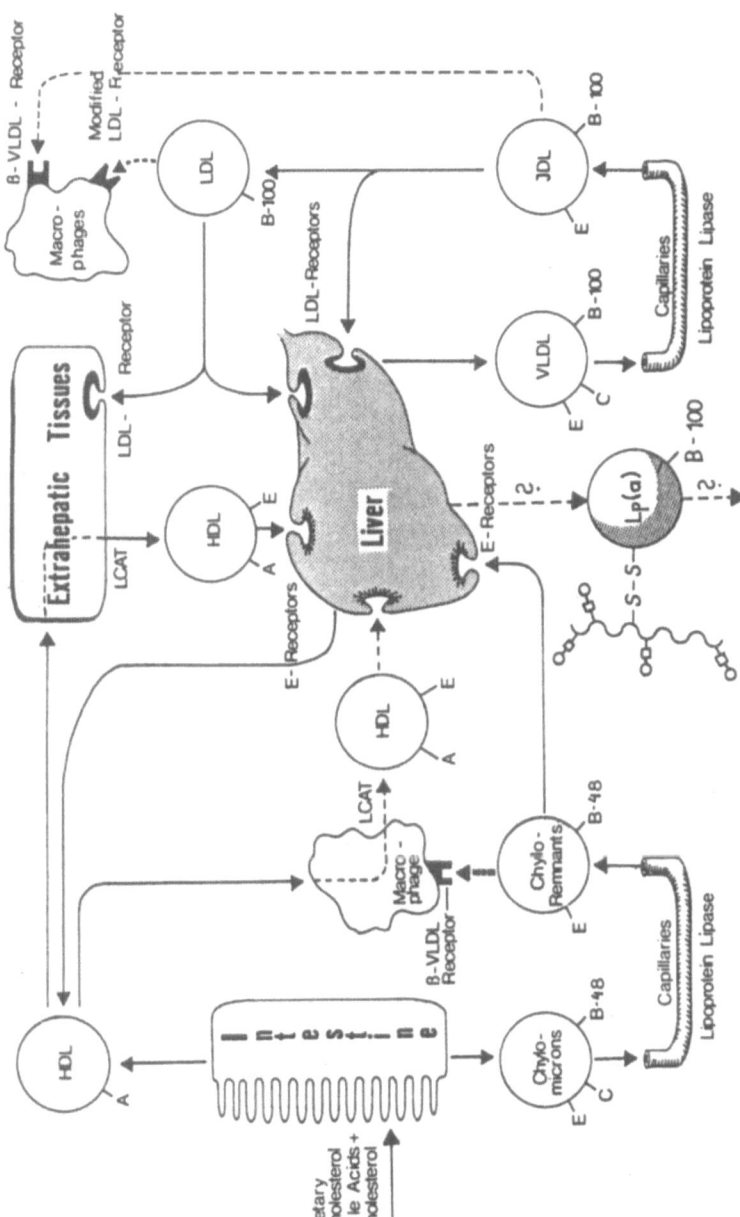

Figure 2 Major pathways of cholesterol transport in men. A, B-100, B-48, C, E denote the respective apolipoproteins. LCAT = lecithin–cholesterol acyltransferase. HDL, LDL, IDL, VLDL = high-, low-, intermediate-, and very low density lipoproteins respectively

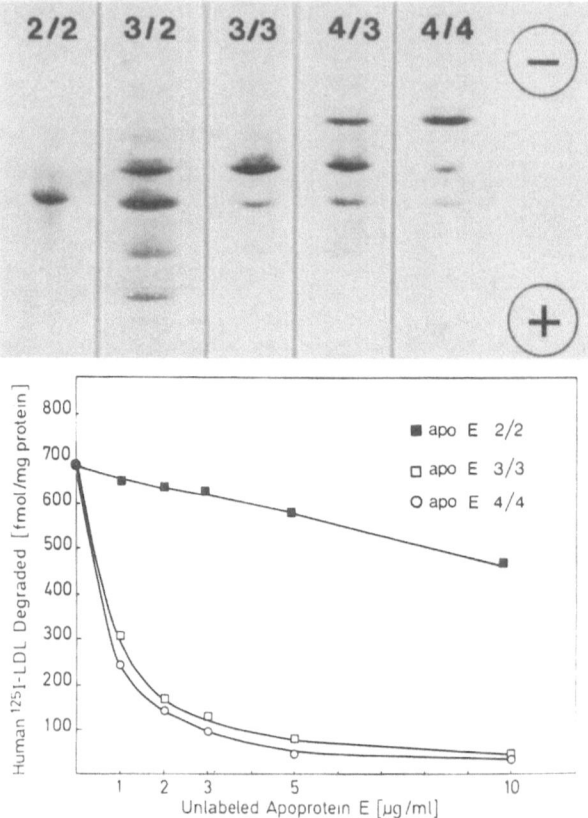

Figure 3 Upper Part: Immunoblots demonstrating five of the common apoE phenotypes. For methodological details see Menzel and Utermann (1986)
Lower Part: Competition by apoE–phospholipid complexes of the specific LDL-mediated degradation of ^{125}I-LDL by Hela cells (G. Utermann, S. Motzny, unpublished result)

to *increase* apoB by 4.92 mg/dl and total cholesterol by 7.09 mg/dl. The apoE gene locus explains roughly 20%, 12% and 4% of the variability in apoE, apoB and total cholesterol in Germans respectively. Similar results have been obtained in all westernized populations studied so far (see Figure 4). However, studies in different ethnic groups indicate that the apoE allele effect may also depend on cultural background (Utermann, 1987). In Indians and Malays the ApoB and cholesterol raising effect of the $\varepsilon 4$ allele was not present (E. Boerwinkle, N. Saha, G. Utermann, unpublished observations) and the effect of the $\varepsilon 2$ allele on cholesterol was only moderate or absent.

The effect of apoE genes on lipid metabolism is most pronounced in E2 homozygotes that present with a rather uniform type of dyslipidaemia. Independent of ethnic origin E2/2 subjects exhibit low total cholesterol, apoB and low LDL concentrations but elevated apoE and on the average mildly elevated VLDL

Table 1 Allele frequencies of the apoE polymorphism among 8 ethnically or geographically distinct populations

Populations	*Alleles*			
	N	*2*	*3*	*4*
Germany[1]	1031	0.077	0.773	0.150
Finland[1]	615	0.040	0.773	0.227
Tyrol[2]	473	0.084	0.798	0.108
Iceland[3]	176	0.063	0.773	0.165
Japan[4]	318	0.079	0.847	0.074
China[5]	175	0.106	0.794	0.100
India[5]	136	0.070	0.875	0.055
Malay[5]	117	0.115	0.770	0.115

[1]Utermann, 1987; [2]H. J. Menzel, S. Schrangl-Will, G. Utermann, unpublished; [3]H. J. Menzel, G. Sigurdsson, G. Utermann, unpublished; [4]T. Sata, G. Utermann, unpublished; [5]N. Saha, G. Utermann, unpublished.

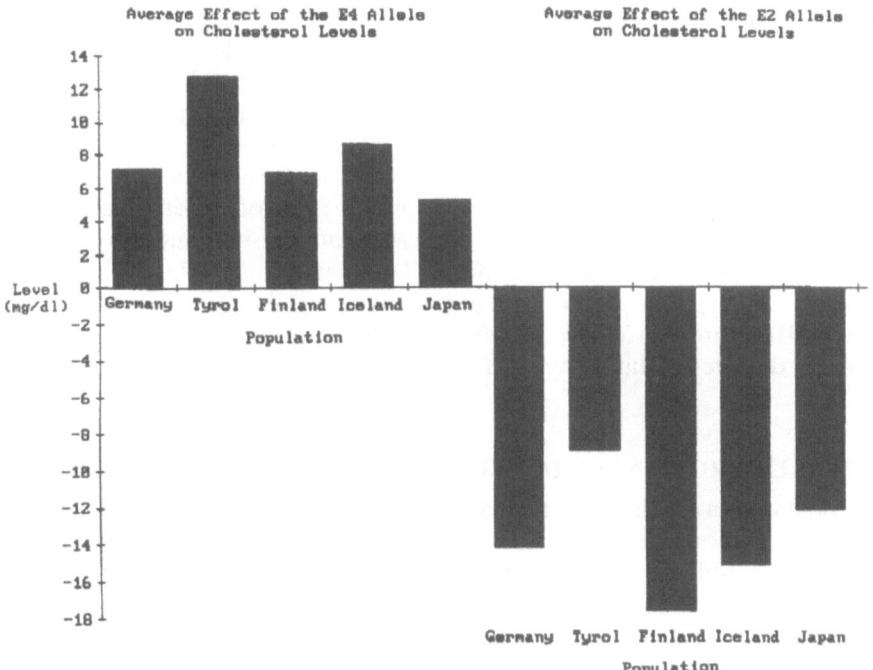

Figure 4 Graphic demonstration of apoE allele effects on total cholesterol levels in different populations (E. Boerwinkle and G. Utermann, 1988 and unpublished observations)

concentrations. All E2/2 subjects exhibit a so-called beta-VLDL subfraction (on agarose gel-electrophoresis) that represents remnants of triglyceride-rich lipoproteins. This characteristic dyslipidaemia is called primary dysbetalipoproteinaemia. It is surprising that apoE2/2 subjects, despite homozygosity for the functionally defective apoE that should result in the accumulation of lipoproteins in plasma, have subnormal cholesterol levels. This question is still unresolved but we have recently proposed a mechanism that may explain the situation in E2 homozygotes (Utermann, 1985, 1987). In brief we have postulated that the defective catabolism of chylomicron remnants that depend on apoE for uptake by the liver secondarily affects LDL-catabolism. The delayed uptake of chylomicron remnants that carry dietary cholesterol not only results in an accumulation of remnants (beta-VLDL) but also may trigger a compensatory upregulation of LDL receptors of the liver and thereby result in an increased catabolism of LDL.

APOLIPOPROTEIN E POLYMORPHISM AND MULTIFACTORIAL HYPERLIPIDAEMIA

The significant effects of the apoE gene locus on plasma lipoprotein metabolism that are reflected in lipid and apolipoprotein levels suggest that genes at this locus are also involved in the pathogenesis of multifactorial hyperlipoproteinaemia. Indeed subjects with an $\varepsilon 2$ allele have a 1.7-fold increased risk of developing hyperlipidaemia. The risk of developing mixed hyperlipidaemia is increased 1.5-fold for a subject with an $\varepsilon 4$ allele and 2.5-fold in one with the $\varepsilon 2$ allele. Subjects with an $\varepsilon 4$ allele have a 2.5-fold increased risk of developing pure hypercholesterolaemia (Utermann *et al.*, 1984). The tightest association, however, is between apoE2 homozygosity and type III hyperlipidaemia reflecting the aetiological role of defects in apolipoprotein E in the pathogenesis of this form of hyperlipidaemia that is characterized by the accumulation in plasma of massive amounts of remnant lipoproteins but low LDL concentrations (Utermann *et al.*, 1975, 1977b). Hence it represents the hyperlipidaemic form of primary dysbetalipoproteinaemia. Clinically the type III disorder is associated with premature coronary and peripheral athero-sclerosis, diabetes mellitus, hyperuricaemia and xanthomatosis.

HYPERLIPOPROTEINAEMIA TYPE III – A MULTIFACTORIAL DISORDER

We are faced with the fact that apoE2 homozygosity is associated with both subnormal cholesterol levels and a specific form of hyperlipidaemia called type III. ApoE2/2 subjects with low cholesterol and with severe type III hyperlipidaemia may coexist in the same family (Utermann *et al.*, 1979b). Careful studies that include protein sequencing and receptor-binding experiments have shown that apoE2 homozygotes with low and high cholesterol have indeed the same defective protein (Rall *et al.*, 1983). Extended family studies have provided evidence that the major difference between the two categories of apoE2 homozygotes is that non-type III forms of hyperlipidaemia segregate independently from apoE types in families of probands with type III hyperlipidaemia but not in those of probands with low cholesterol (Utermann *et al.*, 1979b). In several families identified by a proband with severe type III disorder with xanthomatosis the coexisting hyperlipida-emia had the characteristics of familial combined hyperlipidaemia (Utermann *et al.*, 1979b; Hazzard *et al.*, 1981). This has led to the hypothesis that coinheritance of two copies of the ε2 alleles with 'hyperlipidaemia genes' results in type III hyperlipidaemia (Utermann *et al.*, 1979b).

Lipid concentrations in most type III patients are extremely sensitive to dietary perturbation, hormones and lipid lowering drugs. Xanthomatosis may disappear following therapy. Since the final phenotype is determined by genetic and environmental factors, we consider type III hyperlipidaemia to be a multifactorial disorder (Figure 5).

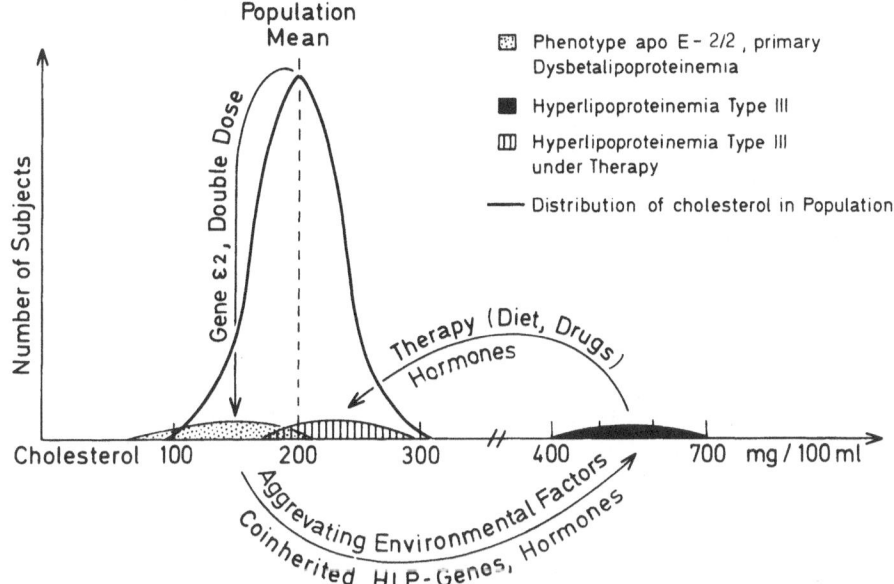

Figure 5 Hyperlipoproteinaemia type III. Model of a multifactorial disorder

One group has recently claimed that the second genetic factor in type III hyperlipidaemics is also in the apoE–CII–CI gene cluster, suggesting the existence of a 'type III haplotype' (Klasen *et al.*, 1987). However, their data do not explain the apoE-independent segregation of other forms of hyperlipidaemia in type III families. Probably final proof of the 'two genetic factor' hypothesis will only be possible when the riddle of familial combined hyperlipidaemia has been solved and when it will be possible directly to demonstrate that individuals with severe type III hyperlipidaemia and xanthomatosis carry both the defective apoE genes and the gene(s) for familial combined hyperlipidaemia.

In a fraction of patients with type III hyperlipidaemia other mutant forms of apoE have been observed, all of which have in common a defect in binding to the LDL-receptor (Rall *et al.*, 1982; Mahley *et al.*, 1984). In at least one family there is evidence that the defective apoE (apoE3-Leiden) alone may be sufficient to cause type III hyperlipidaemia even in heterozygotes and it has been concluded that the disorder is a dominant in this family (Havekes *et al.*, 1986).

POLYMORPHISM OF Lp(a) LIPOPROTEIN

In 1963 Berg demonstrated that a rabbit antibody against human betalipoproteins was able to distinguish two genetic lipoprotein phenotypes that he designated Lp(a+) and Lp(a−). Later it was shown by several groups that the inherited antigenic property resides in a distinct lipoprotein subpopulation called the Lp(a) lipoprotein that has a density of about 1.05–1.1 g/ml (Wiegandt *et al.*, 1968; Ehnholm *et al.*, 1972). The concentration of this lipoprotein varies widely between individuals (from virtually non-detectable to 200 mg/dl) but is very constant in a given individual.

In caucasians the distribution of Lp(a) concentrations is highly skewed with most subjects having low Lp(a) concentrations (Albers and Hazzard, 1974). Several groups have studied the inheritance of the quantitative Lp(a) trait but the genetics remained largely unclear. A major dominant gene for high Lp(a) levels with polygenic background has been postulated (Sing *et al.*, 1974). Recently Hasstedt and Williams (1986) proposed a model where three alleles Lp^A, Lp^a, and Lp^0 determine high, intermediate, and low concentrations. Research on the Lp(a) lipoprotein was greatly stimulated by reports that suggested an implication of Lp(a) lipoprotein in the aetiology of atherosclerosis (Renninger *et al.*, 1965; Berg *et al.*, 1974; Kostner *et al.*, 1981; Rhoads *et al.*, 1986).

Biochemical characterization of the isolated Lp(a) lipoprotein revealed that it is a complex molecule composed of a particle resembling LDL in protein and lipid composition and of a high MW glycoprotein that is unique for Lp(a) lipoprotein. This glycoprotein is bonded to apoB100 in the LDL by a disulphide bridge (Gaubatz *et al.*, 1983; Fless *et al.*, 1984; Utermann *et al.*, 1987). We have recently developed a method of demonstrating the Lp(a) glycoprotein, or apo(a) as it is called by others, directly in human plasma or sera by SDS-gel electrophoresis under reducing conditions followed by immunoblotting with poly- and monoclonal anti-Lp(a)

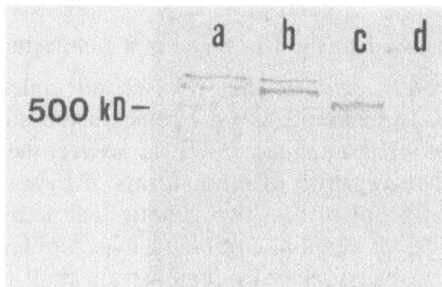

Figure 6 Western blot demonstrating the principal Lp(a) glycoprotein phenotypes. Total delipidated sera were subjected to SDS-polyacrylamide-gel electrophoresis under reducing conditions. Proteins were blotted onto nitrocellulose and Lp(a) glycoprotein was demonstrated with an affinity purified rabbit-anti-Lp(a) antibody followed by a gold-labelled sheep anti-rabbit IgG. Lanes a and b: double-band phenotypes (S1/S2); lane c: single band phenotype (B); lane d: 0-phenotype

antibodies (Utermann *et al.*, 1987, 1988a; Figure 6). Application of this technique in population and family studies has yielded the following results: at least six species of Lp(a) glycoprotein with apparent MW ranging from approximately 400 kD to 700 kD can be distinguished. We have categorized these proteins according to their relative mobility compared to apoB100 (MW~500 kD) into phenotypes F, B, S1, S2, S3 and S4 and the respective doubleband phenotypes. The different Lp(a) species are coded for by alleles designated Lp^F, Lp^B, Lp^{S1}, Lp^{S2}, Lp^{S3} and Lp^{S4} at a single genetic locus (Utermann et al., 1987, 1988b). In addition these studies demonstrated existence of a null allele (Lp^0). Frequencies of Lp(a) alleles are given in Table 2.

Table 2   Lp(a) polymorphism

Isoform	MW(kD)	Allele	Frequency
Lp(a) F	~430	Lp^F	rare
Lp(a) B	~500	Lp^B	0.012
Lp(a) S1	~540	Lp^{S1}	0.042
Lp(a) S2	~600	Lp^{S2}	0.096
Lp(a) S3	~670	Lp^{S3}	0.092
Lp(a) S4	~750	Lp^{S4}	0.105
—	—	Lp^0	0.654

Simultaneous determination of Lp(a) *lipo* protein concentration and Lp(a) *glyco* protein phenotypes in 441 unrelated subjects revealed a highly significant association between both genetic phenomena. Mean and median Lp(a) concentrations are high in phenotypes with Lp B, Lp S1 and Lp S2 isoforms. They are low in phenotypes Lp S3 and Lp S4 and very low or absent in the Lp 0 type (Utermann

Table 3   Lp(a) lipoprotein concentrations (mg/dl) in plasma donors with different single band Lp(a) phenotypes

Lp(a) phenotype	N	Mean	SD	Median
B	7	47.4	20.8	40.0
S1	26	18.5	15.6	13.5
S2	61	20.7	16.7	19.0
S3	58	9.6	5.8	8.5
S4	67	9.42	5.9	8.0

Kruskal–Wallis test $p < 0.001$; from Utermann et al. 1988a

et al., 1987, 1988a; Table 3). Together the population and family data have led us to propose the following working hypothesis (Figure 7). Lp(a) *glyco* protein phenotypes are genetically determined by a series of codominant alleles probably at the Lp(a) structural gene locus. This gene locus (including its regulatory elements) at the same time is a major gene locus determing Lp(a) *lipo* protein concentration in plasma. The higher order distribution of Lp(a) concentrations in the population may then be explained by the assumption that each Lp(a) allele determines its own distribution of Lp(a) concentrations and that those alleles associated with low concentrations are frequent whereas those associated with high Lp(a) are rare.

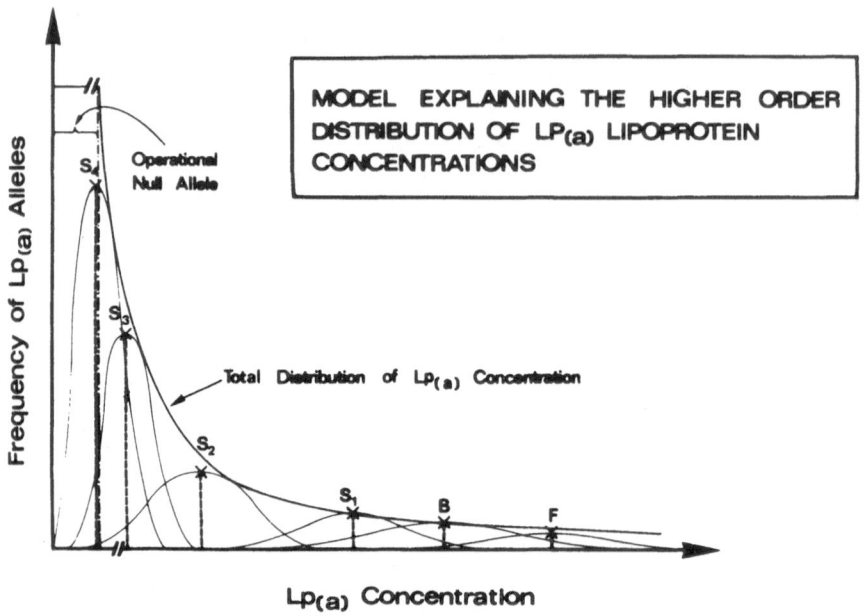

Figure 7 Simplified model explaining the higher order distribution of Lp(a) plasma levels by the hypothesis that variability in Lp(a) concentration in the population is controlled by several apo(a) alleles each of which determines its own mode of concentrations

Preliminary evidence suggests that Lp(a) alleles act additively in determining Lp(a) lipoprotein concentrations (Utermann *et al.*, 1988b). In a given phenotype, however, there is a wide range of Lp(a) lipoprotein concentrations (Utermann *et al.*, 1987, 1988a). Hence the Lp(a) structural gene locus cannot be the only one responsible for Lp(a) lipoprotein levels. These other genes await identification. The fact that extremely high levels of Lp(a) lipoprotein (>100 mg/dl) were seen only in hyperlipidaemic subjects suggests to us that certain genes for hyperlipidaemia also affect Lp(a) lipoprotein levels. Studies in the respective patients hopefully will broaden our understanding of multifactorial hyperlipidaemia and of multifactorial disorders in general.

The basis of the Lp(a) size heterogeneity is presently unclear. Differences in the carbohydrate moiety probably are not primarily involved. Eaton and coworkers (1987) recently demonstrated a high degree of homology of a partial sequence of Lp(a) with plasminogen, e.g. Kringle 4. This raises the intriguing possibility that the size heterogeneity of apo(a) is due to differences in the number of Kringle structures.

ACKNOWLEDGEMENTS

Parts of the work from our laboratory reported in this paper were supported from the Deutsche Forschungsgemeinschaft (DFG), the Österreichischer Fonds zur Förderung der wissenschaftlichen Forschung and by the Thyssen-Stiftung to G.U.

REFERENCES

Albers, J. J. and Hazzard, W. R. Immunochemical quantification of human plasma Lp(a) lipoprotein. *Lipids* 9 (1974) 15–26

Berg, K. A new serum type system in man: the Lp-system. *Acta Pathol. Microbiol. Scand.* 59 (1963) 369–382

Berg, K., Dahlen, G. and Frick, M. H. Lp(a) lipoprotein and pre-β–lipoprotein in patients with coronary heart disease. *Clin. Genet.* 6 (1974) 230–235.

Boerwinkle, E. and Utermann, G. Simultaneous effects of the apolipoprotein E polymorphism on apolipoprotein E, apolipoprotein B and cholesterol metabolism. *Am. J. Hum. Genet.* 78 (1988) 41–46

Breslow, J. L. Human apolipoprotein molecular biology and genetic variation. *Annu. Rev. Biochem.* 54 (1985) 699–727

Brown, M. S. and Goldstein, J. L. A receptor mediated pathway for cholesterol homeostasis. *Science* 232 (1986) 34–47

Dolphin, P. J. Lipoprotein metabolism and the role of apolipoproteins as metabolic programmers. *Can. J. Biochem. Cell. Biol.* 63 (1985) 850–869

Eaton, D. L., Fless, G. M., Kohr, W. J., McLean, J. W., Xu, Q.-T., Miller, C. G., Lawn, R. W. and Scanu, A. M. Partial amino acid sequence of apolipoprotein(a) shows that it is homologous to plasminogen. *Proc. Natl. Acad. Sci. USA* 84 (1987) 3224–3228

Ehnholm, C., Garoff, H., Renkonen, O. and Simons, K. Protein and carbohydrate composition of Lp(a) lipoprotein from human plasma. *Biochemistry* 11 (1972) 3229–3232

Fless, G. M., Rolik, C. A. and Scanu, A. M. Heterogeneity of human plasma lipoprotein(a). Isolation and characterization of the lipoprotein subspecies and their apoproteins. *J. Biol. Chem.* 259 (1984) 11470–11478

Gaubatz, J. W., Heideman, C., Gotto, Jr. A. M., Morrisett, J. D. and Dahlen, G. H. Human plasma lipoprotein(a): structural properties. *J. Biol. Chem.* 258 (1983) 4582–4589

Gregg, R. E., Zech, L. A., Schaefer, E. J., Stark, D., Wilson, D. and Brewer, Jr. H. B. Abnormal *in vivo* metabolism of apolipoprotein E4 in humans. *J. Clin. Invest.* 78 (1986) 815–821

Hasstedt, S. J. and Williams, R. R. Three alleles for quantitative Lp(a). *Genet. Epidemiol.* 3 (1986) 53–55

Havekes, L., De Wit, E., Gevers-Leuven, J., Klasen, E., Utermann, G., Weber, W. and Beisiegel, U. Apolipoprotein E3-Leiden. A new variant of human apolipoprotein E associated with familial type III hyperlipoproteinemia. *Hum. Genet.* 73 (1986) 157–163

Hazzard, W. R., Warnick, G. R., Utermann, G. and Albers, J. J. Genetic transmission of isoapolipoprotein E phenotypes in a large kindred: relationship to dysbetalipoproteinaemia and hyperlipidaemia. *Metabolism* 30 (1981) 79–88

Klasen, E. C., Talmud, P. J., Havekes, L., De Wit, E., Van der Kooij-Meijs, E., Smit, M., Hansson, G. and Humphries, S. E. A common restriction fragment length polymorphism of the human apo E gene and its relationship to type III hyperlipidaemia. *Hum. Genet.* 75 (1987) 244–247

Kostner, G. M., Avorgaro, P., Cazzoloto, G., Marth, E., Bittolo-Bon, G. and Quini, G. B. Lipoprotein Lp(a) and the risk for myocardial infarction. *Atherosclerosis* 38 (1981) 51–61

Mahley, R. W., Innerarity, T. L., Rall, S. C. and Weisgraber, K. H. Plasma lipoproteins: apolipoprotein structure and function. *J. Lipid Res.* 25 (1984) 1277–1294

Menzel, H. J., Kövary, P. M. and Assmann, G. Apolipoprotein A-IV polymorphism in man. *Hum. Genet.* 62 (1982) 349–352

Menzel, H. J. and Utermann, G. Apolipoprotein phenotyping from serum by western blotting. *Electrophoresis* 7 (1986) 492–495

Rall, S. C., Weisgraber, K. H., Innerarity, T. L. and Mahley, R. W. Structural basis for receptor binding heterogeneity of apolipoprotein E from type III hyperlipoprotein subjects. *Proc. Natl. Acad. Sci. USA* 79 (1982) 4696–4700

Rall, S. C., Weisgraber, K. H., Innerarity, T. L., Mahley, R. W. and Assman, G. Identical structural and receptor binding defects in apolipoprotein E2 in hypo- normo- and hypercholesterolemic dysbetalipoproteinia. *J. Clin. Invest.* 71 (1983) 1023–1031

Renninger W., Wendt, G. G., Nawrocki, P. and Weigand, H. Beitrag zur Problematik des Lp-Systems. *Humangenetik* 1 (1965) 658–667

Rhoads, G. G., Dahlen, G., Berg, K., Morton, N. E. and Dannenberg, A. L. Lp(a) lipoprotein as a risk factor for myocardial infarction. *J. Am. Med. Assoc.* 256 (1986) 2540–2544

Robertson, F. W. and Cumming, A. M. Effects of apoprotein E polymorphism of serum lipoprotein concentration. *Arteriosclerosis* 5 (1985) 283–292

Schneider, W. J., Kovanen, P. T., Brown, M. S., Goldstein, J. L., Utermann, G., Weber, W., Havel, R. J., Kotite, L., Kane, J. P., Innerarity, T. L. and Mahley, R. W. Familial dysbetalipoproteinemia. Abnormal binding of mutant apoprotein E to low density lipoprotein receptors of human fibroblasts and membranes from liver and adrenals of rats, rabbits and cows. *J. Clin. Invest.* 68 (1981) 1075–1085

Sing, C. F., Schultz, J. S. and Shreffler, D. C. The genetics of the Lp-antigen II. *Ann. Hum. Genet.* 38 (1974) 47–56

Sing, C. F. and Davignon, J. Role of the apolipoprotein E polymorphism in determining normal plasma lipid and lipoprotein variation. *Ann. Hum. Genet.* 37 (1985) 268–284

Utermann, G. Genetic polymorphism of apolipoprotein E – impact on plasma lipoprotein metabolism. In Crepaldi, G. (ed.) *Diabetes, Obesity and Hyperlipidemias – III*. Elsevier Science Publishers, New York, 1985, pp. 1–28

Utermann, G. Apolipoprotein E polymorphism in health and disease. *Am. Heart J.* 113 (1987) 433–440

Utermann, G., Jaeschke, M. and Menzel, H. J. Familial hyperlipoproteinaemia type III: deficiency of a specific apolipoprotein (apoE III) in the very-low-density lipoproteins. *FEBS Lett.* 56 (1975) 352–355

Utermann, G., Hees, M. and Steinmetz, A. Polymorphism of apolipoprotein E and occurrence of dysbetalipoproteinaemia in man. *Nature* (1977a) 604–607

Utermann, G., Canzler, H., Hees, M., Jaeschke, M., Mühlfellner, G., Schoenborn, W. and Vogelberg, K. H. Studies on the metabolic defect in Broad-β-disease (hyperlipoproteinaemia type III). *Clin. Genet.* 12 (1977b) 139–154

Utermann, G., Pruin, N. and Steinmetz, A. Polymorphism of apolipoprotein E. III: Effect of a single polymorphic gene locus on plasma lipid levels in man. *Clin. Genet.* 15 (1979a) 63–72

Utermann, G., Vogelberg, K. H. and Steinmetz, A. *et al.* Polymorphism of apolipoprotein E: II: Genetics of hyperlipoproteinaemia type III. *Clin. Genet.* 15 (1979b) 37–62

Utermann, G., Feussner, G., Franceschini, G. and Steinmetz, A. Genetic variants of group A apolipoproteins. *J. Biol. Chem.* 257 (1982) 501–507

Utermann, G., Kindermann, I., Kaffarnik, H. and Steinmetz, A. Apolipoprotein E phenotypes and hyperlipidaemia. *Hum. Genet.* 65 (1984) 232–236

Utermann, G., Menzel, H. J., Kraft, H. G., Duba, H. C., Kemmler, H. G. and Seiss, C. Lp(a) glycoprotein in phenotypes. Inheritance and relation to Lp(a)-lipoprotein concentration in plasma. *J. Clin. Invest.* 80 (1987) 458–465

Utermann, G., Kraft, H G., Menzel, H. J., Hopferwieser, T. and Seiss, C. Genetics of the quantitative Lp(a) lipoprotein trait. I. Relation of Lp(a) glycoprotein phenotypes to Lp(a) lipoprotein concentrations in plasma. *Hum. Genet.* 78 (1988a) 41–46

Utermann, G., Duba, C. and Menzel, H. J. Genetics of the quantitative Lp(a) lipoprotein trait. II. Inheritance of Lp(a) glycoprotein phenotypes. *Hum. Genet.* 78 (1988b) 47–50

Weisgraber, K. H., Innerarity, T. L. and Mahley, R. W. Abnormal lipoprotein receptor-binding activity of human E apoprotein due to cysteine arginine interchange at a single site. *J. Biol. Chem.* 257 (1982) 2518-2521

Wiegandt, H., Lipp, K. and Wendt, G. Identifizierung eines Lipoproteins mit Antigenwirksamkeit im Lp-System. *Hoppe-Seyler's Z. Physiol. Chem.* 349 (1968) 489–494

Zannis, V. I. and Breslow, J. L. Human very low density lipoprotein apolipoprotein E isoprotein polymorphism is explained by genetic variation and post-translational modification. *Biochemistry* 20 (1981) 1033–1041

J. Inher. Metab. Dis. 11 Suppl. 1 (1988) 87–90

A Neonatal Screening Approach to the Detection of Familial Hypercholesterolaemia and Family-based Coronary Prevention

D. E. L. Wilcken, B. L. Blades and N. P. B. Dudman
Department of Cardiovascular Medicine, The University of New South Wales, The Prince Henry Hospital, Sydney, New South Wales 2036, Australia

Summary: We measured by radial immunodiffusion apolipoprotein B (ApoB) as a genetic marker for familial hypercholesterolaemia in heel prick blood spot samples on filter paper routinely collected from 5000 3- to 5-day-old neonates for current screening programmes. Dried blood spot ApoB levels were distributed continuously and were 9% higher in female neonates than in males (0.246 ± 0.085 versus 0.225 ± 0.079 g/L of whole blood, mean $\pm$SD, $p < 0.0001$). Neonates of birth weight under 2.5 kg had lower levels than the population mean in both sexes but levels did not change with birth weight within this range; in those with the birth weights greater than 2.5 kg levels increased with increasing birth weight. Sex and birth weight could account for 5.7% of the variability of ApoB. Levels had largely stabilized by day 3 of life. Of parents of neonates whose ApoB levels were among the top 2%, 45 families were available for study when the infants were aged 12.3 ± 3.3 months. In 6 of these families there was a persisting elevation of ApoB both in the infants (levels greater than 0.7 g/L of whole blood) and in one parent who also had elevated ApoB and a lipid profile of familial hypercholesterolaemia phenotype, results indicating an ascertainment rate of between 1 in 365 and 1 in 830 of the screened population. These studies have defined the variables affecting neonatal ApoB levels and establish that neonatal screening for familial hypercholesterolaemia is feasible. Primary prevention of vascular disease in young families with the gene should be possible with this approach.

The management of dyslipoproteinaemia is directed mainly towards the prevention of vascular disease, and the coronary circulation is by far the most common site for this. There is now impressive evidence that lowering serum lipids reduces the risk of coronary heart disease and this has been reviewed recently by Thompson (1987). Coronary prevention programmes are already having some impact in Australia and the USA. In both countries, death rates from coronary heart disease increased steadily during the immediate post-war years and reached a peak in 1968 but since then there has been a progressive decrease. For example, data from the Australian Bureau of Statistics reveal that, for the productive 30–64 year age group, there has been a 41% decline in death rate for men since 1968, and 45% for women. Despite this, coronary artery disease is still the commonest cause of death in this age group.

Journal of Inherited Metabolic Disease. ISSN 0141–8955. Copyright © SSIEM and MTP Press Limited, Queen Square, Lancaster, UK.

These benefits have been achieved largely through small lifestyle changes occurring across a large population. Risk factors for vascular disease have decreased, principally due to dietary changes leading to reductions in serum lipids and to less cigarette-smoking (Heller, 1986; Goldman and Cook, 1984). Lacking at the moment is a comprehensive high-risk strategy to complement this population approach. There is a need to identify groups at special risk who require more definitive regimens to prevent coronary artery disease developing than the population approach can provide. Individuals with familial hypercholesterolaemia form such a high-risk group.

Carl Muller (1939) first comprehensively described familial hypercholesterolaemia 50 years ago and showed that it was dominantly inherited and associated with precocious coronary artery disease. The inborn error occurs in about 1 in 500 of the population and is now known to be due to a marked reduction in low-density lipoprotein receptors on the cell surface (Goldstein and Brown, 1983). This results in increased plasma levels of the atherogenic low-density lipoprotein (LDL) and the LDL carrier protein, apolipoprotein B (ApoB). If familial hypercholesterolaemia could be identified early in the presymptomatic period, the appropriate management could be offered and the prospects for preventing vascular complications improved.

There is now compelling evidence that atherogenesis begins in childhood (McGill, 1984; Newman *et al.*, 1986; Stary, 1987) and that regimens which lower plasma cholesterol to levels effective in preventing adult coronary heart disease are safe in children. Clinics have been established in a number of centres to investigate and, when necessary, treat the children of parents who have evidence of premature vascular disease. If this is useful, we reasoned that it might be helpful to identify such children before the occurrence of a vascular event in one of their parents, so that early intervention could be made available to the whole family in family-based lipid clinics. With these ideas in mind, we have explored a neonatal screening approach to case-finding with a view to implementing a family-based coronary prevention programme for children and their parents who have the gene. If suitable methodology could be incorporated into current neonatal screening programmes for phenylketonuria and hypothyroidism, young high-risk families with the gene for familial hypercholesterolaemia could be identified at very low cost and offered management to postpone or prevent the vascular complications.

Our approach has been to measure ApoB as a genetic marker for familial hypercholesterolaemia in heel prick blood spot samples on filter paper routinely collected from 3–5-day-old neonates for current screening programmes. We first developed a radial immunodiffusion method for measuring ApoB in the dried blood spots and showed that levels in the blood spot and in the serum were closely correlated with plasma LDL cholesterol (Dudman *et al.*, 1985). We then measured dried blood spot ApoB levels in 5000 consecutively born 3–5-day-old neonates. In 4491 of these, we assessed effects of sex, birth weight, gestational age, age at sampling, and delay and temperature of storage before assay on the measured levels of ApoB (Blades *et al.*, 1987).

We found that the dried blood spot ApoB levels were distributed continuously with no distinct population of neonates having abnormally high levels. Analysis of the screened population showed that female neonates had a 9% higher mean ApoB level than males ($p<0.0001$), the mean ± SD being 0.246±0.085 and 0.225±0.079

g/L of whole blood respectively. In neonates of birth weight under 2.5 kg, levels were lower than the population mean in both sexes and did not change appreciably with birth weight within this range. In neonates with birth weight >2.5 kg, ApoB levels increased with increasing birth weight. A similar pattern was found in relation to gestational age for infants ≤36 weeks and those >36 weeks respectively. As expected, gestational age was significantly correlated with birth weight ($r = 0.62$, $p<0.0005$). Heirarchical multiple linear regression analysis revealed that sex, birth weight and gestational age could account for 5.7% of the variability in ApoB, and, after adjustments for these variables, the neonates' age at sampling, ie. 3, 4 or 5 days, did not influence ApoB levels significantly. Thus, levels had largely stabilized by day 3. Assayed ApoB levels decreased with increasing age of sample and temperature of storage of the samples before assay. A further analysis of the effect of diet in a subset of 541 babies revealed no difference in ApoB levels whether the babies were breast fed, formula fed, or breast fed with formula complement. We have reported all these findings in detail recently (Blades *et al.*, 1987).

To investigate the validity of the screening procedure, we arbitrarily recalled infants in the top 2% of values for repeat dried blood spot ApoB determination and measured plasma lipid and ApoB levels in their parents. The recall was undertaken before the above-mentioned variables had been elucidated, and they were not, therefore, taken into account. Forty-five infants of the 103 recalled were available for retesting at a mean age (±SD) of 12.3±3.3 months. These 45 did not differ from the whole group of 103 in relation to the above-mentioned variables. The mean ApoB level (±SD) in these 45 neonates was 0.65±0.20 g/L of whole blood (range 0.30–1.16). In 6 of the infants, there was both a persisting elevation of ApoB (>0.70 g/L of whole blood) and one parent who also had elevated ApoB and hyperlipidaemia of the type II phenotype. Two of these parents, both males, had already had coronary by-pass surgery under the age of 40 years. Their children had the highest ApoB levels among those restudied. Thus, there was good evidence for familial transmission of the disorder in these 6 families.

These results represent an ascertainment rate of 1 in 830 in the total population screened, or, if the 45 neonates retested were truly a representative sample of the top 2%, a rate of 1 in 365. This frequency is of the same order of magnitude as the known frequency of familial hypercholesterolaemia heterozygotes in the general population (Goldstein and Brown, 1983). It is in contrast with the much lower incidence of the disorders currently identified in newborn screening programmes: phenylketonuria (1 in 10000), congenital hypothyroidism (1 in 4000), cystic fibrosis (1 in 2500) and galactosaemia (1 in 40000).

The present studies have defined the variables affecting neonatal ApoB levels and extend appreciably earlier studies by Van Biervliet and colleagues (1986) and Lane and McConathy (1986). With regard to a screening programme, it is clear that the baby's sex, birth weight and gestational age should be taken into account when determining a cut-off point for the recall and when establishing false-positive and false-negative rates. Nevertheless, even without these refinements, the results of the present study are encouraging and suggest that neonatal screening for familial hypercholesterolaemia is feasible. For case finding, it represents an example of extended screening, the infant identifying a young parent, and possibly other family members, with the gene; that is, at least two for the price of one. Although LDL receptor studies have not been performed in the 6 families identified by the screen, the inferential evidence that they do have familial hypercholesterolaemia is nevertheless strong.

At present, young families with a genetically determined predisposition to coronary artery disease are usually identified as a result of one parent having a myocardial infarction. Dietary and other regimens are then established. In the light of evidence that atherosclerosis begins in childhood (McGill, 1984; Newman *et al.*, 1986; Stary, 1987) and that, as reviewed by Thompson (1987), lowering lipid levels reduces cardiovascular risk, it seems logical to establish such regimens early in life in familial hypercholesterolaemia families. It will be of interest to compare compliance and progress in families recruited by this extended screening approach before a vascular episode has occurred in a parent with those recruited after one parent has already sustained a coronary event. We suggest that these two approaches to case ascertainment are complementary and could facilitate the development of family based coronary prevention programmes in young high-risk subjects.

REFERENCES

Blades, B. L., Dudman, N. P. B. and Wilcken, D. E. L. Variables affecting apolipoprotein B measurements in 3- to 5-day-old babies: a study of 4491 neonates. *Pediatr. Res.* 21 (1987) 608–614

Dudman, N. P. B., Blades, B. L., Wilcken, D. E. L. and Aitken, J. M. Radial immunodiffusion assay of apolipoprotein B in blood dried on filter paper – a potential screening method for familial type II hypercholesterolaemia. *Clin. Chim. Acta* 149 (1985) 117–127

Goldman, J. L. and Cook, E. F. The decline in ischemic heart disease mortality rates. *Ann. Intern. Med.* 101 (1984) 825–836

Goldstein, J. L. and Brown, M. S. Familial hypercholesterolaemia. In: Stanbury, J. B., Wyngaarden, J. B., Fredrickson, D. S., Goldstein, J. L. and Brown, M. S. (eds.) *The Metabolic Basis of Inherited Disease*, McGraw-Hill, New York, 5th Edn. (1983) pp. 672–712

Heller, R. F. The rise and fall of cardiovascular disease. *Med. J. Aust.* 144 (1986) 686–688

Lane, D. M. and McConathy, W. J. Changes in the serum lipids and apolipoproteins in the first four weeks of life. *Pediatr. Res.* 20 (1986) 332–337

McGill, H. C. Persistent problems in the pathogenesis of atherosclerosis. *Arteriosclerosis* 4 (1984) 443–451

Muller, C. Angina pectoris in hereditary xanthomatosis. *Arch. Intern. Med.* 64 (1939) 675–700

Newman, W. P., Freedman, D. S., Voors, A. W., Gard, P. D., Srinivasan, S. R., Cresanta, J. L., Williamson, G. D., Webber, L. S. and Berenson, G. S. Relation of serum lipoprotein levels and systolic blood pressure to early atherosclerosis. The Bogalusa Heart Study. *N. Engl. J. Med.* 314 (1986) 138–144

Stary, H. C. Evolution and progression of atherosclerosis in the coronary arteries of children and adults. In: Bates, S. R. and Gangloff, E. C. (eds.) *Atherogenesis and Ageing*, Springer-Verlag, New York (1987) pp. 20–36

Thompson, G. R. Evidence that lowering serum lipids favourably influences coronary heart disease. *Q. J. Med.* New Series 62 (1987) 87–95

Van Biervliet, J. P., Rosseneu, M., Bury, J., Caster, H., Stul, M. S. and Lamote, R. Apolipoprotein and lipid composition of plasma lipoproteins in neonates during the first month of life. *Pediatr. Res.* 20 (1986) 324–328

J. Inher. Metab. Dis. 11 Suppl. 1 (1988) 91–93

The Paediatric Lipid Clinic in Birmingham

M. TARLOW, A. GREEN, D. WORTHINGTON and E. BUCHANAN
Birmingham Children's Hospital, Ladywood Middleway, Ladywood, Birmingham B16 8ET, UK

Public and professional awareness of lipid disorders and their relevance to premature vascular disease is growing rapidly. Many families who have suffered from premature vascular disease now realise that risks can be reduced. The Birmingham paediatric lipid clinic was born over three years ago in an attempt to satisfy this unmet need. This short paper concentrates on two aspects only: the selection of high risk children and the practical organization of the clinic.

SELECTION

From the beginning, the clinic was envisaged as one for high-risk children only. Previous experience with a paediatric lipid clinic had highlighted the major problems of long-term compliance in very many children. It was hoped that by selecting only children at high risk, some of the management problems would be lessened. The clinic is closely linked with an adult clinic at Dudley Road Hospital, a district general hospital in Birmingham, from which many of the patients are referred.

The children in the clinic fall into two broad groups – those with hypercholesterolaemia and those with hypertriglyceridaemia.

Hypercholesterolaemia

These children are detected through family screening. Adults admitted to the coronary care unit at Dudley Road Hospital have serum cholesterol and triglyceride levels routinely measured. Those with raised lipid levels are referred to an adult physician with an interest in lipid disorders. He, in turn, screens all adult members of the family, and, if there is evidence of hypercholesterolaemia in any of these other family members, the children are referred to the paediatric lipid clinic. Thus, all hypercholesterolaemic children seen by this route have a family history of premature vascular disease and have more than one other member of the family with a raised serum cholesterol. In addition, almost all families have skin or tendon xanthomas or a premature corneal arcus.

These children, therefore, all have familial hypercholesterolaemia as defined on clinical grounds and are therefore all high-risk patients.

In addition to these, about 20% of referrals at present are from adult physicians, chemical pathologists or general practitioners for screening of children from families with hypercholesterolaemia and premature vascular disease.

Journal of Inherited Metabolic Disease. ISSN 0141–8955. Copyright © SSIEM and MTP Press Limited, Queen Square, Lancaster, UK.

Hypertriglyceridaemia

The second large group of patients in the children's lipid clinic in Birmingham is that with hypertriglyceridaemia – children with elevated chylomicron levels in the blood in association with actual or functional deficiency of tissue lipoprotein lipase (Frederickson's type I). These patients are said to be very rare, but there are already 11 from 6 families in the clinic. Six of these eleven were detected on the neonatal screening programme in the City of Birmingham. This was established by the late Dr Noel Raine, Chemical Pathologist at Birmingham Children's Hospital, who was one of the founders of the Society for the Study of Inborn Errors of Metabolism. This programme uses blood collected in heparinized capillary tubes for amino acid screening, for PKU, hypothyroidism and sickle cell haemoglobin. A diagnostic bonus of this screening technique is that lipaemic plasma can be detected in the neonatal period and these rare children with hypertriglyceridaemia can therefore be detected very early.

Four of the other five patients were detected on screening the family once the first child in a family had been diagnosed, and the fifth was discovered opportunistically when admitted to hospital for an unrelated problem.

ORGANIZATION

Following initial referral, the child's home diet is assessed by a 3-day dietary diary in which everything consumed over this period is recorded. This record is returned to the clinic dietician, and is assessed using a computer programme so that, when the child attends the clinic, a detailed dietary analysis is available.

The child is brought to the clinic fasting, and capillary blood samples are collected by the laboratory staff between 9 and 9.30 a.m. These samples are assayed for cholesterol and triglyceride and the results are brought to the clinic by the biochemist for 10.30 a.m. For the last 12 months, a few children have been monitored by home capillary blood sampling performed by the parents and posted to the hospital. This enables advice to be given over the telephone and often allows management with fewer clinic visits.

Both the biochemist and the dietician attend the clinic with the paediatrician: this eases any communication problems and ensures each patient gets optimal attention.

On their initial attendance, a full clinical and genetic history is taken and the relevant clinical examination is performed. Since the clinic has selected to see high-risk patients, it is often the case that the child's father has been very ill, or may even have died from coronary heart disease in the preceding months. The children are frequently anxious, and several have had left-sided chest pain of psychological origin. A full cardiological examination and reassurance has cleared this pain rapidly in all instances.

All patients with familial hypercholesterolaemia are managed on diet alone for at least a six-month period. Drug therapy is only used in extremely high-risk families (determined by family history of overt coronary heart disease) and situations where dietary therapy has been unsuccessful. Families are counselled about familial hypercholesterolaemia and management plans are determined in consultation with them. All patients are regularly followed up at intervals varying from 2 weeks to 6 months, depending on the clinical situation; over a four-year period, no patients have been lost to follow-up. It is hoped that this will provide useful information in the future on the effects of intervention in childhood hypercholesterolaemia.

J. Inher. Metab. Dis. 11 Suppl. 1 (1988) 94–109

Recent Advances in Cystic Fibrosis

M. A. McPHERSON

Department of Medical Biochemistry, University of Wales College of Medicine, Heath Park, Cardiff CF4 4XN, UK

Summary: Cystic fibrosis, one of the most common lethal inherited disorders in N. European and N. American populations, is characterized by the production of abnormally viscous mucous secretions in the lungs and digestive tract. The pathophysiological basis of the disease is unknown. However, during the last few years, rapid advances in molecular genetics and biochemical and physiological studies on cystic fibrosis epithelial cells have led to optimism that the cystic fibrosis defect will soon be identified. Current evidence suggests that the basic disturbance lies in altered regulation of protein secretion and electrolyte transport leading to an imbalance in composition of epithelial secretions in cystic fibrosis patients. Increasing knowledge of the mechanisms regulating production and secretion of mucins and movement of electrolytes across the cell membrane should lead to development of pharmacological manipulation(s) to correct the cellular abnormality. Ultimately, it is hoped that this will lead to the development of a rational treatment for cystic fibrosis patients.

INTRODUCTION

Cystic fibrosis is a common life-threatening genetic disease in Caucasian populations. It has an autosomal recessive mode of inheritance with an estimated carrier frequency of approximately 1:20 in the UK. Although molecular genetic studies have mapped the cystic fibrosis locus to the middle of the long arm of chromosome 7 (White *et al.*, 1985; Wainwright *et al.*, 1985), the cystic fibrosis gene has not yet been identified. However, there is optimism, as reviewed below, that complementary molecular genetic and cellular biochemical approaches will soon lead, not only to identification of the defective gene, but also to knowledge of how an abnormality in a single protein results in the complex pathophysiology of the disease.

Cystic fibrosis is characterized as an exocrinopathy, with clinical features of chronic lung disease and pancreatic insufficiency (Goodchild and Dodge, 1985). Pilocarpine-stimulated sweat electrolytes are elevated in cystic fibrosis patients (di Sant Agnese *et al.*, 1953) and this is the most widely used diagnostic test (Gibson and Cooke, 1959). Recent studies have also evaluated measurements of nasal transepithelial potential difference *in vivo* as a means of distinguishing controls and cystic fibrosis patients (Knowles *et al.*, 1981; Alton *et al.*, 1987b; Sauder *et al.*,

Journal of Inherited Metabolic Disease. ISSN 0141–8955. Copyright © SSIEM and MTP Press Limited, Queen Square, Lancaster, UK.

1987). The symptoms of cystic fibrosis result from the production, by epithelial cells, of abnormally viscous mucous secretions which occlude the lungs, pancreatic ducts and vas deferens (Goodchild and Dodge, 1985). It is not clear whether the increased viscosity of the secretions is due to excessive mucus production or alteration in physicochemical properties, caused, for example, by alteration in the water content of the secretions. An investigation of the biochemical mechanisms leading to the production of abnormal mucus and how this relates to the elevated sodium chloride content of cystic fibrosis sweat is fundamental to understanding the basic biochemical disturbance in cystic fibrosis. Within the last few years, studies on affected epithelial cell types have defined a common abnormality which suggests that the cystic fibrosis defect lies in a regulator protein involved in stimulus–response coupling (McPherson and Dormer, 1987, 1988; McPherson and Goodchild, 1988). This would lead to inappropriate amounts of electrolytes and proteins being secreted in response to a stimulus and might, therefore, explain the characteristic abnormalities seen in cystic fibrosis.

MAPPING THE CYSTIC FIBROSIS GENE

Studies aimed at identifying the cystic fibrosis gene in the absence of knowledge of the protein defect have looked for protein or DNA markers linked to cystic fibrosis. A breakthrough came in 1985 when, after approximately 40% of the genome had been excluded for cystic fibrosis, a linkage was established. Initially, biochemical studies showed that a polymorphic activity of the serum enzyme, paraoxonase, segregated with cystic fibrosis in family linkage studies (Eiberg *et al.*, 1985); but paraoxonase had not been assigned to a particular chromosome. The discovery that an anonymous polymorphic DNA probe, DoCRI-917 (D7S15), was linked to both cystic fibrosis and paraoxanase (Tsui *et al.*, 1985) soon led to the localization of the cystic fibrosis gene to chromosome 7 (Knowlton *et al.*, 1985). Other markers much closer to the cystic fibrosis locus: pJ3.11 (D7S8), an anonymous DNA sequence and met, a met oncogene probe were identified and the cystic fibrosis gene was thus mapped to 7q2.2–7q3.1 (White *et al.*, 1985; Wainwright *et al.*, 1985). Several other probes, including the T-cell receptor β-chain gene (TCRB) and a collagen gene probe (COL1A2), have also been shown to be linked to cystic fibrosis, and, on the basis of recombination data, an order of 7qter-TCRB-J3.11-CF-met-7C22-B79a-COL1A2-7cen has been assigned (Beaudet *et al.*, 1986; Estivill *et al.*, 1987; Zengerling *et al.*, 1987; Lathrop *et al.*, 1988). At least 95% of families studied segregate with the same locus, indicating that cystic fibrosis is, in the majority of cases, a single gene disorder (Beaudet *et al.*, 1986) as was suggested from previous studies based on the incidence of first cousin marriages among cystic fibrosis parents (Romeo *et al.*, 1985).

Genetic data indicate that pJ3.11 (D7S8) and met flank the cystic fibrosis gene and are within 2–3 000 kbases. Because of their close proximity to the cystic fibrosis locus, the probes have been used successfully in prenatal diagnosis and carrier detection in informative families with an affected child (Farrall *et al.*, 1986a,b). Prenatal diagnosis based on measurements of activity of microvillar enzymes in amniotic fluid (Brock *et al.*, 1985) is still carried out, however, in families at risk where the DNA probes are uninformative or where DNA from an affected child is not available for analysis. Strategies aimed at identifying the CF gene (Estivill *et al.*, 1987) have used chromosome-mediated gene transfer to construct mouse/human hybrid cell lines which contain both met and J3.11 and, by inference, the cystic fibrosis locus. These have been analysed for coding sequences by identifying HTF islands, using a methylation-sensitive restriction enzyme, since high-frequency non-methylated CpG regions are known to precede many vertebrate genes. Two new probes, CS.7 and XV-2c, which define one such coding sequence map closer to cystic fibrosis than either J3.11 or met and are within 40 k bases of the cystic fibrosis locus (Estivill *et al.*, 1987). The probes are in such strong disequilibrium with cystic fibrosis (approximately 95% of cystic fibrosis chromosomes have the same haplotype) that they can be used potentially for carrier exclusion in the general population. In addition, data argue against genetic heterogeneity at the cystic fibrosis locus and it has been suggested that a single mutational event accounts for most of the cases of cystic fibrosis (Estivill *et al.*, 1987).

The coding sequence defined by the markers CS.7 and XV-2c has recently been sequenced from normal and cystic fibrosis individuals and shows no mutation, suggesting that, although very close to the cystic fibrosis locus, this is not the cystic fibrosis gene. Recently, chromosome walking techniques have isolated another candidate gene sequence (B. Wainwright, personal communication). Northern blot analysis is being used to determine whether the gene is expressed in epithelial cell types affected in cystic fibrosis and whether the sequence is mutated in cystic fibrosis individuals. Ultimately, it may also be necessary to show that insertion of mRNA or antibody against a putative protein coded for by the candidate gene modifies epithelial cell function in a manner predicted from cellular physiological and biochemical studies. As reviewed below, these studies suggest a defect in a regulator of ion transport and protein secretion in cystic fibrosis epithelial cells.

REGULATION OF SECRETION IN EPITHELIAL CELLS

Type of stimuli

Epithelial cells are polarized such that secretion of proteins and fluids across the apical (or luminal) membrane is controlled by neurotransmitters of the autonomic nervous system and by circulating hormones which act at the basolateral (or serosal) membrane. The type of agonist determines the nature of the response. Thus, stimulation by cholinergic agonists or cholecystokinin evokes both enzyme and fluid secretion from exocrine pancreas (Case, 1978), whereas cholinergic stimulation of salivary and sweat glands results in copious fluid secretion containing small amounts of protein (Schneyer *et al.*, 1972). Noradrenaline primarily increases protein output by its β-effect and fluid secretion by its α-effect in salivary glands. β-Adrenergic stimulation, as in sweat glands, is thus a minor fluid-secreting stimulus (Schneyer *et al.*, 1972; Sato and Sato, 1984). By contrast, in respiratory airways epithelial cells adrenaline, acting by its β-effect, is a major stimulus for fluid secretion (Welsh, 1986). The primary fluid secreted by epithelial cells is a sodium chloride-rich isotonic fluid which, in exocrine glands, is modified as it passes down the ducts. Pancreatic duct cells secrete bicarbonate and water in response to the gut hormone, secretin (Hadorn *et al.*, 1968); whereas salivary and sweat duct cells reabsorb sodium and chloride in excess of water and secrete small amounts of potassium and bicarbonate, resulting in a final saliva or sweat which is hypotonic. There is evidence that ductal reabsorption of electrolytes in sweat (Schwarz and Simpson, 1985; Pedersen *et al.*, 1986) and salivary glands (Schneyer and Thavornthon, 1973) is under autonomic control.

Studies on human tissues are not as complete as those on animals where detailed mechanisms for secretion of proteins and fluids and reabsorption of electrolytes have been proposed (Petersen, 1986; Welsh and Liedtke, 1986; McPherson and Dormer, 1987). However, key experiments on scarce human epithelial and exocrine tissues suggest that they have similar mechanisms for producing secretions and that they are regulated in a similar way (McPherson and Dormer, 1987).

Intracellular mechanisms controlling protein and fluid secretion

Secretory proteins, which include digestive enzymes and mucins, are packaged in a highly condensed form in zymogen granules. They are secreted by fusion of granule membrane to apical cell membrane and release of contents: an event known as exocytosis. The final events explaining how membrane fusion occurs on stimulation of exocytosis (Zimmerberg, 1987) are not well understood; but there is evidence for involvement of Ca^{2+} and GTP-binding proteins (Gomperts, 1986). The trigger for protein secretion must also involve movement of the secretory granule to the cell membrane (Burgoyne and Cheek, 1987) such that fusion can occur. Morphological studies have shown that human salivary glands contain zymogen granules similar to those seen in animal cells (Mangos and Donnelly, 1981). It has also been shown that cholinergic stimulation induces amylase release from human exocrine pancreatic acinar cells (Petersen *et al.*, 1985) and that β-adrenergic stimulation is the primary stimulus for amylase secretion from human parotid acinar cells (Mangos, 1981) and for amylase and mucin secretion from human submandibular acinar cells (McPherson *et al.*, 1985, 1986a).

The current model proposed for fluid secretion from exocrine acinar cells and respiratory airways epithelial cells incorporates a Na^+–K^+-ATPase in the basolateral membrane as the driving force (Petersen, 1986; Welsh and Liedtke, 1986; McPherson and Dormer, 1987). Na^+ is pumped out of the cell in exchange for K^+ in a ratio of $3Na^+:2K^+$. Na^+ re-enters on a basolateral Na^+–K^+–$2Cl^-$ cotransporter, which leads to a concentration of intracellular Cl^- above its equilibrium potential. Opening of an apical membrane Cl^- channel allows Cl^- to exit passively from the cell. An important feature of the model, which has been demonstrated in human exocrine pancreatic acinar cells (Petersen *et al.*, 1985) and tracheal cells (Welsh and Liedtke, 1986) is a Ca^{2+}-activated basolateral K^+ channel, which, when open, allows K^+ to exit across the basolateral membrane, thus preventing depolarization. This allows further entry of K^+ on the cotransporter and results in increased Cl^-

secretion. Na^+ is thought to follow by a paracellular route and water by osmosis, leading to secretion of an isotonic primary fluid. In this model, maximum fluid secretion is evoked by a stimulus which opens both apical membrane Cl^- channels and basolateral K^+ channels, and a stimulus which opens only apical Cl^- channels evokes a small amount of fluid secretion.

A link has been proposed between regulation of protein secretion by exocytosis and fluid secretion triggered by opening of apical membrane Cl^- channels (McPherson *et al.*, 1986a,b; McPherson and Dormer, 1987; McPherson and Goodchild, 1988). A mechanism for such a link has been suggested by recent evidence indicating that Cl^- channels are present on secretory granule membranes (DeLisle and Hopfer, 1986). Thus, fusion of zymogen granule membrane with apical membrane during exocytosis could lead to insertion of apical membrane Cl^- channels. During stimulation of protein secretion, opening of an apical Cl^- channel at the site of exocytosis might aid in wash out of granule contents. However, under non-physiological stimulatory conditions, as are evoked by a pure α- or β-adrenergic stimulus, stimulation of fluid and protein secretion can be dissociated in salivary acinar cells (Mangos *et al.*, 1975; Martinez *et al.*, 1976; Quissell and Barzen, 1980; McPherson and Dormer, 1984a; Doughney *et al.*, 1987b).

The main intracellular messengers controlling secretion of fluids and proteins in epithelial cells are cyclic AMP and Ca^{2+} (Dormer *et al.*, 1987; McPherson and Dormer, 1987). There is evidence that β-adrenergic agonists and the gut hormone, secretin, act by increasing cyclic AMP concentrations and that α-adrenergic and cholinergic stimuli act by increasing intracellular Ca^{2+}. In exocrine pancreatic acinar cells, direct evidence has accumulated that Ca^{2+} is a key regulator of secretion. Thus, increases in intracellular free Ca^{2+}, in response to carbamylcholine, precede enzyme secretion (Dormer, 1983; Ochs *et al.*, 1985) and incorporation of Ca^{2+} chelators into the cells inhibits stimulation of secretion (Dormer, 1984). In exocrine pancreas and sweat glands, cholinergic stimulation is initially independent of extracellular Ca^{2+} (Dormer *et al.*, 1987; Pedersen, 1987). Evidence suggests that Ca^{2+} is mobilized from an intracellular store in the rough endoplasmic reticulum by the action of inositol 1,4,5-trisphosphate (Streb *et al.*, 1984; Brown *et al.*, 1987), formed as a result of cholinergic stimulation of phospholipase C in the basolateral membrane (Doughney *et al.*, 1987a,c). The other product of phospholipase C action, diacylglycerol, activates protein kinase C, which is also thought to play a role in exocrine secretion (DePont and Fleurens-Jakobs, 1984). Noradrenaline has been shown to cause rapid increases in Ins 1,4,5-P_3 and Ins P_4 in submandibular acinar cells, which are mediated by an α-effect and are likely to be related to Ca^{2+}-mediated fluid secretion (Doughney *et al.*, 1987b).

Although second messengers regulating secretion have been identified, the mechanisms of how they act to trigger secretory events are undefined. Current evidence suggests that cyclic AMP, acting via cyclic AMP-dependent protein kinase (A-kinase), and Ca^{2+}, acting via calmodulin to stimulate Ca^{2+}/calmodulin-dependent protein kinase, stimulates phosphorylation of specific protein(s) and that this triggers apical membrane Cl^- secretion and exocytosis. Mechanisms controlling reabsorption of electrolytes in the sweat duct are less well understood, although progress in understanding the actions of autonomic agonists (Lee *et al.*, 1984; Pedersen *et al.*, 1986; Pedersen, 1987; Doughney *et al.*, 1987c; Kealey, 1987) and kinins (Brayden, Cuthbert and Lee, unpublished) is being made, using cultured cells.

Interactions between messenger pathways have been demonstrated in submandibular acinar cells where β-adrenergic stimulation of mucin release, although accompanied by increases in cyclic AMP concentration, is partially dependent on extracellular Ca^{2+} (Quissell and Barzen, 1980; McPherson and Dormer, 1984a) and where it has been shown that β-agonists increase Ca^{2+}-mobilization (McPherson and Dormer, 1984a,b). There is increasing evidence for 'cross-talk' between messenger pathways: for example, it was recently demonstrated in an erythrocyte model that protein kinase C can phosphorylate adenylate cyclase and enhance its activity (Yoshimasa *et al.*, 1987) and it is well known that Ca^{2+}/calmodulin can activate adenylate cyclase and cyclic nucleotide phosphodiesterase in different cell types (Klee and Newton, 1985).

DEFECTIVE β-ADRENERGIC RESPONSES IN CYSTIC FIBROSIS

Studies on a variety of affected epithelial cell types *in vitro* have shown a marked defect in β-adrenergic responsiveness of cystic fibrosis cells. Responses to other agonists which increase cyclic AMP levels are also markedly reduced in cystic fibrosis, suggesting a generalized regulatory defect beyond the level of the receptor.

Table 1   Defective cyclic AMP-mediated responses in cystic fibrosis epithelial cells

Agonist	Decreased response	Reference
Isoproterenol Noradrenaline	Secretion of mucins and amylase from submandibular acinar cells	McPherson *et al.*, 1984, 1985, 1986a, 1987
Isoproterenol	Fluid secretion from sweat glands	Sato and Sato, 1984
Isoproterenol	Cl^- reabsorption in cultured sweat duct cells	Pedersen *et al.*, 1986
Isoproterenol Adrenaline Forskolin 8-Bromo-cyclic AMP	Apical membrane Cl^- secretion in respiratory epithelial cells	Welsh and Liedtke, 1986 Frizzell *et al.*, 1986 Widdicombe, 1986 Boucher *et al.*, 1986 Cotton *et al.*, 1987
Prostaglandin E_2	Cl^- secretion in jejunal tissues	Taylor *et al.*, 1987

The studies leading to this consensus, which are summarized in Table 1, are described below.

Secretion of mucins and amylase in submandibular acinar cells

Submandibular glands from cystic fibrosis patients show dilated ducts, similar to those in the pancreas, with inspissations in the secretions which have an elevated electrolyte content (Goodchild and Dodge, 1985). However, the glands do not have the severe fibrosis and acinar cell degeneration seen in the pancreas and are not chronically infected; thus they provide a good model for investigating the primary cause of abnormalities in cystic fibrosis. *In vitro* studies on small numbers of cystic fibrosis glands have shown a consistent, markedly defective response to β-adrenergic stimulation (McPherson *et al.*, 1985, 1986a). The morphology of control and cystic fibrosis tissues was similar and *in vitro* preparations were shown to be viable by measurements of ATP content and lactate dehydrogenase release (McPherson *et al.*, 1985, 1986a). In addition, secretory responses of control human submandibular acini and tissue fragments incubated *in vitro* were similar to those of rat in terms of maximum secretion rates and dose dependence of agonist stimulation (McPherson and Dormer, 1984b; McPherson *et al.*, 1986a, 1987).

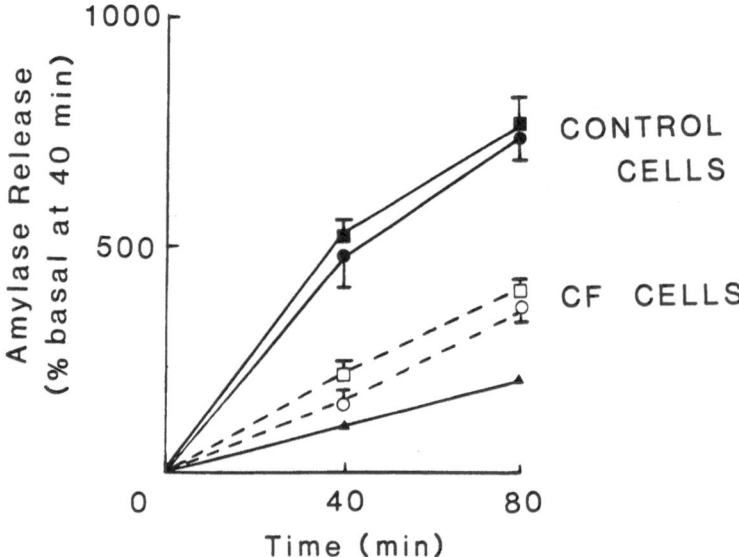

Figure 1 Amylase secretion in response to isoproterenol and noradrenaline in control and cystic fibrosis submandibular acinar cells. Amylase release measured as described (McPherson *et al.*, 1985, 1986a) in response to 10μmol$\,$L^{-1} isoproterenol (●) or 10μmol$\,$L^{-1} noradrenaline (■) in control cells, and to 10μmol$\,$L^{-1} isoproterenol (○) or 10μmol$\,$L^{-1} noradrenaline (□) in cystic fibrosis cells. ▲ basal (unstimulated) release. Data are mean±SEM for at least three observations

Figure 1 shows a time course of amylase release from control and cystic fibrosis submandibular acinar cells in response to a maximally effective concentration of either the β-agonist, isoproterenol, or the physiological neurotransmitter, noradrenaline. In control cells isoproterenol and noradrenaline stimulated secretion to the same extent, approximately four-fold above basal. In contrast, amylase secretion was markedly reduced in cystic fibrosis cells ($p<0.01$ for difference from secretion in control cells); although increases above basal release in response to both isoproterenol and noradrenaline were seen (Figure 1). In two of the cystic fibrosis glands, noradrenaline increased amylase secretion to a greater degree than did isoproterenol, but the overall effect was not significant. In terms of mucin release, an increase approximately two-fold above basal level was seen in response to isoproterenol ($10\,\mu$mol L^{-1}) in control cells; but no response to isoproterenol ($10\,\mu$mol L^{-1}) was observed in cystic fibrosis cells ($108\pm9\%$ basal, $n = 3$).

Table 2 Protein secretion in human submandibular acinar cells is mediated by stimulation of β-adrenergic receptors

Experimental conditions		Amylase release (% basal)	
		Control cells	Cystic fibrosis cells
Isoproterenol	($10\,\mu$mol L^{-1})	386 ± 43	$143\pm\ 2$
+ propranolol	($30\,\mu$mol L^{-1})	110 ± 12	–
+ phentolamine	($30\,\mu$mol L^{-1})	421 ± 48	–
Noradrenaline	($10\,\mu$mol L^{-1})	327 ± 36	219 ± 42
+ propranolol	($30\,\mu$mol L^{-1})	148 ± 17	123 ± 22
+ phentolamine	($30\,\mu$mol L^{-1})	352 ± 38	248 ± 38

Amylase release from tissue fragments incubated *in vitro* for 40 min under the experimental conditions shown was measured as described (McPherson *et al.*, 1985, 1986a). Data are mean±SEM for four observations from control glands and mean±range for two observations from cystic fibrosis glands

Table 2 shows that the actions of either isoproterenol or noradrenaline in control and cystic fibrosis cells were blocked by the β-blocker, propranolol, but not by the α-blocker, phentolamine. As in rat cells (McPherson and Dormer, 1984a), noradrenaline stimulation was not completely inhibited by the β-blocker, suggesting a non-α, non-β component of noradrenaline action. In cystic fibrosis cells, the small amount of secretion induced by noradrenaline was mediated by stimulation of β- but not α-receptors (Figure 1), ruling out a switch in function at the receptor level, but not necessarily at an intracellular level (McPherson *et al.*, 1986a,b, 1987). Mucin secretion in control cells evoked by isoproterenol or noradrenaline also gave the same pattern of responses to adrenergic blockers (McPherson *et al.*, 1984a, 1987).

Thus, secretion of mucins and amylase in response to isoproterenol and noradrenaline was mediated by stimulation of β-adrenergic receptors and this was markedly reduced in cystic fibrosis submandibular acinar cells.

Fluid secretion and electrolyte transport in sweat and respiratory epithelial cells

Experiments by Quinton (1983) showed that isolated perfused cystic fibrosis sweat ducts were markedly impermeable to Cl$^-$. The finding of an increased bioelectric potential difference across cystic fibrosis respiratory epithelia (Knowles *et al.*, 1981) led to the hypothesis (Quinton, 1983) that there was a common defect in Cl$^-$ permeability in all affected epithelia in cystic fibrosis. More recently, defective β-adrenergic regulation of Cl$^-$ transport and fluid secretion has been demonstrated in cystic fibrosis sweat and respiratory airway epithelial cells.

Thus, Sato and Sato (1984) showed that sweating *in vivo* and fluid secretion from isolated sweat glands *in vitro* was not increased in response to isoproterenol in cystic fibrosis patients, but that most of the control subjects responded. However, only small volumes of sweat were produced since β-adrenergic stimulation is not the primary stimulus for fluid secretion in these cells. The results have been substantiated by a recent independent study (Behm *et al.*, 1987), where a decreased sweat response to isoproterenol in cystic fibrosis heterozygotes (Sato and Sato, 1984) has also been confirmed. In tracheal and nasal airways epithelial cells, β-adrenergic stimulation is a main stimulator of fluid secretion and opens apical Cl^- channels and basolateral K^+ channels (Welsh and Liedtke, 1986).

Several groups of investigators have shown, using Ussing chambers, intracellular microelectrodes (Widdicombe, 1986; Cotton *et al.*, 1987) and patch clamp techniques (Welsh and Liedtke, 1986; Frizzell *et al.*, 1986), that β-adrenergic stimulation of apical membrane Cl^- transport is defective in cystic fibrosis. In cultured tracheal cells, patch clamped in the cell attached mode, isoproterenol opened apical membrane Cl^- channels in 50–60% of control cells but not in cystic fibrosis cells. Cystic fibrosis cells were also unresponsive to 8-bromo-cyclic AMP and forskolin (Table 1). In addition, coupled to the decreased Cl^- transport, was an observed increase in Na^+ transport in cystic fibrosis cells (Boucher *et al.*, 1986), which suggested a generalized regulatory defect, rather than a defect only at the site of the Cl^- channel. This was in agreement with studies on salivary acinar cells (McPherson *et al.*, 1985, 1986a,b) and the finding that bicarbonate secretion from pancreatic ducts in response to the gut hormone, secretin, was also defective in cystic fibrosis patients (Hadorn *et al.*, 1968). Elevated electrolytes in cystic fibrosis sweat result from a decreased ability of the sweat duct to reabsorb sodium chloride. Quinton (1983) showed that this was due to a marked decrease in chloride permeability of isolated perfused cystic fibrosis sweat ducts and a similar defect has also been demonstrated in cultured cystic fibrosis sweat duct cells (Pedersen *et al.*, 1986). In addition, Cl^- conductance, in the presence of amiloride, was increased in response to isoproterenol in control but not in cystic fibrosis sweat duct cells (Pedersen *et al.*, 1986). Very recent data indicate that Cl^- transport in cystic fibrosis jejunal tissues is not increased by agonists which are thought to act by increasing cyclic AMP concentrations (Taylor *et al.*, 1987).

The β-adrenergic and Cl^- permeability defect was expressed in epithelial cells in the absence of cystic fibrosis serum or sweat (McPherson *et al.*, 1985, 1986a; Sato and Sato, 1984; Bijman and Quinton, 1984; Welsh and Liedtke, 1986; Frizzell *et al.*, 1986; Cotton *et al.*, 1987) and in cells maintained in culture for a few weeks (Yankaskas *et al.*, 1985; Pedersen *et al.*, 1986), suggesting an intrinsic cellular abnormality, rather than an abnormality resulting from influence of an abnormal circulating serum factor (McPherson *et al.*, 1983; Dorin *et al.*, 1987). The finding that the transepithelial potential difference of transplanted trachea was normal in a cystic fibrosis patient one year after heart-lung transplant (Alton *et al.*, 1987c) also supports this view.

A consensus has thus emerged for a defect in a regulator of protein secretion and Cl^- transport in affected epithelial cells in cystic fibrosis.

THE NATURE OF THE REGULATORY DEFECT

In all of the affected tissues studied, results suggest that the decreased responsiveness in cystic fibrosis is not due to a receptor abnormality or a defect in receptor–adenylate cyclase coupling. Thus, maximum cyclic AMP increases in response to β-agonists have been shown to be normal in cystic fibrosis submandibular (McPherson *et al.*, 1985, 1987), sweat (Sato and Sato, 1984; T. Kealey, personal communication) and respiratory epithelial cells (Welsh and Liedtke, 1986; Widdicombe, 1986; Boucher *et al.*, 1986) and non-receptor-mediated responses, induced by cyclic AMP analogues or forskolin, are also decreased in cystic fibrosis cells (Table 1).

The finding that decreased β-responses of cystic fibrosis submandibular acinar cells were partially restored by a cyclic nucleotide phosphodiesterase inhibitor (McPherson *et al.*, 1985, 1986a, 1987) that β-responses in the presence of theophylline were abnormally increased in cystic fibrosis parotid acinar cells (Mangos, 1981) and that cyclic AMP levels were slower to reach maximum in cystic fibrosis sweat glands than controls (Sato and Sato, 1984) suggested overactivity of a cyclic nucleotide phosphodiesterase in cystic fibrosis. This led to the proposal that alteration in activity of calmodulin, which is known to activate cyclic nucleotide phosphodiesterase (Klee and Newton, 1985), would disturb, not only cyclic AMP metabolism, but also Ca^{2+} homeostasis in cystic fibrosis, and, thus, might be directly related to the genetic defect (McPherson *et al.*, 1985, 1986a, b, 1987). Calmodulin itself has been excluded as the site of the basic defect in cystic fibrosis (Scambler *et al.*, 1987). However, an increased calmodulin-like activation of cyclic AMP phosphodiesterase has been demonstrated in cystic fibrosis submandibular extracts; this correlates with increased calmodulin binding to a 61 000 molecular weight protein (McPherson *et al.*, 1986b, 1987; Shori *et al.*, 1988).

The altered calmodulin-binding (acceptor) protein demonstrated in cystic fibrosis submandibular glands could be one of the target enzymes of calmodulin or a modulator protein which interacts to modify activation by calmodulin of one of its target enzymes. A calmodulin-acceptor protein, calcinerurin, first identified in brain tissues, inhibits Ca^{2+}/calmodulin-dependent activation of cyclic nucleotide phosphodiesterase and also acts as a Ca^{2+}/calmodulin-dependent phosphatase (Pallen and Wang, 1985). Alteration in a calmodulin-acceptor protein with similar dual functions in cystic fibrosis exocrine and epithelial cells (Figure 2) might neutralize the actions of cyclic AMP by (a) increasing cyclic AMP breakdown due to alteration in its inhibitory action on calmodulin-dependent phosphodiesterase, and (b) dephosphorylating protein(s) phosphorylated by A-kinase. This would provide a link between altered calmodulin function in cystic fibrosis (McPherson *et al.*, 1986a,b, 1987; Shori *et al.*, 1988) and the finding that A-kinase was ineffective in opening apical membrane Cl^- channels when added to excised membrane patches from cystic fibrosis respiratory cells (Schoumacher *et al.*, 1987; Li *et al.*, 1988). Alternatively, an inappropriate substrate for phosphorylation by A-kinase (Schoumacher *et al.*, 1987) and Ca^{2+}/calmodulin-dependent protein kinase might underlie the cystic fibrosis defect, particularly as a role for Ca^{2+}/calmodulin and calmodulin acceptor protein(s) in stimulating Cl^- transport in bovine tracheal apical membranes vesicles has been directly demonstrated (Dubinsky, 1987). The hypothetical

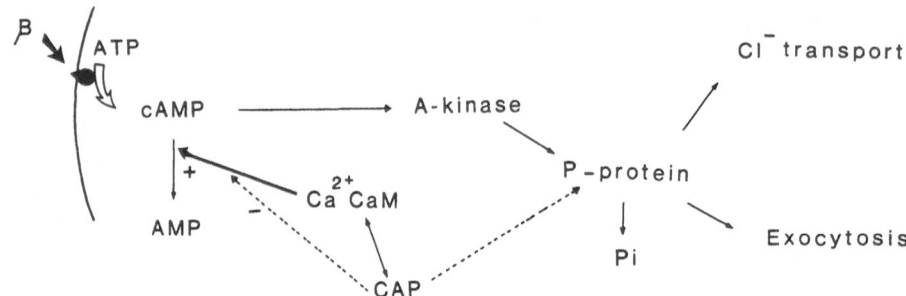

Figure 2 A putative link between an altered calmodulin-acceptor protein and defective regulation of apical membrane transport in cystic fibrosis. A putative alteration in a calmodulin-acceptor protein (CAP) in cystic fibrosis could modify both activation of cyclic nucleotide phosphodiesterase by Ca^{2+}/calmodulin and protein phosphorylation mediated by A-kinase or Ca^{2+}/calmodulin-dependent protein kinase.
Abbreviations: cAMP – cyclic 3′ 5′-adenosine monophosphate; A-kinase – cyclic AMP-dependent protein kinase; CaM – calmodulin; CAP – calmodulin-acceptor protein; P-protein – phosphorylated protein(s); Pi – inorganic phosphate

scheme (Figure 2) illustrates how alteration in a Ca^{2+}-dependent regulator protein might modify cyclic AMP-mediated events. Calmodulin function in cystic fibrosis cells could also be disturbed as a result of altered Ca^{2+} entry, mediated by alteration in pathways of inositol phosphate formation (Pedersen *et al.*, 1987; Doughney *et al.*, 1987c). In either case, selective actions of Ca^{2+} would also be predicted to be altered in cystic fibrosis. Altered Ca^{2+}-mediated responses have been reported in cystic fibrosis patients (Davis *et al.*, 1980; Katz *et al.*, 1984) and there is increasing evidence for altered Ca^{2+}-related events in cystic fibrosis epithelial cells (Mangos, 1981; Pedersen, 1987; Pedersen *et al.*, 1987; Doughney *et al.*, 1987c; Taylor *et al.*, 1987).

It is thus apparent that the cyclic AMP and Ca^{2+} pathways regulating apical membrane events of secretion of proteins and fluid and reabsorption of electrolytes are interlinked. It is important to determine the site(s) of interaction (cross-talk) between the messenger pathways operative during physiological stimulation to elucidiate where the primary defect in cystic fibrosis lies and to determine which pathways are likely to be modified as a consequence of the defect.

PERSPECTIVES

There is optimism that complementary molecular genetic and cellular biochemical studies will soon lead to knowledge of the basic defect in cystic fibrosis. Molecular genetic approaches have already provided DNA probes which give reliable prenatal diagnosis in families at risk. It is likely that identification of the mutation(s) in the cystic fibrosis gene will make it theoretically possible for carrier detection and prenatal screening for cystic fibrosis to be extended to the general population.

In terms of devising a rational treatment for cystic fibrosis patients, pharmacological manipulation of defective epithelial cell responses at present affords a more realistic possibility than does gene therapy. In this regard, it will be necessary to understand how the cystic fibrosis defect leads to abnormalities in secretory function which are seen in cystic fibrosis epithelial cells. The importance of understanding interactions in messenger pathways controlling secretion is apparent since it might be possible to intervene pharmacologically at a functionally relevant site beyond the cystic fibrosis protein defect. For example, pharmacological stimulation of compensatory pathways which can initiate secretion of proteins and electrolytes might lead to correction of abnormalities in cystic fibrosis secretions which are thought to cause most of the clinical manifestations of the disease.

ACKNOWLEDGEMENTS

I am very grateful to Dr R. L. Dormer for many helpful discussions. I thank Dr M. C. Goodchild, Professor J. A. Dodge, Dr M. Alfaham, Dr P. K. Jeffery and Mr R. Sharma for their invaluable help in providing cystic fibrosis submandibular material. I also thank the Cystic Fibrosis Research Trust, UK for generous support.

REFERENCES

Alton, E. W. F. W., Batten, J., Hodson, M., Wallwork, J., Higenbottam, T. and Geddes, D. M. Absence of electrochemical defect of cystic fibrosis in transplanted lung. *Lancet* 1 (1987a) 1026

Alton, E. W. F. W., Hay, J. G., Munro, C. and Geddes, D. M. Measurement of nasal potential difference in adult cystic fibrosis, Young's syndrome and bronchiectasis. *Thorax* (1987b) in press

Beaudet, A., Bowcock, A., Buchwald, M., Cavalli-Sforza, L., Farrall, M., King, M.-C., Klinger, K., Lalouel, J.-M., Lathrop, G., Naylor, S., Ott, J., Tsui, L.-C., Wainwright, B., Watkins, P., White, R. and Williamson, R. Linkage of cystic fibrosis to two tightly linked DNA markers: joint report from a collaborative study. *Am. J. Hum. Genet.* 39 (1986) 681-693

Behm, J. K., Hagiwara, G., Lewiston, N. J., Quinton, P. M. and Wine, J. J. Hyposecretion of β-adrenergically induced sweating in cystic fibrosis heterozygotes. *Pediatr. Res.* 22 (1987) 271-276

Bijman, J. and Quinton, P. M. Apparent absence of cystic fibrosis sweat factor on ion-selective and transport properties of the perfused human sweat duct. *Pediatr. Res.* 18 (1984) 1292-1296

Boucher, R. C., Stutts, M. J., Knowles, M. R., Cantley, L. and Gatzy, J. T. Na$^+$ transport in cystic fibrosis respiratory epithelia. Abnormal basal rate and response to adenylate cyclase activation. *J. Clin. Invest.* 78 (1986) 1245-1252

Brock, D. J. H., Bedgood, D., Barron, L. and Hayward, C. Prospective prenatal diagnosis of cystic fibrosis. *Lancet* 2 (1985) 1175-1178

Brown, G. R., Richardson, A. E. and Dormer, R. L. The role of a Ca^{2+}, Mg-ATPase of the rough endoplasmic reticulum in regulating intracellular Ca^{2+} during cholinergic stimulation of rat pancreatic acini. *Biochim. Biophys. Acta* 902 (1987) 87-92

Burgoyne, R. D. and Cheek, T. R. Reorganisation of peripheral actin filaments as a prelude to exocytosis. *Biosci. Rep.* 7 (1987) 281-288

Case, R. M. Synthesis, intracellular transport and discharge of exportable proteins in the pancreatic acinar cell and other cells. *Biolog. Rev.* 53 (1978) 211-354

Cotton, C. U., Stutts, M. J., Knowles, M. R., Gatzy, J. T. and Boucher, R. C. Abnormal apical cell membrane in cystic fibrosis respiratory epithelium. *J. Clin. Invest.* 79 (1987) 80–85

Davis, P. B., Shelhamer, J. R. and Kaliner, M. Abnormal adrenergic and cholinergic sensitivity in cystic fibrosis. *N. Engl. J. Med.* 302 (1980) 1453–1456

De Lisle, R. C. and Hopfer, U. Electrolyte permeabilities of pancreatic zymogen granules: implications for pancreatic secretion. *Am. J. Physiol.* 250 (1986) G489–G496

De Pont, J. J. H. M. and Fleurens-Jakobs, A. M. M. Synergistic effect of A23187 and a phorbol ester on amylase secretion from rabbit pancreatic acini. *FEBS Lett.* 170 (1984) 64–68

Di Sant'Agenese, P. A., Darling, G. A. and Shea, E. Abnormal electrolyte composition of sweat in cystic fibrosis of the pancreas. *Pediatrics* 12 (1953) 549–557

Dorin, J. R., Novak, M., Hill, R. E., Brock, D. J. H., Secher, D. S. and van Heyningen, V. A clue to the basic defect in cystic fibrosis from cloning the cystic fibrosis antigen gene. *Nature (London)* 326 (1987) 614–617

Dormer, R. L. Direct demonstration of increases in cytosolic free Ca^{2+} during stimulation of pancreatic enzyme secretion. *Biosci. Rep.* 3 (1983) 233–240

Dormer, R. L. Introduction of calcium chelators into isolated rat pancreatic acini inhibits amylase release in response to carbamylcholine. *Biochem. Biophys. Res. Commun.* 119 (1984) 876–883

Dormer, R. L., Brown, G. R. Doughney, C. and McPherson, M. A. Intracellular Ca^{2+} in pancreatic acinar cells: regulation and role in stimulation of enzyme secretion. *Biosci. Rep.* 7 (1987) 333–344

Doughney, C., Brown, G. R., McPherson, M. A. and Dormer, R. L. Rapid formation of inositol-1,4,5, P_3 in response to carbachol in rat pancreatic acini. *Biochim. Biophys. Acta* 928 (1987a) 341–347

Doughney, C., Dormer, R. L. and McPherson, M. A. Adrenergic regulation of formation of inositol phosphates in rat submandibular acini. *Biochem. J.* 241 (1987b) 705–709

Doughney, C., Pedersen, P. S., McPherson, M. A. and Dormer, R. L. Autonomic regulation of inositol phosphate formation in cultured human sweat duct cells. *Pediatr. Pulmonol. Suppl.* 1 (1987c) 119

Dubinsky, W. P. Reconstitution of chloride channels from tracheal epithelia. *Pediatr. Pulmonol*, Suppl. 1 (1987) 51–52

Eiberg, H., Mohr, J., Schmiegelow, K., Nielsen, L. S. and Williamson, R. Linkage relationships of paraoxonase (PON) with other markers: indication of PON-cystic fibrosis synteny. *Clin. Genet.* 28 (1985) 265–271

Estivill, X., Farrall, M., Scambler, P. J., Bell, G. M., Hawley, K. M. F., Lench, N. J., Bates, G. P., Kruger, H. C., Frederick, P. A., Stanier, P., Watson, E. K., Williamson, R. and Wainwright, B. J. A candidate for the cystic fibrosis locus isolated by selection for methylation-free islands. *Nature (London)* 326 (1987) 840–845

Farrall, M., Law, H.-Y., Rodeck, C. H., Warren, R., Stanier, P., Super, M., Lissens, W., Scambler, P., Watson, E., Wainwright, B. and Williamson, R. First-trimester prenatal diagnosis of cystic fibrosis with linked DNA probes. *Lancet* 2 (1986a) 1402–1405

Farrall, M., Scambler, P. J. and Klinger, K. W. Cystic fibrosis carrier detection using a linked gene probe. *J. Med. Genet.* 23 (1986b) 295–299

Frizzell, R. A., Rechkemmer, G. and Shoemaker, R. L. Altered regulation of airway epithelial cell chloride channels in cystic fibrosis. *Science* 233 (1986) 558–560

Gibson, L. E. and Cooke, R. E. A test for concentration of electrolytes in cystic fibrosis of the pancreas utilizing pilocarpine by iontophoresis. *Pediatrics* 23 (1959) 545–552

Gomperts, B. D. Calcium shares the limelight in stimulus-secretion coupling. *TIBS* 11 (1986) 290–292

Goodchild, M. C. and Dodge, J. A. *Cystic Fibrosis Manual of Diagnosis and Management*, 2nd Edn., Balliere Tindall Press, Eastbourne, 1985

Hadorn, B., Zoppi, G., Shmerling, D. H., Prader, A., McIntyre, I. and Anderson, C. M. Quantitative assessment of exocrine pancreatic function in infants and children. *J. Pediatr.* 73 (1968) 39–50

Katz, S., Schoni, M. H. and Bridges, M. A. The calcium hypothesis of cystic fibrosis. *Cell Calcium* 5 (1984) 421–440

Kealey, T. Phosphorylation studies in isolated human eccrine sweat glands. In: Mastella, G. and Quinton, P. M. (eds.) *Cellular and Molecular Basis of Cystic Fibrosis*, San Francisco Press, San Francisco (1987), pp. 150–154

Klee, C. B. and Newton, D. L. Calmodulin: an overview. In: Parratt, J. R. (ed.) *Control and Manipulation of Calcium Movement*, Raven Press, NY (1985) 131–146

Knowles, M. R., Gatzy, J. T. and Boucher, R. C. Increased bioelectric potential difference across respiratory epithelia in cystic fibrosis. *N. Engl. J. Med.* 305 (1981) 1489–1495

Knowlton, R. G., Cohen-Haguenauer, O., Van Cong, N., Frezal, J., Brown, V. A., Barker, D., Braman, J. C., Schumm, J. W., Tsui, L.-C., Buchwald, M. and Donis-Keller, H. A polymorphic DNA marker linked to cystic fibrosis is located on chromosome 7. *Nature (London)* 318 (1985) 380–382

Lathrop, G. M., Carroll, M., O'Connell, P., Wainwright, B., Leppert, M., Nakamura, Y., Lench, N., Kruyer, H., Dean, M., Park, M., Vande Woude, G., Lalouel, J. M., Williamson, R. and White, R. Refined linkage map of chromosome 7 in the region of the cystic fibrosis gene. *Am. J. Hum. Genet.* (1988) in press

Lee, C. M., Jones, C. J. and Kealey, T. Biochemical and ultrastructural studies of human eccrine sweat glands isolated by shearing and maintained for seven days. *J. Cell. Sci.* 72 (1984) 259–274

Li, M., McCann, J. D., Liedtke, C. M., Nairn, A. C., Greengard, P. and Welsh, M. J. Cyclic AMP-dependent protein kinase opens chloride channels in normal but not cystic fibrosis airway epithelium. *Nature (London)* 331 (1988) 358–360

McPherson, M. A. and Dormer, R. L. Mucin release and calcium fluxes in isolated rat submandibular acini. *Biochem. J.* 224 (1984a) 473–481

McPherson, M. A. and Dormer, R. L. Control of secretion of mucin-type glycoproteins from human and rat submandibular acini. *Biochem. Soc. Trans.* 12 (1984b) 652–653

McPherson, M. A. and Dormer, R. L. The molecular and biochemical basis of cystic fibrosis. *Biosci. Rep.* 7 (1987) 167–185

McPherson, M. A. and Dormer, R. L. Cystic fibrosis: a disease of stimulus-response coupling. *TIBS* 13 (1988) 10–13

McPherson, M. A. and Goodchild, M. C. The biochemical defect in cystic fibrosis. *Clin. Sci.* 74 (1988) 337–345

McPherson, M. A., Dodge, J. A. and Goodchild, M. C. Cystic fibrosis serum stimulates mucin secretion but not calcium efflux from rat submandibular acini. *Clin. Chim. Acta* 135 (1983) 181–188

McPherson, M. A., Dormer, R. L., Dodge, J. A. and Goodchild, M. C. Autonomic secretory responses of control and cystic fibrosis submandibular acinar cells. In: Lawson, D. (ed.) *Cystic Fibrosis: Horizons*, John Wiley & Sons, Chichester (1984) p. 51

McPherson, M. A., Dormer, R. L., Dodge, J. A. and Goodchild, M. C. Adrenergic secretory responses of submandibular tissues from control subjects and cystic fibrosis patients. *Clin. Chim. Acta* 148 (1985) 229–237

McPherson, M. A., Dormer, R. L., Bradbury, N. A., Dodge, J. A. and Goodchild, M. C. Defective β-adrenergic secretory responses in submandibular acinar cells from cystic fibrosis patients. *Lancet* 2 (1986a) 1007–1008

McPherson, M. A., Dormer, R. L., Dodge, J. A. and Goodchild, M. C. Biochemical basis of cystic fibrosis. *Nature (London)* 323 (1986b) 400

McPherson, M. A., Dormer, R. L., Bradbury, N. A., Shori, D. K. and Goodchild, M. C. Regulation of secretion of amylase and mucins in control and cystic fibrosis submandibular acinar cells: role of cyclic AMP and Ca^{2+}. In: Mastella, G. and Quinton, P. M. (eds.) *Cellular and Molecular Basis of Cystic Fibrosis*, San Francisco Press, San Francisco (1988), pp. 343–354

Mangos, J. A. Isolated parotid acinar cells from patients with cystic fibrosis. Functional characterization. *J. Dent. Res.* 60 (1981) 797–804

Mangos, J. A. and Donnelly, W. H. Isolated parotid acinar cells from patients with cystic fibrosis. Morphology and composition. *J. Dent. Res.* 60 (1981) 19–25

Mangos, J. A., McSherry, N. R., Barber, T., Arvanitakis, S. N. and Wagner, V. Dispersed rat parotid acinar cells. II Characterisation of adrenergic receptors. *Am. J. Physiol.* 229 (1975) 560–565

Martinez, J. R., Quissell, D. O. and Giles, M. Potassium release from the rat submaxillary gland *in vitro*. I. Induction by catecholamines. *J. Pharm. Exp. Ther.* 198 (1976) 385–394

Ochs, D. L., Korenbrot, J. I. and Williams, J. A. Relation between free cytosolic calcium and amylase release by pancreatic acini. *Am. J. Physiol.* 249 (1985) G389–G398

Pallen, C. J. and Wang, J. H. A multifunctional calmodulin-stimulated phosphatase. *Arch. Biochem. Biophys.* 237 (1985) 281–291

Pedersen, P. S. Cholinergic influence on chloride permeability in sweat duct cell cultures from normals and patients with cystic fibrosis. *Med. Sci. Res.* 15 (1987) 769–770

Pedersen, P. S., Brandt, N. J. and Larsen, E. H. Qualitatively abnormal beta-adrenergic response in cystic fibrosis sweat duct cell culture. *IRCS Med. Sci.* 14 (1986) 701–702

Pedersen, P. S., Larsen, E. H. and Brandt, N. J. Restitution of chloride permeability in cystic fibrosis. *Med. Sci. Res.* 15 (1987) 151–152

Petersen, O. H. Calcium-activated potassium channels and fluid secretion by exocrine glands. *Am. J. Physiol* 251 (1986) G1–G13

Petersen, O. H., Findlay, I., Iwatsuki, N., Singh, J., Gallacher, D. V., Fuller, C. M., Pearson, G. T., Dunne, M. J. and Morris, A. P. Human pancreatic acinar cells: studies on stimulus–secretion coupling. *Gastroenterology* 89 (1985) 109–117

Quinton, P. M. Chloride impermeability in cystic fibrosis. *Nature (London)* 301 (1983) 421–422

Quissell, D. O. and Barzen, K. A. Secretory responses of dispersed rat submandibular cells. *Am. J. Physiol.* 238 (1980) C99–C106

Romeo, G., Bianco, M., Devoto, M., Menozzi, P., Mastella, G., Giunta, A. M., Micalizzi, C., Antonelli, M., Battistini, A., Santamaria, F., Castello, D., Marianelli, A., Marchi, A. G. and Manca, A. Incidence in Italy, genetic heterogeneity and segregation analysis in cystic fibrosis. *Am. J. Hum. Genet.* 37 (1985) 388–394

Sato, K. and Sato, F. Defective β-adrenergic response of cystic fibrosis sweat glands *in vivo* and *in vitro. J. Clin. Invest.* 73 (1984) 1763–1771

Sauder, R. A., Chesrown, S. E. and Loughlin, G. M. Clinical application of transepithelial potential difference measurements in cystic fibrosis. *J. Pediatr.* 111 (1987) 353–358

Scambler, P. J., McPherson, M. A., Bates, G., Bradbury, N. A., Dormer, R. L. and Williamson, R. Biochemical and genetic exclusion of calmodulin as the site of the basic defect in cystic fibrosis. *Hum. Genet.* 76 (1987) 278–284

Schwarz, V. and Simpson, I. M. N. Is salt reabsorption in the human sweat duct subject to control? *Clin. Sci.* 68 (1985) 441–447

Schneyer, L. H. and Thavornthon, T. Isoproteronol-induced stimulation of sodium absorption in perfused salivary duct. *Am. J. Physiol.* 224 (1973) 136–139

Schneyer, L. H., Young, J. A. and Schneyer, C. A. Salivary secretion of electrolytes. *Physiol. Rev.* 52 (1972) 720–777

Schoumacher, R. A., Shoemaker, R. L., Halm, D. R., Tallant, E. A., Wallace, R. W. and Frizzell, R. A. Phosphorylation fails to activate chloride channels from cystic fibrosis airway cells. *Nature (London)* 330 (1987) 752–754

Shori, D. K., Bradbury, N. A., Goodchild, M. C., Dormer, R. L. and McPherson, M. A. Altered calmodulin function in cystic fibrosis. *Biochem. Soc. Trans.* 16 (1988) 345–346

Streb, H., Bayerdorffer, E., Haase, W., Irvine, R. F. and Schulz, I. Effect of inositol-1,4,5-triphosphate on isolated subcellular fractions of rat pancreas. *J. Memb. Biol.* 81 (1984) 241–253

Taylor, C. J., Baxter, P. S., Hardcastle, J. and Hardcastle, P. T. Absence of secretory response in jejunal biopsy samples from children with cystic fibrosis. *Lancet* 2 (1987) 107–108

Tsui, L.-C., Buchwald, M., Barker, D., Braman, J. C., Knowlton, R., Schumm, J. W., Eiberg, H., Mohr, J., Kennedy, D., Plavsic, N., Zsiga, M., Markiewicz, D., Akoto, G., Brown, V., Helms, C., Gravins, T., Parker, C., Rediker, K. and Donis-Keller, H. Cystic fibrosis locus defined by a genetically linked polymorphic DNA marker. *Science* 230 (1985) 1054–1057

Wainwright, B. J., Scambler, P. J., Schmidtke, J., Watson, E. A., Law, H.-Y., Farrall, M., Cooke, H. J., Eiberg, H. and Williamson, R. Localization of cystic fibrosis locus to human chromosome 7cen-q22. *Nature (London)* 318 (1985) 384–385

Welsh, M. J. The respiratory epithelium. In: Andreoli, T. E., Hoffman, J. F., Fanestil, D. D. and Schultz, S. G. (eds.) *Physiology of Membrane Disorders.* Plenum, NY (1986) pp. 751–766

Welsh, M. J. and Liedtke, C. M. Chloride and potassium channels in cystic fibrosis airway epithelia. *Nature (London)* 322 (1986) 467–470

White, R., Woodward, S., Leppert, M., O'Connell, P., Hoff, M., Herbst, J., Lalouel, J.-M., Dean, M., and Vande Woude, G. A closely linked genetic marker for cystic fibrosis. *Nature (London)* 318 (1985) 382–384

Widdicombe, J. H. Cystic fibrosis and β-adrenergic response of airway epithelial cell cultures. *Am. J. Physiol.* 251 (1986) R818–R822

Yankaskas, J. R., Knowles, M. R., Gatzy, J. T. and Boucher, R. C. Persistence of abnormal chloride ion permeability in cystic fibrosis nasal epithelial cells in heterologous culture. *Lancet* 1 (1985) 954–956

Yoshimasa, T., Sibley, D. R., Bouvier, M., Lefkowitz, R. J. and Caron, M. G. Cross-talk between cellular signalling pathways suggested by phorbol-ester-induced adenylate cyclase phosphorylation. *Nature (London)* 327 (1987) 67–70

Zengerling, S., Olek, K., Tsui, L.-C., Grzeschik, K.-H., Riordan, R. and Buchwald, M. Mapping of DNA markers linked to the cystic fibrosis locus on the long arm of chromosome 7. *Am. J. Hum. Genet.* 40 (1987) 228–236

Zimmerberg, J. Molecular mechanisms of membrane fusion: steps during phospholipid and exocytotic membrane fusion. *Biosci. Rep.* 7 (1987) 251–268

J. Inher. Metab. Dis. 11 Suppl. 1 (1988) 110

SYMPOSIUM – THE ETHICS OF ANTENATAL DIAGNOSIS AND THE TERMINATION OF PREGNANCY

Introduction

Editors: D. P. BRENTON[1] and J. W. SEAKINS[2]
[1]*University College and Middlesex School of Medicine, London* and [2]*Institute of Child Health, London, UK*

The suggestion for this symposium came to the Council of the SSIEM from one of the editors (J.W.S.) who pointed out that the Society's last considerations in this area had been 10 years previously in a rather medico-legal context (see the paper of G. A. Ratner, 13th Annual Symposium of the SSIEM: *Medico-social management of inherited metabolic disease*). The emphasis of the present symposium was planned to be quite different. It was concerned with ethical issues primarily and only in the limited context of inherited disorders or congenital abnormalities, not in any other context. Similarly, it was not concerned with *in vitro* fertilization, embryo research or other matters of that kind, although some contributors did make reference to them. The content of the symposium was based on three invited speakers, Professor Marcus Pembrey, Rev. Nicholas R. Holtam and Dr Raanan Gillon, who were asked to deal respectively with the medical and scientific aspects of the problem, the specific Christian viewpoints and the broader ethical issues. These papers were followed by a discussion.

The discussion took place under the Chairmanship of Professor Forrester Cockburn with all of the three main speakers present, as well as the conference participants and a number of parents with contacts in several parent support groups. A verbatim report of the discussion would have been excessively long and the editors have tried, under a number of headings, to represent fairly the views expressed. We are particularly grateful for the contributions made to the discussion by the parents who made the issues real from their own experience and the shared experiences of others. These included Mrs. Margaret Hall, Mrs. Christine Lavery for the Society for Mucopolysaccharide Diseases in Children, Mrs. Lesley Greene for The Research Trust in Metabolic Diseases in Children, Mr. Bob Glover for the Muscular Dystrophy Group and Dr. Paul Dakin. Without them, the discussion would have been immeasurably poorer and their contributions changed an 'armchair' debate into a meaningful interchange of views. The editors hope that all views expressed at the time have been fairly represented and accept responsibility with apologies if that is not the case.

The symposium was supported by a grant from The Nuffield Trust and the expenses of publication were defrayed by donations from the Medical Protection Society and the Medical and Dental Defence Union of Scotland.

Journal of Inherited Metabolic Disease. ISSN 0141–8955. Copyright © SSIEM and MTP Press Limited, Queen Square, Lancaster, UK.

J. Inher. Metab. Dis. 11 Suppl. 1 (1988) 111–119

Antenatal Diagnosis and the Termination of Pregnancy – What the Churches have to Say

REV. N. R. HOLTAM

Lincoln Theological College, Drury Lane, Lincoln, LN1 3BP, UK

INTRODUCTION

This paper attempts to describe what the churches have to say about developments in antenatal diagnosis. It is not, primarily, an analysis of the arguments. Although the arguments are presented denominationally (principally Roman Catholic and Anglican), it is important to note that disagreement about morality within the churches is not straightforwardly denominational. Conservative tends to speak easily with conservative, liberal with liberal, radical with radical, across denominational boundaries (Gustafson, 1978, p. 30). In addition it needs to be noted that the churches speak in different ways and that their statements have differing authority.

Christians have a longstanding moral concern with the beginnings of life. Although there is almost nothing in the scriptures dealing directly with the matter, in the early Church there was considerable reaction against the widespread practices of infanticide and abortion. In the UK, the recent debate in the churches has been focussed by the 1967 Abortion Act, and, currently, by the Warnock Committee's report on *Human Fertilization and Embryology* (DHSS, 1984). The lines of argument developed in these discussions can then be applied to the particular recent developments in antenatal diagnosis. Central to these debates has been the discussion of the status of the embryo.

THE STATUS OF THE EMBRYO

In their official statements, the churches have made two apparently contradictory statements about the moral status of the embryo. Roman Catholic teaching has stressed the protection due to the embryo from conception. Summarizing the Roman Catholic position, Fr Kevin Kelly says:

> "For all practical purposes the position of the Roman Catholic Church equates the status of the embryo with the status of any other living human being and determines its dignity accordingly". (Kelly, 1987, p. 65)

It has been suggested that the Roman Catholic position is the 'safest' interpretation of the facts. It does not imply certainty about the status of the embryo but:

Journal of Inherited Metabolic Disease. ISSN 0141–8955. Copyright © SSIEM and MTP Press Limited, Queen Square, Lancaster, UK.

"As long as we are not 100 per cent certain . . . we must play safe and avoid all risk of directly killing a human being. And since it [i.e. the official Roman Catholic position] accepts that 100% certainty on this issue is humanly impossible, it regards its position on abortion as definitive and unchangeable". (Kelly, 1987, p. 52)

Consequently, any activity destructive of the individual human embryo is considered to be immoral. The Roman Catholic position may, therefore, be described as 'absolutist'. Only one exception is admitted: the indirect termination caused by medical intervention necessary to save the life of a pregnant woman.

All the other main denominations in the UK have expressed a variety of positions which accept that the gradual formation of the embryo into a person is reflected in its changing moral status. Human from the beginning, the embryo is not properly described as a person. A working party of the Church of England's Board for Social Responsibility summarizes this position as follows:

"Respect is due to an embryo at all stages, but protection of life in the sense that a post-natal child would have it, is afforded only after some particular threshold during pregnancy. This has variously been identified with animation, quickening, sentience or viability. The Warnock Inquiry recommended that protection should be afforded at the time of the formation of the primitive streak since 'this marks the beginning of individual development' (paragraph 11.22), and the Board for Social Responsibility in its response to the Inquiry agreed with this position. . . . However, other points of embryonic develop-ment can be argued to be crucially significant. For example, the establishing of a functioning nerve-net at around 40 days after conception can be regarded as a necessary criterion for the beginning of personal life, paralleling the common acceptance of brain-death (as distinct from, say, heart failure) as the mark of the end of physical life." (BSR, 1985, paragraph 88)

Such a gradualist approach is to be found strongly represented in the Christian tradition. In making a distinction between the seriousness of causing an abortion very early in pregnancy and the greater seriousness of causing one later, Christians were simply accepting the contemporary scientific and philosophical theories that said formation or ensoulment took place several weeks after conception. Indeed, a 1965 Anglican Report suggested that the absolute principle that the fetus was inviolable was 'a novel departure' from the Christian moral and legal tradition. This Anglican Report pointed out that in Catholic teaching the killing of the fetus prior to animation was not always regarded as murder and that it was only in 1869 that Pius IX issued a Bull which ended a distinction previously made in canon law by pronouncing excommunication on all who procured abortion without mention of time (BSR, 1965, p. 28).

The abolutist and gradualist approaches within the Christian churches appear to be totally contradictory. However, if the debate is viewed in terms of the underlying principles of the argument, considerable common ground is to be found. Kelly shows how there is agreement that 'reverence and respect is to be owed to the human embryo from the moment of conception'. The embryo has intrinsic value, i.e. value which is recognized, not conferred.

"This means that its good (its life, growth and development) must be respected and can only be put in second place or even sacrificed if it is in direct conflict with some higher good related to the dignity of human beings". (Kelly, 1987, p. 75)

This respect due to the embryo results in a very protective attitude towards it. The protectiveness of the absolutist is such that no activity destructive of the human embryo is permitted. Prenatal diagnosis, therapeutic procedures and experiments are only acceptable when they respect the life and integrity of the embryo and are directed towards its safeguard and healing (John Paul II, 1987, p. 298). The absolutist position looks logically incoherent at this point because research and therapy in this area inevitably progress hand in hand. Even so, this position retains popular emotive force.

The protectiveness of the gradualists means that some strictly limited destructive intervention may be permissible in the life of the embryo and, possibly, fetus. So, the Church of England's Board for Social Responsibility accepted the limit recommended by Warnock that experiments on human embryos be confined to the period of 14 days after conception (BSR, 1984, 10.3).

ABORTION

The attitude to the status of the embryo is indicative of attitudes to abortion. All the churches in their official statements agree that abortion is not an alternative to contraception. For example, the 1965 Anglican Report asserts that the general inviolability of the fetus is normative and that its right to live and develop should be defended as a first principle (BSR, 1965, p. 31f).

A harder case is presented by the potentially handicapped child. In 1980, the Roman Catholic Archibishops of Great Britain argued that to say anyone was better off dead was tantamount to asserting the rightness of euthanasia. They said that such an attitude was at odds with 'the inner riches of human experience' and they pointed out that 'many handicapped people (are) themselves deeply opposed to abortion' (Roman Catholic Archbishops, 1980, para 20). Their case clearly rests on a regard for the embryo and fetus as a person.

By contrast, the 1965 Anglican Report argued that the justification of abortion always depended on the effects the potential child would have on the mother. Compassion for the mother has to be balanced with responsibility to the fetus. All justification for abortion should be dealt with under the category of risk to the life or mental or physical health of the mother, and this should be seen as integrally connected to the life and well-being of her family (BSR, 1965, p. 61). Potential handicap is therefore not singled out as a particular case, though it is clearly more likely to have a deleterious effect on both mother and family. The authors of this report hoped that each decision for abortion would be the result of intraprofessional consultation and that this would result in the development of a 'genuine moral tradition in medical practice' (BSR, 1965, p. 43).

The 1967 Abortion Act

The 1967 Act was at variance with the 1965 Anglican Report in two crucial respects. First, it made specific provision for the termination of the potentially seriously handicapped independently of any consideration of the circumstances of the mother. Second, by attaching the environmental clause (taken from the Anglican Report) to the determining of risk to the health of the mother without the condition of her carrying a potentially seriously handicapped child, the way was opened for doctors to perform abortions because the risk to the life of the mother is statistically smaller if the pregnancy is terminated in the early months than if it is allowed to go to term (Dunstan, 1974, p. 86f).

All the churches have expressed concern about the implementation of the Act. In these expressions, a third area of concern has emerged. The 'conscience clause', intended to protect the place of medical personnel conscientiously opposed to abortion, has proved to be inadequate. All the churches have argued that this clause needs strengthening (Boyd *et al.*, 1986, p. 31). However, to argue for reform of the Abortion Act may underestimate the seriousness of the changes taking place in our society. G. R. Dunstan has commented on the widespread acceptance in society at large of medical intervention for the destruction of life, something which he says is previously unknown in our ethics (Dunstan, 1974, p. 87).

Oliver O'Donovan sees the failure of the Act's conscience clause as having far-reaching implications. The freedom of self-determination for the pregnant woman, made possible by increased technological skill and made available by the 1967 Act, has been at the expense of the physician's freedom. For O'Donovan, the Act's conscience clause inevitably failed because:

> "The organisation of mass medicine requires predictability of performance. A hospital schedule cannot be planned around individuals who may, or may not, when it comes to, assist at an abortion. The rule must be that if they can't stand the heat they must get out of the kitchen, and the best that can be said for the conscience clause is that it has sometimes provided a graceful mode of exit". (O'Donovan, 1984, p. 7)

In these deplorable circumstances, the implementation of the Act raises a question as to whether the patient and the doctor can still be involved in joint conscientious moral decision making. The technological revolution has its own momentum. Crudely, if it is possible and the patients want it, the doctors will have to do it. The moralist in these circumstances cannot advise on the rights and wrongs of particularly difficult moral decisions, for the expression of opinion has no power. Rather, the moralist must attempt to be an interpreter, explaining "how and why these decisions now come to us in these forms and present these difficulties" (O'Donovan, 1984, p. viii).

MORAL DECISION-MAKING

Of those who think there are still significant moral choices to be made, none think that the moralist is able to provide clear and unequivocal advice to those facing the moral dilemmas of modern medicine. At best, what can be offered are interim judgements made in rapidly changing circumstances. Such judgements must always be contingent upon changing circumstances and, therefore, provisional. Even so, a Roman Catholic theologian, Karl Rahner, has said that "a particular contingent judgement of this kind can still be the only correct one *in its situation*" (Boyd *et al.*, 1986, p. 146.)

It is important to note that none of the churches seeks to remove the burden of individual decision making from the actors immediately involved; that is, in the case of abortion, the pregnant woman and her doctors. The churches see their task as assisting people in the interpretation of complex circumstances (as O'Donovan above) and helping to inform people's consciences. No church considers itself the sole supplier of either the correct interpretation or all the relevant information.

Perhaps this needs to be qualified with regard to Roman Catholicism. According to the major teaching documents of that Church, an individual's properly informed conscience is sovereign. Even if a person is mistaken, (s)he cannot be blamed for obeying their conscience (e.g. *'Guadium et Spes'*, 1965, p. 16). However, the treatment of the American moral theologian, Fr Charles Curran, presents a different approach. Declaring Fr Curran to be "not suited and not qualified" to teach Catholic theology, Cardinal Ratzinger (the head of the Curia's Doctrinal Office) said "the faithful are not required to obey only the infallible teachings" (always matters of doctrine, never of morals) "but also to submit to the intelligence and will of the supreme pontiff or the college of bishops . . . in matters of faith or morals" (*The Guardian* 20 August 1986). In this, the Cardinal appears to be saying that only a conscience in agreement with the teachings of the Pope is properly informed and correct. Such a view has caused widespread concern among Roman Catholics and in ecumenical debate.

In dealing with genuine moral dilemmas, moral decision making cannot be a matter of strict logic for, if it were, the dilemma would no longer exist. Morality is, therefore, the product of the best possible reasons *and* the best possible feelings. It is a matter of almost aesthetic judgement in which all the ingredients, some of which will be in conflict, are brought to the best possible outcome in the particular case.

PASTORAL PRACTICE

In pastoral practice, it is quite common to meet people who have been made to feel guilty by the moral perfectionism they have perceived in the Church's teaching. This is particularly true of people who have had abortions. I have tried to indicate that this is a misapprehension of the Church's teaching. Perfection does not exist this side of heaven and, in God's good creation, we have to live with the provisional and the contingent. The single criterion that matters is whether a decision was made conscientiously or not. The recognition of this can still leave people with feelings of guilt. Their decision may have been rushed or pressured by circumstances. Again, I have tried to indicate that what matters is that it was the best possible decision in the circumstances. Even where it was not, there is meant to be a pastoral practice of loving the sinner whilst hating the sin.

IMPROVEMENTS IN ANTENATAL DIAGNOSIS

The earlier and more accurate identification of serious handicap in antenatal diagnosis is likely to be welcomed by both gradualist and absolutist alike. The fetus will be given greater protection because of a reduction in the number of abortions carried out on the grounds of risk of serious handicap rather than of firm diagnosis.

It is important that the mother is told the truth clearly about the accuracy of the particular diagnostic process. In the weighing of complex evidence, it is essential that adequate counselling be made available.

It is helpful to the pregnant woman that diagnosis can now take place in the first trimester before she need tell people that she is pregnant. This may make it easier for her should she decide to have the pregnancy terminated. Whilst acknowledging this gain for the woman, the absolutist will be doubtful about making termination any easier.

In the case of the preimplantation diagnosis of genetic disease in IVF embryos the gradualist will see no particular moral problem. However, the absolutist will deplore the simultaneous generation of multiple embryos and their consequent wastage. However, even the production of a single embryo would not solve the absolutist's moral problem. Presumably, on this view, even a genetically defective embryo ought to be given the opportunity to develop and every effort be made to implant it.

The main problems of improved antenatal diagnosis of serious handicap seem to me to be more to do with the operation of the 1967 Act than with the techniques themselves. The provision for abortion on the grounds 'that there is a substantial risk that if the child was born it would suffer from such physical or mental abnormalities as to be seriously handicapped' looks extremely worrying on two counts.

First, who is to decide that the potential handicap of the child is sufficiently serious to justify abortion? If the decision remains, as at present, with the pregnant woman and the two doctors, the law is likely to continue to prove itself permissive in such a way as to allow 'a woman's right to choose'. In retaining a say in a medical judgement, the medical profession might limit the information made available to the pregnant woman. This would be in keeping with the current practice of not disclosing the sex of a fetus after chorionic villus sampling. But, in the increasingly significant grey area between 'serious' and 'not serious', it is not clear what criteria would be determinative nor whether the restriction of information should be the result of legislation or professional agreement.

Second, to permit the abortion of the seriously handicapped simply because they will be seriously handicapped is to say something very worrying about the values of our society. The matter was well put by the present Pope in a homily about the sick, elderly, handicapped and dying:

"Without the presence of these people in your midst, you will be tempted to think of health, strength and power as the only important values to be pursued in life". (Boyd *et al.*, 1986, p. 43)

Consequently, I am inclined to think that the 1965 Anglican Report was right in seeking to limit the justification for abortion to the effect the child would have on the mental and physical health of the mother. The principal decision is then not about the seriousness of the handicap or the intrinsic worth of the handicapped person but about the capacity of those who will carry the greatest responsibility for them. It is not a decision which can be made by the pregnant woman and the two doctors alone. It has to be the product of wider consultation in order to make sure that the decision is well informed. Such provision would be subject to exactly the same sort of abuse as is presently experienced with the abortion of normal fetuses. This seems unavoidable given that law and morality are not identical. However, the particularly difficult circumstances surrounding the weighing of evidence about potential handicap could result in the need for a greatly improved counselling service and the establishment of that 'genuine moral tradition in medical practice' hoped for by the 1965 Anglican Report. Limiting the grounds for abortion in this way would require a firm commitment from society that handicapped people have a proper and safeguarded place within it. Handicapped people, their parents and families require extra resources to be made available to them. There is a danger that because the decision to carry a handicapped child to term increasingly becomes the conscious decision of the mother, the family will be expected to bear the entire burden consequent on having the child. This must not be allowed to happen.

RULES FOR CONVERSATION

It must be obvious that the debate of abortion within the churches has been a highly contentious one, particularly in recent years. The strong division of opinion is much the same in our wider society. Consequently it may be helpful to conclude this paper with a summary of Richard McCormick's ten 'rules for conversation' for those who differ on the abortion issue (McCormick, 1981, pp. 176–188). They are the product of painful experience within the churches and deserve to be better known.

(1) Attempt to identify areas of agreement.

(2) Avoid slogans.

(3) Represent the opposing position accurately and fairly.

(4) Distinguish the pairs right–wrong, good–bad. For McCormick, good–bad refers to the intention of the person doing the action. "One's action can therefore be morally good, but still morally wrong. It can be morally right, but morally bad". Such a distinction "allows one to disagree agreeably – that is without implying, suggesting or predicting moral evil of the person one believes to be morally wrong. This would be a precious

gain in a discussion that often witnesses this particular and serious collapse of courtesy".

(5) Try to identify the core issue at stake. For McCormick, "the core issue . . . concerns the moral claims the nascent human being . . . makes on us. Do these frequently or only very rarely yield to what appear to be extremely difficult alternatives? And above all, why or why not? . . . It is illumined neither by flat statements about the inviolable rights of fetuses nor assertions about a woman's freedom of choice. These promulgate conclusions. They do not share with us how one arrived at it".

(6) Admit doubts, difficulties and weaknesses in one's own position.

(7) Distinguish the formation and the substance of moral conviction. According to McCormick, "ethical formations, being the product of human language, philosophy and imperfection, are only more or less adequate to the substance of our moral convictions at a given time. . . . The formation can easily betray the substance . . . it is frequently difficult to know just what is changeable, what permanent . . . (To) conduct discussion as if substance and formation were identical is to get enslaved in formulations. Such captivity forecloses conversations".

(8) Distinguish morality and public policy.

(9) Distinguish morality and pastoral care or practice. McCormick explains this distinction as follows: "A moral statement is . . . an abstract statement not in the sense that is has nothing to do with real life, or with particular decisions, but in the sense that it abstracts or prescends from the ability of this or that person to understand it and live it.

Pastoral care (and pastoral statements), by contrast, looks to the art of the possible. It deals with an individual where that person is in terms of his or her strengths, perceptions, biography, circumstances (financial, medical, educational, familial, psychological). Although pastoral care attempts to expand perspectives and maximize strengths, it recognizes at times the limits of these attempts".

(10) Incorporate the woman's perspective, or women's perspectives. Maybe the church has been worse at this than is general in society, but it is widely felt that women have lacked the influential voice in this discussion that they are entitled to expect.

REFERENCES

BSR (Church of England Board for Social Responsibility): A Committee appointed by the Church Assembly BSR, *Abortion: an Ethical Discussion*, CIO, 1965

BSR, *Human Fertilization and Embryology*, BSR, 1984

BSR, The Report of a Working Party on Human Fertilization and Embryology of the BSR, *Personal Origins*, CIO, 1985

Boyd, K., Callaghan, B. and Shotter, E. *Life Before Birth: A Search for Consensus on Abortion and the Treatment of Infertility*, SPCK, 1986

DHSS, *Report of a Committee of Inquiry into Human Fertilization and Embryology* ('The Warnock Report'), HMSO, 1984

Dunstan, G. R. *The Artifice of Ethics*, SCM, 1974

Guadium et Spes. Catholic Truth Society, 1965

Gustafson, J. *Protestant and Roman Catholic Ethics: Prospects for Rapprochement*, SCM, 1978

John Paul II. Instruction on respect for life and its origin and on the dignity of procreation. *The Tablet*, 14.3.87, pp. 293–298

Kelly, K. T. *Life and Love: Towards a Christian Dialogue on Bioethical Questions*, Collins, 1987

McCormick, R. A. *How Brave a New World?*, SCM, 1981

O'Donovan, O. *Begotten or Made?*, Clarendon, 1984

Roman Catholic Archbishops of Great Britain, *Abortion and the Right to Live*, CIS, 1980

J. Inher. Metab. Dis. 11 Suppl. 1 (1988) 120–124

Ethics and Clinical Practice

R. Gillon
Imperial College Health Service, London, UK

I suppose it is worth starting by saying that I am a general practitioner also trained in philosophy. I am interested in today's subject of medical ethics from two perspectives, first at a more philosophical level and second as a general practitioner with wide clinical responsibilities for my patients. Secondly, I ought to apologize in advance to anyone who is philosophically trained for the very skimpy nature of what is to follow, but, in the time available, I can only outline some basic ethical issues relevant to medicine in general and to problems associated with inborn errors of metabolism in particular.

In the minds of many people there are two moral theories, or types of moral theory, that often seem to conflict. On the one hand, we have Utilitarianism, which sets out to maximize happiness or welfare for as many as possible within society, and, on the other hand, Deontology which emphasizes duty and rules: don't lie, don't kill, do this, don't do that, etc. It often seems that these two types of theory are separated by an unbridgeable gulf. On the contrary, I think that sophisticated exponents of both types of theories can, if they wish to, join in recommending for practical ethics, four basic *prima facie* moral principles. The first one is respect for autonomy, and the others beneficence, non-maleficence and justice.

Autonomy is something that characterizes the whole of our moral life, our specialness as it were. As people we have autonomy, or the ability to rule ourselves in the context of rationality, i.e. in the context of thinking about, and reflecting on, how we should behave. Respect for people's autonomy is one of the fundamental principles in all moral theories. However, respect for autonomy also requires respect for the autonomy of *all* affected by any proposed action. It must be emphasized that respect for autonomy does not just mean that, if somebody wants to do something, that is the end of the matter, he or she must be allowed to do it. On the contrary, it's a matter of assuming that, if people have thought out and worked out what they want to do, then one should not oppose them *provided* that the decision is compatible with respect for the autonomy of *all* involved. Now, such respect is a basic presupposition (although it is often not explicit) in any medical relationship. After all, it is only because people come to us, asking for our help, that we are allowed to begin the business of helping them in the first place and thus respect for their autonomy is implicit from the beginning.

What in practice tends to happen is that that the second and third principles, those of beneficence and non-maleficence, take precedence in many doctors' minds because that's what most patients *want* us to be concerned with. But it is important to realise that the principles of beneficence (of helping others) and non-maleficence

Journal of Inherited Metabolic Disease. ISSN 0141-8955. Copyright © SSIEM and MTP Press Limited, Queen Square, Lancaster, UK.

(of not harming others) are implicitly conditional, in health care relationships, upon respect for the patient's autonomy in the first place.

Regarding beneficence and non-maleficence, the next two of these four principles, there are clear differences in the scope of obligation for each. Thus, even if you do not have an obligation to help a person, you still have a *(prima facie)* obligation not to harm that person. It is clear there *are* circumstances when you don't have any particular obligation to benefit people. Indeed, I don't suppose I have any particular obligation to benefit the vast majority of people in this world. If I do, then it is impossible to fulfill, and one of the characteristics of moral obligations is that they should be in principle fulfillable. However, once you do have some obligation to benefit others, then you always have to take into account, when trying to do so, the general obligation not to harm them. Thus, to benefit someone's health with minimal risk of harm is our aim in health care. While the principle of non-maleficence doesn't require any presuppositions about beneficence, the principle of beneficence, of helping others, always requires consideration of non-maleficence.

Finally, there is the fourth principle, justice, the most difficult to summarize, but, essentially, the need to be fair. There are two major areas where this applies, first, in the area of distribution of resources, and, second, in the context of respecting people's rights. Those four principles, autonomy, beneficence, non-maleficence and justice, are, I believe, basic to all moral theories, and thus to all theories of medical morality for I don't think there is any separate entity, 'medical ethics', which is separate from 'ethics' in general. It's just 'ethics' applied to medical care.

As well as the four principles outlined, there is also the question of their scope to be considered. To whom or to what do we owe these various obligations? As indicated above, I believe the scope of beneficence is limited to a subclass of people whom we have an obligation to benefit. Non-maleficence, on the other hand, not harming others, is surely universal; we owe it, *prima facie*, to everybody in the present and in the future. The principle of autonomy is again universal. We have a *prima facie* obligation to respect the autonomy of other people, whoever they are, at least by not interfering to overrule that autonomy, provided such respect extends to the autonomy of all affected. It is much more difficult to decide what should be the scope of justice. Whom do we have to be fair to? Surely we have to respect people's rights. But do we have to be fair to everybody in the distribution of resources? There are questions about the scope of distributive justice that I cannot explore here.

Consider now our particular context of prenatal diagnosis and termination of pregnancy. I have indicated that the obligation to benefit people, with minimal harm, and to do so because they wish us to do so is a sort of central moral nexus for medical care. Once people come to us and want us to do something to help them medically, then we begin to act in a professional relationship to try to meet those desires and to try to help them as far as that is acceptable. In responding to their request, we are respecting their autonomy. In trying to benefit them with minimal harm (beneficence and non-maleficence), we have to start by isolating what are the proposed benefits and what are the possible harms. Some sort of a 'weighing-up' is required to achieve the optimal medical benefit required with the minimal harm. In considering our various courses of action, some possibilities are potentially more harmful than others. It may be that the probabilities – although they are much more precisely predictable in your field of genetic disorders – are still by no means certain. We are always going to have to live with uncertainties. The whole art, as it were, of predicting benefits and harms is going to be probabilistic, even if we have got very finely tuned systems for assessing the benefits and harms.

And then the question of scope arises – *whose* benefits and harms are we to consider, a particularly difficult question in medical genetics. In the normal context, the health care professions consider primarily the benefits and harms to the individual patient. But we also have to consider the benefits and harms to others, at least in so far as we may be proposing something that might harm others, because we have accepted a general universal *prima facie* obligation not to harm others. We cannot actually get away from this obligation not to harm others (non-maleficence) even though our special obligation is to consider the benefit and harms to the actual patient. This is one of the great tensions that exist in medical ethics. We need to acknowledge that it exists. Concern, on the one hand, for the benefits and harms to society as a whole – the Utilitarian objective of maximizing the welfare of all – must be reconciled, on the other hand, with the particular interests of our individual patients, to whom we have a special obligation. I think that it is possible to justify *having* those special obligations, but they cannot be absolute obligations.

How can we apply all this to the questions of antenatal diagnosis and the termination of pregnancy? I am fairly sure that it is the question of scope that is the central moral issue here. If you believe that the embryo, from the moment of conception, is equivalent to one of us, then I think your assessment of what to do, of what your moral obligations are going to be, will be quite different from your assessment if you don't. If you believe that the embryo and the fetus do not come into as important a moral category as we do, then, obviously, the question arises as to when you can kill it in the interests of people. Now I take it that people at this symposium, by virtue of the sorts of interests they have, are likely to fall into the latter camp, which, as a matter of fact, so do I. This view says that the embryo is not in the same moral category as you and I are, so does not have the same rights, and, in particular, that it does not have the *prima facie* right to life which we accord each other. But, obviously, if you don't believe that and take the view that the embryo is in the same moral category as you or I, then you would not be able to accept the kind of interventions which require the possibility of a deliberate abortion or killing of an embryo or fetus. Thus, the first and crunch issue for anyone looking into the ethics of a prenatal diagnosis is to decide whether it is going to be possible for that person, whether a health care professional or a putative parent, to countenance abortion. That will depend on his or her answer to this question about the moral status of the embryo. Does the person believe that the embryo and the fetus are in the same camp as we are, morally speaking, or not.

Even if we decide, along with most Roman Catholics and the Right to Life groups, that embryos are in the same camp as you and I, it doesn't follow that no work on embryos should be countenanced. What I think comes up here is the moral differentiation between therapeutic and non-therapeutic interventions. The objective of therapeutic intervention is basically to benefit the patient. The objective of non-therapeutic intervention is to benefit others. Now, even if you believe that the embryo is one of us, I can see no argument that stems from this belief which could justify banning interventions to benefit the embryo or fetus, for example by gene transfer, if it were possible, in order to replace the deleterious gene. Quite on the contrary, it seems to me that if you believe that the embryo is morally speaking one of us, then you would actually have a moral obligation to try to help that embryo, to try to benefit it therapeutically. Thus, I don't think that one should assume there is a total gulf, even between those who believe in the embryo as a person from the moment of conception and those who don't. The gulf does remain, however, in the context of termination.

The question which clearly remains is what course of action should be followed, and why, when differences of moral opinions are present. It seems to me that here we actually have to think about the mechanisms we have in society for resolving such differences. As I see it, there are always going to be differences of moral opinion at some level or other and, if that is true, second order moral questions arise about how best to act when such differences cannot be resolved. My own view, for what it is worth, is that what any democratic system tries to do is to allow the autonomy of a whole community of people to be reflected in the structures that are represented in their laws. In such a society, what we try to do when we have conflicting moral views is to look through the debates, the conflicting justifications as it were for opposing positions and then, by use of an agreed system within the law, to accept a legal resolution of the conflict when it is not possible to allow both sides totally to have their own way. The nearest I think we get to that in the case of abortion is that it is strictly controlled under the law and that no one who believes in the moral equivalence of fetuses is required to do abortions. At the same time, there is considered to be sufficient justification available on the other side (with sufficient numbers of people able rationally to defend their position so far as the moral status of embryos is concerned) to make it legitimate for the law to provide options. Thus, it is possible for people who believe in abortion and the termination of embryonic life to do so, under the strict control of society via its laws. But no one who rejects the morality of abortion is required to have one or do one. That is the situation we have in this country at the moment. It is always open for opponents of abortion and embryo research and all associated issues to try to persuade by reason through the parliamentary process the majority in the country who take the opposite view and that is exactly what they continue to do. As I see it, it is a matter of trying to respect the autonomy of a whole range of conflicting opinions, and then to have a system for decision making, when this proves impossible, which itself is morally defensible. Given such a system and given its justifiability, then I think that so far as the law is concerned we have to accept that the fetus has been accorded a lower moral status than that of a person – a status that allows us after due deliberation to kill it for the (sufficient) benefit of people. The same, presumably, should apply to the embryo.

I have rather rushed through what I wanted to say, but let me just approach one of the issues that I was specifically asked to address. What is the role of the medical profession in influencing the moral climate? It seems to me that what I have just said indicates an answer to that. As far as influencing the moral climate is concerned, we have as much right or role or prerogative of doing so as any other member of society, no more, no less. In a democratic society, no one's moral judgements should be given more weight than those of any other person and I don't think that we, as doctors or health care workers, should arrogate to ourselves any such right. What we do have is an individual right as citizens to try to influence. In addition, we also have, as doctors, a special sort of expertise which can inform people in their decision making about the likely outcomes of different courses of action.

With regard to this, knowledge and understanding are always changing. Justification for our actions depends upon what we have learned from experience or research. Research may find ways of benefiting certain abnormal embryos. In the case of embryo research, new and perhaps earlier methods of removing the disabilities of certain sorts of embryos may be discovered, and, if not, identification of major abnormalities followed by abortion may, where the mother so desires, prevent the development of an abnormal embryo or fetus into a handicapped person, with all the trauma that this can entail. In these areas, I think people in the health care professions do have an obligation to discover impartially what are the probable outcomes of the alternative decisions which we might make and then to try to feed that knowledge into the moral decision-making processes both of the individuals in their care and of the decision makers for society.

J. Inher. Metab. Dis. 11 Suppl. 1 (1988) 125–129

Discussion

THE STATUS OF THE EMBRYO AND FETUS

The centrality of this issue has been made clear in the first and second papers. In a sense, viewpoints on the moral status of fetus and embryo were often derived from the religious viewpoints. The absolute sanctity of life was expressed by one contributor very clearly on these grounds (see below). The absolute position was questioned on the grounds that, among spontaneously aborted fetuses, there is a high incidence of chromosomal or other abnormalities. For this contributor, the natural order of things seems to favour the spontaneous abortion of the abnormal fetus. From that viewpoint, antenatal diagnosis and termination of abnormal pregnancies is an extension of a natural process. Most contributors did not hold the absolutist position. Termination was, for them, a less distressing option than continuation of a pregnancy leading to a very handicapped child. It was also clear that the earlier prenatal diagnosis and termination are possible, the less distressing the decision-making process can be. No attempt was made in discussion to define the age at which the fetus is deemed to have the same moral equivalence as an adult by those who are prepared to accept early termination.

THE EFFECTS OF DIFFERENT RELIGIOUS BELIEFS AND THE PREVAILING VIEWS IN SOCIETY

Many contributors were anxious to emphasize the importance of religious beliefs, not simply Christianity (and its subdivisions) or Judaism, but also other world religions. Parents with religious beliefs containing a large fatalistic element, for example, may not wish to consider antenatal diagnosis or termination. These religious viewpoints frequently carry much more weight with parents than national laws which may be more permissive than the religious beliefs. There seemed to be general agreement amongst participants of the need for sensitive understanding of these viewpoints.

In some countries, religious belief is so widely shared in the community that the prevailing view of society is essentially the religious viewpoint. Historically, societies have taken different attitudes even to newborn children – witness, for example, the habit of exposing weak or abnormal babies in classical times. There are also societies which still value male offspring more than female and some anxieties were expressed in discussion that, when an antenatal diagnosis reveals the sex of a child incidentally, that, rather than other issues, may influence a parental decision for or against termination. Laws tend to reflect the majority viewpoint in society and, for most Western countries now, the laws are more liberal than the religious views of minorities.

One contributor who had lost both a brother and a sister with cystic fibrosis and had shared all the traumas of their young lives and eventual deaths was able to say, from the Christian viewpoint of that family, that man is made in the image of

Journal of Inherited Metabolic Disease. ISSN 0141-8955. Copyright © SSIEM and MTP Press Limited, Queen Square, Lancaster, UK.

God and that all life has sanctity even when impaired by handicap or disability. For this family, antenatal diagnosis which could help early medical care would be helpful but termination would conflict with their view on the sanctity of life. Others, because of their personal experience of affected children, have rejected the strongly stated position of their church and terminated an affected pregnancy.

INTRINSIC AND INSTITUTIONAL RIGHTS

Phrases which recur in discussion refer to 'rights' – the 'rights of parents' to a normal child, for example, or the 'right to life' of the fetus. The question of rights is fundamental and was clarified by distinguishing two sorts of rights. There are rights that are ascribed to people by virtue of being people – they are 'intrinsic rights' and the right to life comes into that category. It must be assumed that everyone would recognize the right to life, i.e. that we cannot kill someone once certain attributes of being a person are present. The problem for those who do not hold the absolutist position is to define what those attributes are and when, in fetal life, they can be said to be present *in reality*. For those holding an absolutist view of the right to life, it is not *the reality* of the attributes of being a person which matter but the *potentiality* for their development. Other rights are institutional – that is the rights of all members of society to equal opportunities in sharing the benefits of that society. For example, parents do have a right to share in the efficient application of current knowledge to ensure, as far as possible, a normal child.

ATTITUDES TO TERMINATION. WHAT INFLUENCES THE DECISION?

The importance of language and terminology and the emotional connotations of the various terms was seen by one parent as an important facet of all discussions on antenatal diagnosis and termination. Scientific detachment with its own terminology is useful in debate, but runs the risk of undervaluing the emotional context. A welter of emotion, on the other hand, prevents clear thinking about the issues.

Among the parents present, it rapidly emerged that a decision to terminate was an emotionally very distressing experience. Mothers think, not in terms of 'embryo' or 'fetus', but in terms of a 'baby'. The distress of having to decide whether to have the baby or not is very major indeed. Several factors seem to enter the decision making apart from those of religious belief.

(1) The quality of life for the baby or child. Diseases which lead to death in the first year or two of life, such as Tay–Sachs' disease, would perhaps be most readily accepted as grounds for termination. The idea of the baby slowly dying, from intrauterine life onwards as the abnormal storage material accumulates, was quite real to some people. For diseases which allowed a relatively normal early childhood but later progressive handicap and early death, such as muscular dystrophy, there was great concern for the burden placed upon the individual sufferer of coming to terms with such a distressing outcome physically and mentally. In parental surveys, diseases which can be

effectively treated by diet, for example phenylketonuria, with the promise of a reasonable quality of life may not be seen as meriting termination. Where treatment is expensive, there may be economic pressures to terminate a pregnancy. Some parents would still like antenatal diagnosis even if they do not wish to terminate, partly because they feel it aids prompt treatment but partly because parents seem to need a period of mental preparation before the birth – a sort of getting used to the idea.

Several contributors sounded a cautionary warning about too easily making assumptions about the quality of life. How far are we entitled to go in stating that 'someone's life is not worth living'?

(2) Previous experience of a handicapped child. It was clear from the parents present that previous experience altered their viewpoint. First-hand experience of the quality of life for the affected child had repercussions for the next decision. Parents who had suffered with their children and found caring for them in many ways a fulfilling and rewarding experience in spite of the stress nevertheless frequently were not prepared to have another handicapped child, not because of their own (parental) suffering, but because of their first-hand knowledge of the child's distress. This was found to be true in parental surveys even for Catholic parents who would sometimes set aside their religious views and seek termination.

(3) The effects of a handicapped child on normal siblings in the family. It was felt that the time and effort which had to be put into caring for the affected family member was often detrimental to the upbringing of the normal siblings who became relatively deprived. The effect of a handicapped child on the relationship between the parents was not discussed but, clearly, it can present a marital strain.

(4) One mother described very clearly her own changing feelings. After her first child was diagnosed with a serious genetic illness, the knowledge and experience of that left her in no doubt at her second pregnancy that she would terminate it if the child proved abnormal. In this, she probably felt more strongly than her husband because the burden of caring for the first child had fallen on her. Fortunately, the second child was normal as was the third. At the time of her fourth pregnancy, the first child had been dead for 6 years. The memories of hard work, strain, stress and suffering had subsided leaving only the fonder memories. The decision to terminate then after 16 weeks of pregnancy would have been difficult. For herself, she would have accepted another handicapped child but not for the sake of her two normal children. Fortunately, the fourth child was normal also. The essential point was that she felt somewhat differently on each occasion she had to make the decision.

COUNSELLING

For the parents present, this meant much more than spelling out the numerical risks of having an affected child, fundamental though that obviously is. A single interview with a clinical geneticist was felt to be insufficient for a decision on termination in some cases. The difficulties of obtaining an accurate picture of what to expect with regard to the quality of life in an affected child was a major problem. Even for experienced doctors, that is difficult – one doctor described, in detail, the difficulty of explaining to a mother carrying a fetus with Down's syndrome the full implications of a diagnosis which itself shows clinical variation in severity. Hunter's syndrome was another example used where clinical variation is appreciable and the need for prospective parents not only to appreciate that range but to understand that the apparently fairly normal child may become bedbound, deaf and incontinent by the second decade. A plea was made for a wider roler for parental groups in offering this kind of advice – see below. Another plea was made for post-termination counselling. The need to talk through the problem does not finish for many parents with the termination.

Parental support groups

A wider role for these groups was discussed in relationship to patient counselling. The knowledge embodied within these groups and the first-hand experience of the disease by the members of the group could be invaluable to many. The knowledge of developments within the limited medical field with which any group is concerned is often good and the literature produced by them most helpful. Doctors could do more in ensuring that parents or prospective parents are put in touch with such groups. This point was not developed further in discussion.

MATERNAL ENVIRONMENT AND FETAL DAMAGE

The moral dilemma was discussed which might arise if a mother is neglectful of her own condition when this is damaging to the fetus. Maternal phenylketonuria was taken as an example of a maternal state requiring a strict diet to prevent fetal damage. It was argued that, if a decision was taken by the mother not to terminate the pregnancy, then she has a moral obligation to comply with treatment in the interests of her baby. Her intention that the fetus should develop (which a decision not to terminate the pregnancy clearly implies) brings the moral obligation to treat it with the same care as any other child. There is, apparently, no law in the United Kingdom which could enforce treatment on the mother although, apparently, in the USA, the position may be different, with examples of Courts trying to order treatment on the grounds of fetal abuse. Even with legal enforcement, maternal compliance could not be assured and, since diet should start pre-conception for the best outcome, the moral position of the mother is more complex than first appears.

ATTITUDES WITHIN SOCIETY TO THE HANDICAPPED

It was suggested that, as antenatal diagnosis improves for a wide range of disorders, the burden of being the unfortunate parents with a handicapped child may seem to be even greater. A decision not to terminate a pregnancy at risk of handicap might be regarded by many in society as irresponsible and such attitudes towards loving and caring parents may be very hurtful. These points and others indicate the urgent need for a caring and compassionate society, willing to accept the handicapped and to provide for them the very substantial help which they and their families need. There must be total respect for the position of those who decide not to terminate for fetal abnormality.

PRE-IMPLANTATION DIAGNOSIS

It was suggested that some potential parents might actually prefer '*in vitro*' fertilization and the selection of genetically normal fertilized ova for implantation rather than early prenatal diagnosis and termination which is still very distressing for some women. Professor Pembrey had already spoken against genetically manipulating fertilized ova as opposed to the selection of normal ova referred to here.

GUIDELINES AND LEGAL RESTRICTIONS

There was no general desire to see a legal framework, which could create more anomalies than it would solve, other than the existing one which would, in any case, probably be modified in the UK to restrict termination to before 24 weeks in the not too distant future. There was much greater interest in some sort of framework of guidelines. Suggestions for guidelines were fewer, however. There was agreement that termination of pregnancy simply to ensure a child of chosen sex was unacceptable. Termination should be for serious handicap only. But who should define serious handicap? If respect for autonomy of the parents involved is accepted as a key guideline, as was generally agreed, does that mean simply that a serious handicap is whatever the parents define as a serious handicap? For example, would a disease like phenylketonuria, which is treatable but with a very serious potential handicap if untreated or very poorly treated, be appropriate grounds for termination? If, in general, it is not good grounds for termination, what extenuating circumstances might make it so? Could any system of guidelines take account of all possible extenuating circumstances? Regrettably, there was no opportunity to discuss these matters further.

SOCIETY FOR THE STUDY OF INBORN ERRORS OF METABOLISM

The SSIEM was founded in 1963 by a small group in the North of England but now has more than 70% of its members outside the UK. The aim of the Society is to promote the exchange of ideas between professional workers in different disciplines who are interested in inherited metabolic disorders. This aim is pursued in scientific meetings and publications.

The Society holds an annual symposium concentrating on different topics each year with facilities for poster presentations. There is always a clinical aspect as well as a laboratory component. The meeting is organized so that there is ample time for informal discussion; this feature has allowed the formation of a network of contacts throughout the world. The international and multidisciplinary approach is also reflected in the *Journal of Inherited Metabolic Disease*.

If you are interested in joining the SSIEM then contact the Treasurer: Dr. I. B. Sardharwalla, Willink Biochemical Genetics Unit, Royal Manchester Children's Hospital, Pendlebury, Manchester M27 1HA, UK. The current subscription is £25 per year payable January 1st each year. This subscription includes 4 issues and 2 supplements of the *Journal of Inherited Metabolic Disease* as well as the regular circulation of a newsletter.

J. Inher. Metab. Dis. 11 Suppl. 2 (1988) 131

Preface to Short Communications

This issue is devoted to selected short communications based on oral and poster presentations at the free sessions of the Annual Meeting of the Society for the Study of Inborn Errors of Metabolism held in Sheffield, 22–25th September 1987. As well as reflecting the topic of the main symposium, lipoprotein disorders, the oral and poster free communications covered a wide variety of other aspects of the study of inherited metabolic disease. Those presentations not reported elsewhere in this issue are listed below. The book of abstracts of all free communications may be purchased from the Society for the Study of Inborn Errors of Metabolism at 15 Saint Thomas' Drive, Hatch End, Pinner, Middlesex HA5 4SX, UK, price £5, including postage.

Many of the short communications were submitted for the SSIEM Award, which is judged on the basis of the manuscript prepared for the *Journal* rather than on the material presented at the meeting. This year the prize was awarded to R. J. A. Wanders, C. W. T. van Roermund, M. J. A. van Wijland, R. H. B. Schutgens, A. W. Schram, J. M. Tager, H. van den Bosch and C. Schalkwijk for their paper "X-Linked adrenoleukodystrophy: Identification of the primary defect at the level of a deficient peroxisomal very long chain fatty acyl-CoA synthetase using a newly developed method for the isolation of peroxisomes from skin fibroblasts".

With pressure on space in all scientific journals we hope that contributors and users will accept our suggestion that these papers be generated and used as short communications rather than as preliminary abstracts, at least in part. This year we have been able to accommodate only 68% of those offered for publication and thus an element of appraisal is inherent in their selection. It is clear to the editors that some are preliminary communications which allow priority to be established. However, others are worthwhile additional records which are adequate in themselves as contributions to our accumulated experience and do not require additional recording.

R. J. Pollitt
R. A. Harkness
G. M. Addison

Journal of Inherited Metabolic Disease. ISSN 0141–8955. Copyright © SSIEM and MTP Press Limited, Queen Square, Lancaster, UK.

J. Inher. Metab. Dis. 11 Suppl. 2 (1988) 132–134

Free Communications

Studies of lipoprotein synthesis and clearance in patients with familial endogenous hypertri-glyceridaemia and/or hyperapobetalipoproteinaemia. *J. A. Cortner, D. R. Cryer and P. M. Coates*

A genetic disorder with elongated peroxisomes due to a deficiency in the peroxisomal β-oxidation system. *B. T. Poll-The, J. P. Bonnefont, J. Scotto, M. Fardeau, H. Ogier, R. J. A. Wanders, R. B. H. Schutgens, A. W. Schram and J. M. Saudubray*

Propionate and protein turnover in methylmalonic and propionic acidaemia. *J. H. Walters, G. N. Thompson, J. V. Leonard, C. Hetherington, K. Bartlett, D. Halliday and A. A. Green*

Apoprotein assays: a cautionary note on exposing hidden antigenic sites. *M. S. Billingham, A. Green, R. A. Hall and C. J. Bailey*

Metabolism of vitamin A-labelled intestinal lipoprotein in man: a multi-compartmental analysis. *P. M. Coates, N-A. Le, P. R. Gallagher and J. A. Cortner*

The application of new DNA polymorphisms in the low density lipoprotein receptor gene for diagnosis of familial hypercholesterolaemia. *R. Taylor, M. S. Jeenah and S. E. Humphries*

A probe for lipoprotein lipase detects a polymorphism with stu1. *P. J. Bell, S. Erenbeck, K. Darnfors, G. Bjursell and S. E. Humphries*

Abetalipoproteinaemia. An Italian case. *R. Parini, B. Bardelli, A. M. Giunta and G. Vergani*

Therapeutic modification of abnormal lipids in erythrocyte membranes in juvenile neuronal lipofuscinosis (Batten's disease, JNCL). *M. J. Bennett, G. P. Hosking, R. Gayton, G. Thompson, J. H. Galloway and I. J. Cartwright*

Recurrent encephalopathy in a child of normal growth with a defect of long-chain fatty acid oxidation. *E. M. Layward, S. Hodges, P. G. F. Swift, R. J. Pollitt, M. J. Bennett and K. Bartlett*

Carnitine deficiency with cardiomyopathy presenting as neonatal hydrops: successful response to carnitine therapy. *P. Steenhout, D. Blum, A. Clercx, C. Elmer, F. Vertongen and E. Vamos*

Familial carnitine deficiency with myocardial fibroelastosis in a 4-year-old girl. *S. Lindstedt, I. Nordin, B. Eriksson and A. Oldfors*

3-Hydroxydecanedioic aciduria – a new inborn error of metabolism? *O. Stokke, E. Jellum, E. A. Kvittingen and B. F. Kase*

Acyl-CoA dehydrogenase deficiency in heart tissue from infants who died suddenly and unexpectedly and who demonstrated fatty change in liver. *F. Allison, M. J. Bennett, S. Variend and P. C. Engel*

The diagnosis of fatty acid oxidation defects by the measurement of CO_2 release from radiolabelled fatty acids in lymphocytes. *R. G. F. Gray, S. Simpson, L. Jenkinson and A. Green*

A comparative analysis for a sensitive and accurate determination of metabolites in organic acidurias. *E. Erasmus, L. J. Mienie and C. J. Reinecke*

A new case of 3-hydroxy-3-methylglutaryl-CoA lyase deficiency with severe physical and mental handicap. *H. C. Losty, J. Allen and S. J. Wallace*

Mild clinical presentation in a new case of 3-hydroxy-3-methylglutaryl-CoA lyase deficiency. *M. Orera, B. Merinero, C. Perez-Cerda, A. Jimenez, P. Sanz, M. J. Garcia, M. Martinez-Pardo and M. Ugarte*

Treatment of propionic acidaemia with carnitine and special feeds. *D. M. Isherwood, D. C. Davidson, R. Diplexcito, E. Kerr, J. A. West-Jordan, R. J. Abraham, R. D. Griffiths and R. H. T. Edwards*

A case of propionic acidaemia: unusual presentation and biochemical response to treatment. *B. Fowler, A. K. Holmes, I. B. Sardharwalla and R. W. Newton*

Excretion of propan-1,3-diol and 3-hydroxypropionic and malonic acids apparently caused

Journal of Inherited Metabolic Disease. ISSN 0141-8955. Copyright © SSIEM and MTP Press Limited, Queen Square, Lancaster, UK.

by abnormal bacterial metabolism in the gut. *B. Fowler, I. B. Sardharwalla, M. A. Edwards, R. G. F. Gray and R. J. Pollitt*

Chronic renal disease in methylmalonic acidaemia. *J. H. Walters, A. Michaelski, W. Wilson and J. V. Leonard*

Screening for biotinidase deficiency in the north-east of Italy. *A. Burlina, L. Carcereri and D. Gaburro*

Severe alopecia in a girl with partial biotinidase deficiency. *E. Ruidor, A. Ribes, S. Uriz, J. Brunet, F. Giribert, A. Ballabriga and A. Saez*

Isolation of cDNA clones for the human pyruvate dehydrogenase E1a subunit and characterisation of the mRNA. *H-H. M. Dahl, S. M. Hunt, W. M. Hutchinson and G. K. Brown*

N-Acetyl aspartic aciduria: 2 cases in sibs with progressive cerebral atrophy. *P. Divry, C. Vianey Liaud, M. Mathieu and Macabeo*

A new case of combined homocystinuria and methylmalonic aciduria. *E. Holme, C-E. Jacobsson, R. Jagenburg and B. Kristiansson*

5,10-Methylenetetrahydrofolate reductase deficiency – diagnosis and therapy in a teenage boy. *E. Zammarchi, S. Falorni, G. Arnetoli, M. Paganini and U. Wendel*

Determination of various folate forms in human cerebrospinal fluid, plasma and red blood cells. *Y. S. Shin, O. Zelgar, B. Wartner, S. Kruz and W. Endres*

Autoimmune thyroiditis in a girl with persistent tyrosinaemia. *P. D'Eufemia, A. Cantani, U. Ruberto, F. Martino, L. Boniforti and O. Giardini*

Type 1 hereditary tyrosinaemia: purification using immunoaffinity chromatography and characterization of fumarylacetoacetase from human liver. *H. van Faassen and R. Berger*

Type 1 hereditary tyrosinaemia: isolation and characterization of cDNA for human fumarylacetoacetase. *R. Berger, H. van Faassen, T. Reversma, J. W. Taanman and E. Agsteribbe*

Clinical and biochemical findings in eight cases of tyrosinaemia type 1 (fumarylacetoacetase deficiency). *M. A. Edwards, G. W. Rylance and A. Green*

A new variant of leucinosis. *P. C. Clemens*

Metabolism of L-isoleucine and L-allo-isoleucine in patients with maple syrup urine disease and healthy subjects. *U. Wendel, P. Schadewaldt and U. Langenbeck*

Heterozygote detection of arginase deficiency and argininosuccinic aciduria by assay of erythrocyte enzymes. *Y. S. Shin, B. Kruis, M. Brockstedt and W. Endres*

Study of citrulline levels in a family with a citrullinaemia patient. *P. C. Clemens and C. Plettner*

Aspartame intake and its effect on phenylalanine and the phenylalanine metabolites. *R. Matalon, K. Michaels, D. Sullivan and P. Levy*

Report of a multicentre trial of a new product for the treatment of maternal phenylketonuria. *J. T. Ireland*

Plasma values of phenylacetic acid in phenylacetic acid in phenylketonuria. *P. C. Clemens, A. Kohlschutter and M. Schuenemann*

Visual and brainstem auditory evoked potentials in patients with hyperphenylalaninemia. *R. Korinthenberg, F. Fullenkemper and K. Ullrich*

Maternal phenylketonuria: effects of dietary treatment and phenylalanine control in early gestation. *M. G. Beasley and I. Smith*

Maternal phenylketonuria collaborative study: USA and Canada. *R. Koch, C. Azen, E. G. Friedman, F. dela Cruz, H. Levy, R. Matalon, B. Rouse and W. B. Hanley*

RFLP of the phenylalanine hydroxylase gene in the Spanish population. Haplotypes of three PKU-families. *G. Morales, M. C. Lopez, C. Alonso and M. Ugarte*

Phenylketonuria in Eastern Canada: an anomalously high prevalence rate in the maritime region. *D. E. C. Cole, C. M. Veinot, A. T. J. McDonald and P. Ferreira*

PKU and glycogen storage disease type III in sibs of one family. *H. D. Bakker, A. H. van Gennip, N. G. G. M. Abeling and A. Hammond*

Dihydropteridine reductase activity in eluates from dried blood spots: automation of an assay for a national screening service. *I. M. Surplice, P. D. Griffiths, A. Green and R. J. Leeming*

An evaluation of Becton Dickinson's neonatal blood spot Simultrac TSH/T4. *S. Thompson and R. Kennedy*

A simple concept for the screening of amino- and organic acidurias. *S. Stockler, M. Klopf, B. Pokits, H. Schinagl, W. Erwa and E. Paschke*

HPLC analysis of physiological amino acids as their phenylthiocarbamyl derivatives. *E. Christensen*

Separation of amino acids by high performance liquid chromatography using precolumn derivatisation with dansyl chloride. *H. Schierbeek and R. Berger*

Quantitative amino acid analysis: reversed-phase HPLC compared to IEC. *J. M. Leah, T. Palmer and M. Griffin*

The effect of light on diurnal variation of blood amino acids in neonates. *S. Mantagos, A. Moustogianni, A. Varvarigou and C. Frimas*

Experience in prenatal diagnosis in peroxisomal disorders. *R. B. H. Schutgens, R. J. A. Wanders, J. M. Tager, G. Schrakamp and H. van der Bosch*

X-linked adrenoleukodystrophy: alternative metabolic reactions for the accumulated very long chain fatty acids. *C. J. Reinecke and S. I. Potgieter*

Are the proteinase inhibitors CBZ-PHE-ALA-CHN$_2$ and leupeptin useful in the treatment of patients with metachromatic leukodystrophy? *K. Ullrich, St. Schmiereck, R. Korinthenberg and K. von Figura*

Egasyn affects the processing, intracellular transport and turnover of β-glucuronidase in mouse liver. *K. Pfister, N. Bosshard, M. Zopfi and R. Gitzelmann*

Bone marrow transplantation in a patient with α-mannosidosis. *A. Cooper, A. Will, C. Hatton, I. Sardharwalla, D. Evans and R. Stevens*

β-Glucuronidase deficiency in an adult – clinical and biochemical findings. *R. G. F. Gray, A. Green, T. J. Constable and S. Bundey*

Steroid sulphatase (STS) deficiency in a family with an X/Y translocation: molecular studies using a STS code cDNA probe. *A. Ballabio, G. Parenti, R. Carrozzo, G. Coppa, L. Felici, V. Migliori, M. Silengo, P. Franceschini and G. Andria*

Glucose and palmitate metabolism in patients with glycogenosis type III. *G. P. A. Smit, T. E. Chapman, I. E. Mulder, D-J. Reijngoud, F. A. J. Muskiet and R. Berger*

Determination of branching enzyme in erythrocytes: a case of prenatal exclusion of type IV glycogenosis. *Y. S. Shin, H. Steiguber, P. Klemm, W. Endres, O. Schwab and G. Wolff*

Glycogen storage disease in 4 consecutive generations with demonstrated erythrocyte phosphorylase β-kinase deficiency. *M. Blaskovics, S. Allen, Y-K. Xu, T. Roe, W. G. Ng*

Acute renal failure as a presenting symptom in hypoxanthine–guanine and adenine phosphoribosyltransferase deficiency. *H. A. Simmonds, J. S. Cameron, T. M. Barrett, M. J. Dillon, S. R. Meadow and R. S. Trompeter*

Increased urinary β-aminoisobutyric acid after thymine loading in a patient with undetectable dihydropyrimidine dehydrogenase activity in leukocytes. *A. H. van Gennip, H. D. Baker, A. Zoetekouw and N. G. G. M. Abeling*

A new case of molybdenum cofactor defect with subsequent prenatal exclusion of an affected fetus chorionic villus assay. *A. Aukett, M. J. Bennett, G. P. Hosking, P. Divry and P. Baltassat*

Interpretation of sweat test results – the effect of patient age. *J. M. Kirk and A. Westwood*

Effect of sweat rate on electrolyte concentration in children with and without cystic fibrosis. *K. Carpenter and E. Worthy*

Regional experience of cystic fibrosis prenatal diagnosis. *D. Bozon, I. Maire, A. Vialle, G. Mandon, P. Guibaud and R. Gilly*

Linkage analysis in cystic fibrosis families from Italy. *M. Seia, M. Ferrari, L. Piceni Sereni and A. Giunta*

J. Inher. Metab. Dis. 11 Suppl. 2 (1988) 135–138

Short Communication

An Erroneous Apolipoprotein E-3 Band in High Density Lipoprotein Fractions

A. V. RAWLINGS and T. DEEGAN

Cardiothoracic Unit, Broadgreen Hospital, Liverpool L14 3LB, UK

As a result of genetic polymorphism and post-translational glycosylation with sialic acid, apolipoprotein E (apoE), which is distributed mainly between the triglyceride-carrying lipoproteins and the high density lipoproteins (HDL), shows heterogeneity when submitted to isoelectric focusing (Zannis and Breslow, 1981). The three common genetic isoforms of apoE are designated E-4, E-3 and E-2 and contain zero, one and two cysteine residues per molecule, respectively (Weisgraber *et al.*, 1981). These isoforms can be easily separated by isoelectric focusing, but co-migration of sialoforms with the true genetic isoforms complicates the patterns. According to the currently accepted genetic model, three apoE alleles ($\varepsilon 4$, $\varepsilon 3$ and $\varepsilon 2$) can occur at a single gene locus, and specify six apoE phenotypes: E4/4, E3/3, E2/2, E4/3, E4/2 and E3/2 (Zannis *et al.*, 1982).

Fredrickson Type III hyperlipoproteinaemia (Type III HLP), characterised by hypertriglyceridaemia, hypercholesterolaemia and dys-β-lipoproteinaemia, is generally considered to be primarily associated with the E2/2 phenotype (Brewer *et al.*, 1983). However, the detection of an apoE-3 band in the HDL fraction in this condition has led to the proposition that Type III HLP can arise from the abnormal composition of the very low density lipoprotein (VLDL) particles rather than from the presence of apoE-2 homozygosity (Holmquist, 1984). This paper reports a study of the apolipoprotein patterns in the VLDL and HDL fractions of Type III HLP subjects and other normo- and hyperlipidaemic subjects with various apoE phenotypes, undertaken in an attempt to resolve these differing opinions.

METHODS

Blood samples, withdrawn by antecubital venepuncture in the fasting state, were collected in ice-cold tubes containing EDTA ($1.5 \, \text{mg} \, \text{mL}^{-1}$) and serine protease inhibitors (PPACK; $1 \, \mu\text{mol} \, \text{L}^{-1}$; Calbiochem). Plasma was obtained by centrifugation at $2000 \, g$ for $10 \, \text{min}$ at $4°\text{C}$. VLDL and HDL, separated by preparative ultracentrifugation at $10°\text{C}$ using standard procedures (Lindgren *et al.*, 1972) and delipidated by acetone/alcohol followed by ether, were examined by polyacrylamide rod gel isoelectric focusing (Warnick *et al.*, 1979). Gels were stained with Coomassie Blue R250. ApoE phenotypes were confirmed by selective chemical modification of the cysteine residues present with cysteamine (Weisgraber *et al.*, 1981). Two-dimensional polyacrylamide gel electrophoresis (2D-PAGE) was ap-

135

plied essentially according to Zannis and Breslow (1981). Isoelectric focusing was followed by sodium dodecyl sulphate–polyacrylamide gel electrophoresis in 12.5% polyacrylamide gels.

RESULTS AND DISCUSSION

In accordance with the genetic model for Type III HLP, Holmquist (1984) demonstrated a lack of apoE-3 and apoE-4 in VLDL fractions of Type III HLP subjects, but detected a protein focusing in the E-3 position in the HDL fractions. These results were confirmed in two subjects with Type III HLP using a single-dimensional isoelectric focusing method. Figure 1 shows the patterns of the apolipoproteins in

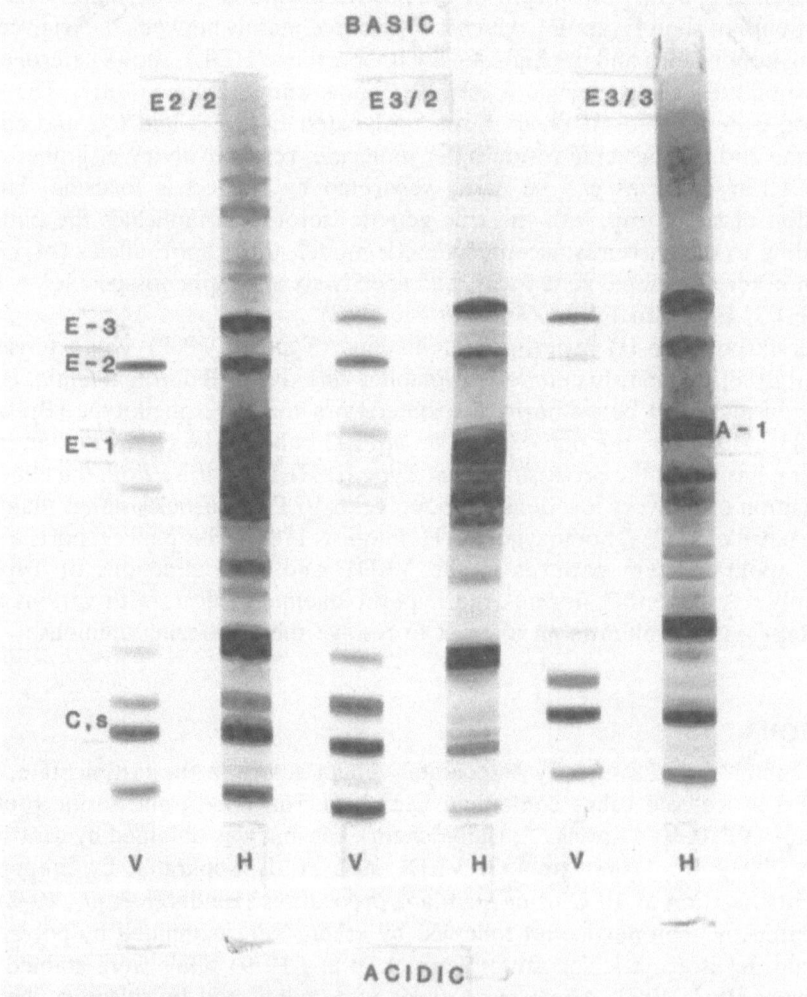

Figure 1 Single-dimensional isoelectric focusing patterns of apolipoproteins from VLDL (V) and HDL (H) fractions from E2/2, E3/2 and E3/3 phenotypes

J. Inher. Metab. Dis. 11 Suppl. 2 (1988)

VLDL and HDL fractions from one Type III HLP subject with the VLDL E2/2 phenotype and two normolipidaemics with the VLDL E3/2 and E3/3 phenotypes, respectively. Comparison of the VLDL and HDL isoelectric focusing patterns showed apparent changes in the apoE phenotypes with emergence of apoE-3 dominance in the HDL fractions. This was especially marked in the subject with Type III HLP, in whom there was an apparent change from the E2/2 phenotype in VLDL to the E3/3 phenotype in HDL.

To define the apoE-3 band in the HDL fractions further, the HDL proteins were examined using 2D-PAGE. The section of the gel containing the apoE and apoA-1 isoforms in the HDL fraction of the Type HLP subject is shown in Figure 2. It is evident that the protein focusing in the E-3 position has approximately the same molecular weight as apoA-1. Moreover, HDL fractions from five normolipid-aemics and five subjects with different hyperlipidaemias, encompassing all the apoE phenotypes, also showed a protein focusing in the E-3 position with the same mass as apoA-1. Upon further examination by two-dimensional immunoelectrophoresis against anti-apoA-1 (Boehringer, Mannheim), the erroneous apoE-3 band demonstrated immunological identity with apoA-1 (not shown). These preliminary results accord with the reported electrophoretic characteristics of pro-apoA-1, an isoform of apoA-1 (Ghiselli *et al.*, 1983), and suggest that the erroneous apoE-3 band described is the same protein.

Although Type III HLP has been reported with phenotypes other than the E2/2 condition, the majority of Type III subjects possess this characteristic (Brewer *et al.*, 1983). The results presented here support this evidence and are consistent with the genetic model of inheritance of the apoE alleles. They also confirm a total lack

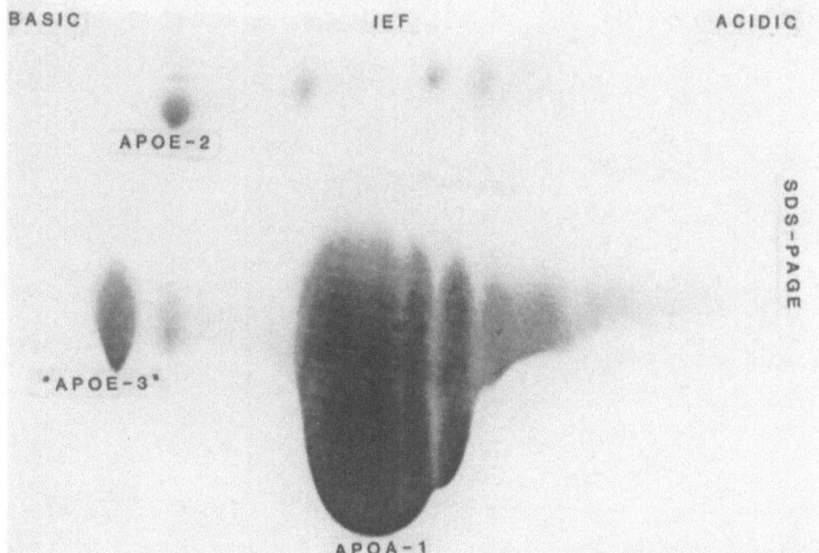

Figure 2 Two-dimensional pattern of HDL apolipoproteins from an E2/2 phenotype. (Only the section of the gel containing the apoE and apoA-1 isoforms is shown

of apoE-3 in the lipoprotein fractions of Type III HLP subjects with the E2/2 phenotype and demonstrate that a protein with the same immunochemical and electrophoretic characteristics as pro-apoA-1 can mask the apoE proteins and result in erroneous phenotyping when using HDL fractions.

REFERENCES

Brewer, H. B., Jr., Zech, L. A., Gregg, R. E., Schwartz, D. and Schaefer, E. J. Type III hyperlipoproteinaemia: Diagnosis, molecular defects, pathology and treatment. *Ann. Int. Med.* 98 (1983) 623–640

Ghiselli, G., Schaefer, E. J., Light, J. A. and Brewer, H. B., Jr. Apolipoprotein A-1 isoforms in human lymph: effect of fat absorption. *J. Lipid Res.* 24 (1983) 731-736

Holmquist, L. Distribution of apolipoprotein E isoforms between very low and high density lipoproteins of normal subjects and hyperlipidaemic patients with special reference to Type III hyperlipidaemia. *Acta Med. Scand.* 215 (1984) 113–120

Lindgren, F. T., Jensen, L. C. and Hatch, F. T. The isolation and quantitative analysis of serum lipoproteins. In Nelson, G. J. (ed.) *Blood Lipids and Lipoproteins: Quantitation, Composition and Metabolism.* Wiley-Interscience (1972) pp. 181-276

Warnick, G. R., Mayfield, C., Albers, J. J. and Hazzard, W. R. Gel isoelectric focussing method for specific diagnosis of familial hyperlipoproteinaemia Type 3. *Clin. Chem.* 25 (1979) 279–284

Weisgraber, K. H., Rall, C. R., Jr. and Mahley, R. W. Human E apoprotein heterogeneity. *J. Biol. Chem.* 256 (1981) 9077–9083

Zannis, V. I. and Breslow, J. L. Human very low density lipoprotein apolipoprotein E isoprotein polymorphism is explained by genetic variation and post-translational modification. *Biochemistry* 20 (1981) 1033–1041

Zannis, V. I., Breslow, J. L., Utermann, G., Mahley, R. W., Weisgraber, K. H., Havel, R. J., Goldstein, J. L., Brown, M. S., Schonfeld, G., Hazzard, W. R. and Blum, C. Proposed nomenclature of apoE isoproteins, apoE genotypes and phenotypes. *J. Lipid Res.* 23 (1982) 911–914

J. Inher. Metab. Dis. 11 Suppl. 2 (1988) 139–142

Short Communication

Histochemical Abnormalities in Liver and Jejunal Biopsies from a Case of Cholesterol Ester Storage Disease

A. Lageron[1] and J. Polonovski[2]

[1]*INSERM U.9 184, rue du Fg Saint Antoine 75012 Paris, France; and* [2]*CNRS UA 524 CHU Saint Antoine 27, rue de Chaligny 75012, Paris, France*

Cholesterol ester storage disease (CESD) (McKusick 21500) is an uncommon inborn error of metabolism due to an acid lipase (EC 3.1.1.1) deficiency. Among the nearly 20 cases reported in the literature, most data concerned the liver and only two works, those of Partin and Schubert (1969) and Dincsoy *et al.* (1984), reported studies of small intestine. The present study demonstrates the injuries previously described and other abnormalities which give rise to numerous problems.

PATIENT AND METHODS

An 11-year-old girl (C.N.), an only child with no relevant family history, suffering from CESD, underwent percutaneous liver and peroral jejunal biopsies after at least 12 h of fasting, to confirm the diagnosis and to assess the disease impact. The diagnosis was suggested by isolated unexplained persistent hepatomegaly without digestive disorder, steatorrhoea or retarded growth. Abnormal blood biological data consistent with this diagnosis were found including: total lipids $10.2\,g\,L^{-1}$, total cholesterol $9.2\,mmol\,L^{-1}$ and triglycerides $1.5\,mmol\,L^{-1}$. Separated lipoproteins by electrophoresis were as follows with staining appreciation: α-lipoproteins 9%, pre-β- and β-lipoproteins 91%.

Histoenzymological studies of frozen unfixed material from liver and small intestine were carried out using classical techniques previously reported (Lageron, 1978).

Some assays of immunological reactions were performed employing monoclonal antibodies developed in mouse by Sanofi (Montpellier, France) against the human apoproteins B-100 and B-48: L_7 reacting with apo-B-100, and L_9 with apo-B-100+B-48, respectively (Salmon *et al.*, 1984). After L_7 or L_9 incubation, an anti-mouse IgG-labelled by horse-radish peroxidase was applied and peroxidase activity revealed by 3,3'-diaminobenzidine and hydrogen peroxide. Sections were counterstained by Oil Red and haematein.

Specificity was checked by omitting each step of the reaction and control sections, from normal subjects, were studied in parallel. Each antibody was tried on paraffin-embedded and frozen material unfixed or postfixed in 10% formaldehyde, endogen-

Journal of Inherited Metabolic Disease. ISSN 0141–8955. Copyright © SSIEM and MTP Press Limited, Queen Square, Lancaster, UK.

ous peroxidase activity being inhibited in paraffin sections by hydrogen peroxide and in frozen material by periodic acid followed by potassium borohydride.

RESULTS

Liver

The histochemical profile is definitely characteristic of CESD (Lageron, 1978) highlighting esterified cholesterol storage of all hepatocytes while macrophages exhibit a more complex storage including free and esterified cholesterol, free fatty acids and phospholipids. With respect to other cases, the liver of C.N. revealed two unusual features: a cirrhotic pattern and numerous macrophages stained as sea-blue histiocytes after using the May-Grünwald-Giemsa technique, these latter appearing as autofluorescent under ultraviolet light.

L_7 visualized apo-B better than L_9, as fine granulations scattered throughout the cytoplasm of hepatocytes.

Jejunal biopsy

Absence of acid lipase activity was ascertained by both histoenzymological techniques in liver and jejunum and biological ones in leukocytes. Though villi were normal in size and shape, they exhibited two striking features: very enlarged lacteals occupying the major axis of the villus and clusters of macrophages caping mainly the upper part of lacteals; other macrophages being located within lacteals or in deep parts of the lamina propria. All these macrophages were sea blue histiocytes displaying the same histochemical profile as those of the liver.

Absorbtive cells contained small Oil Red-stained vacuoles and the same material was seen, extracellular, throughout the lamina propria. The same staining occurred in large patches within lacteals.

Nearly all the material coloured by Oil Red reacted with lipase–lead sulphide, indicating triglycerides; the other part, Schultz positive, was constituted by cholesterol in which free and esterified forms took an equal part as demonstrated by variants of Emeis' reaction. The phospholipids were located by Baker's reaction at the same place as triglycerides, in lesser amount in epithelial cells but greater in extracellular material where they took an elongated shape. In lacteals they appeared as a wide-mesh net. OTAN visualized mixed hydrophobic lipids and phospholipids in all mucosa. Moreover triglycerides, cholesterol and phospholipids lined the lateral space between absorptive cells as flows. All these flows joined together along their basement membrane.

Visualization of apo-B-100 and B-48, located at the same place, was more marked with L_9 than with L_7. Each supranuclear part of epithelial cells from paraffin-embedded sections exhibited a positive reaction while almost all cells remained negative or slightly positive in frozen material. In all sections, macrophages contained apoproteins-B but it is difficult to identify them individually around extracellular lipid material. At the level of the lacteal, a fine net was seen in paraffin sections, while a large spot filled the lacteal in frozen sections. In this brown spot, counterstaining with Oil Red revealed large lipidic vacuoles among apoproteins. Lacteals exhibited individualized round 'structures' without nuclei with a large triglyceride core, sometimes containing cholesterol crystal, surrounded by a thin, irregular, wavy apo-B ring but this was only in some sections, probably because of their instability. Phospholipids were also identified at the same site (Figure 1).

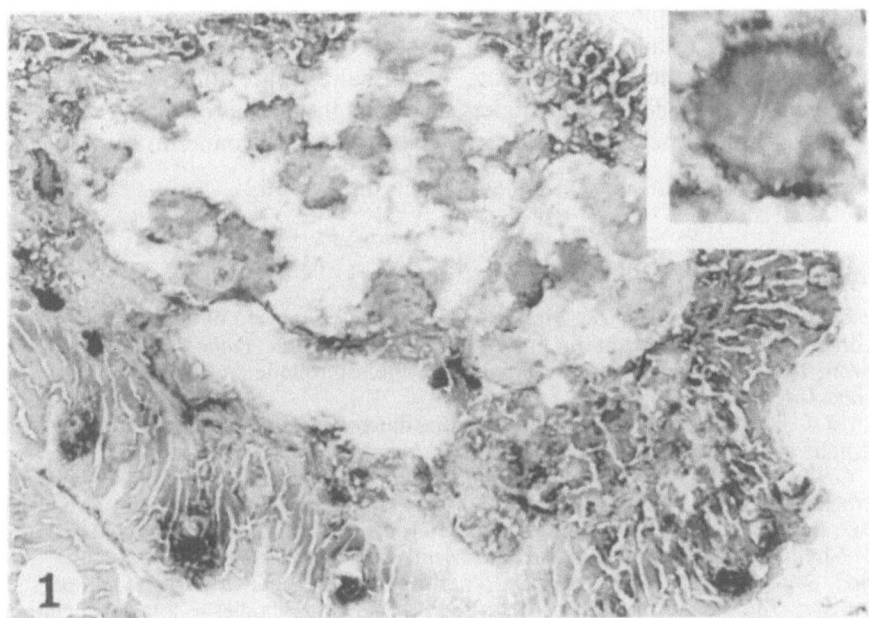

Figure 1 Section of jejunal biopsy from C.N.: L₇ antibody counterstained by Oil Red exhibiting villus with enlarged lacteal filled by chylomicron-like 'structures' (×225). Inset shows detail of the 'structure' with an irregular ring of apo-B, lipidic core and cholesterol crystals (×900).

COMMENTS

Three points might be emphasized:

(1) Are the round 'structures' seen in lacteals true chylomicrons? Though their constituents are exactly those described for chylomicrons, two features count against this hypothesis: their presence after more than 12 h of fasting and their size (200 000 Å). The problem of size is not a real one, enlargement has been reported during triglyceride or cholesterol increase or during intestinal protein synthesis inhibition; in the same field, enlarged chylomicron remnants are seen in Tangier's disease and LCAT deficiency.

Abnormal intestinal retention of chylomicrons is reported only by Roy *et al.* (1987) but this occurred within absorptive cells while, in C.N, retention was in the lacteals. As nothing is known about the movement of normal chylomicrons from lacteals to blood capillaries it is difficult to speculate on abnormal conditions.

(2) The presence of extracellular mixed lipid with an elongated shape (also reported by Partin and Schubert, 1969), which differs from chylomicrons also remained unexplained.

(3) The complex lipid storage within absorptive cells seems to be induced by mechanisms totally different from those observed in abetalipoproteinaemia or protein synthesis inhibition. As intestinal cells are affected by the genetic defect, a total acid lipase deficiency, why might not this defect be responsible for dyslipidic metabolism of these cells and for chylomicron retention?

The way to resolve this puzzle remains to be discovered.

REFERENCES

Dincsoy, H. P., Rolfes, D. B., McGraw, C. A. and Schubert, W. K. Cholesterol ester storage disease and mesenteric lipodystrophy. *Am. J. Clin. Pathol.* 81 (1984) 263–269

Lageron, A. Histoenzymologie de la polycorie cholestérolique. A propos de 5 cas. *Med. Chir. Dig.* 7 (1978) 155–159

Partin, J. C. and Schubert, W. K. Small intestine mucosa in cholesterol ester storage disease. A light and electron microscope study. *Gastroenterology* 57 (1969) 542–558

Roy, C. C., Levy, E., Green, P. H. R., Sniderman, A., Letarte, J., Buts, J. P., Orquin, J., Brochu, P., Weber, A. M., Morin, C. L., Marcel, Y. and Deckelbaum, R. J. Malabsorption, hypocholesterolemia, and fat-filled enterocytes with increased intestinal apoprotein B. Chylomicron retention disease. *Gastroenterology* 92 (1987) 390–399

Salmon, S., Goldstein, S., Pastier, D., Theron, L., Berthelier, M., Ayrault-Jarrier, M., Dubarry, M., Rebourcet, R. and Pau, B. Monoclonal antibodies to low density lipoprotein used for the study of low- and very-low-density lipoproteins, in "ELISA" and immunoprecipitation techniques. *Biochem. Biophys. Res. Commun.* 125 (1984) 704–711

J. Inher. Metab. Dis. 11 Suppl. 2 (1988) 143–145

Short Communication

Cholesteryl Ester Storage Disease: Risk Factors for Atherosclerosis in a 15-Year-Old Boy

R. Longhi[1], C. Vergani[2], R. Valsasina[1], E. Riva[1], C. Galluzzo[1], C. Agostoni[1] and M. Giovannini[1]

[1]5th Department of Pediatrics, Institute of Biomedical Sciences 'Ospedale S. Paolo', University of Milan, and [2]3rd Department of Internal Medicine, Centre for the Study and the Therapy of Metabolic Diseases, University of Milan, Italy

Cholesteryl ester storage disease (CESD; McKusick 21500), an autosomal recessive inborn error of metabolism, is characterized by deficient activity of lysosomal acid lipase. This disorder, unlike the more dangerous expression of the deficiency state, Wolman's disease, can allow patients to live a long time and may not be detected until adulthood, in spite of important biochemical abnormalities. One of these concerns the lipidaemic pattern and the related risk of developing early atherosclerosis due to a high serum total cholesterol and triglyceride levels coupled with very low high-density lipoprotein (HDL) values. As a consequence, a wide range of clinical pictures of the disease is possible, ranging from mild clinical expressions to life threatening ones. We describe here the case of a young subject affected by CESD, emphasizing the risk factors for the development of early atherosclerosis.

CASE REPORT

L.C., a boy, was born in 1971 following a normal pregnancy. On examination, he was found to have hypoplasia of the right pinna. Subsequently, he progressed satisfactorily except for a slight psychomotor delay. At 5 years, hepatosplenomegaly was found. An exhaustive laboratory investigation did not show any abnormality in plasma and urine biochemistry. A subsequent liver biopsy showed considerable fatty infiltration with a modest increase in portal tract fibrosis. He was diagnosed to have CESD at 8 years of age by very low acid esterase activity in peripheral leukocytes ($10 \, nmol \, (mg \, protein)^{-1} \, min^{-1}$) and cultured fibroblasts ($2.9 \, nmol \, (mg \, protein)^{-1} \, min^{-1}$; normal range 76 ± 38). Enzymatic diagnosis was made by A. Mowat and D. Patrick, London. One year later hypercholesterolaemia (with low HDL) and hypertriglyceridaemia developed, and a diet low in saturated fats and animal proteins was started. A complete blood and urine chemistry, including a wide set of lipid-related serum parameters was performed when he was 15 years of age (Table 1). On echography, the adrenals were both found to be mildly enlarged without calcifications. During this clinical 'check-up' asymptomatic hypertension was first noted. On successive checks the finding was confirmed, with diastolic

Journal of Inherited Metabolic Disease. ISSN 0141-8955. Copyright © SSIEM and MTP Press Limited, Queen Square, Lancaster, UK.

Table 1 Biochemical parameters of the patient

Plasma parameters	Value	Normal range
Lipid-related substances[a]		
Total cholesterol	249	152–206
Triglycerides	151	56–120
VLDL	30	12–24
LDL	199	86–142
HDL	20	38–56
HDL-2	8	10–26
HDL-3	12	25–40
Apo AI	64	135–156
Apo AII	22.6	36–42
Apo B	163	94–113
Apo CII	3.7	2.8–4
Apo CIII	6.4	7.3–8.8
Apo E	7.7	3.1–4
Enzymes[b]		
LCAT	20.7	13.3–26.3
SGOT	49	≤37
SGPT	70	≤40
Alkaline phosphatase	1108	≤760
Creatinine	0.83	≤1.2

[a]All values are in $mg\,dL^{-1}$
[b]Values are: LCAT, $nmol\,mL^{-1}h^{-1}$; SGOT and SGPT, $U\,L^{-1}$; alkaline phosphatase, $mg\,dL^{-1}$

pressure ranging from 95 to 105 mmHg, above the 97th percentile for age and height. Systolic arterial pressure was repeatedly found at the upper limits of the normal range (135–140 mmHg). Endocrine investigations on blood (circadian rhythm of cortisol, testosterone, 17β-estradiol, dehydroepiandrosterone sulphate, aldosterone, plasma renin activity in clinostatism and orthostatism) and urine (vanillylmandelic/creatinine ratio), performed to search for a secondary cause of the high blood pressure, showed no abnormality.

The boy is now well, presenting only mild hepatosplenomegaly. His family history showed neither coronary heart disease nor essential hypertension recorded in first-grade relatives; he is not obese and weight and height are appropriate for age.

DISCUSSION

In CESD the impaired function of acid lipase is responsible for the blockade, at the lysosomal step, of the Goldstein and Brown pathway (Brown and Goldstein, 1986). Cholesteryl esters cannot be hydrolysed, and the very low levels of intracellular free cholesterol cannot suppress the endogenous cholesterol synthesis by inhibition of hydroxymethylglutaryl coenzyme A reductase. For the same reason, membrane receptors are not down regulated, resulting in further intracellular storage of cholesteryl esters derived from very-low-density lipoprotein (VLDL) remnants and low-density lipoproteins (LDL). Furthermore, the catabolic rate attributable to the non-receptor pathway (Grundy *et al.*, 1986) could theoretically be impaired. The metabolic fate of cholesterol in CESD appears unbalanced, due to an encumbered centrifugal efflux (liver→ VLDL→ remnants→ LDL→cell) and a low centripetal one (cell→HDL→liver). Our patient has high apolipoprotein (Apo) B (LDL) and Apo E (VLDL) values, while Apo AI–II (HDL) levels are well below the normal range. An increased Apo B/Apo AI ratio appears highly predictive of the development of vascular lesions (Freedman *et al.*, 1986). However, the normal lecithin–cholesterol acyltransferase (LCAT) activity together with low HDL levels raises the possibility of an increased catabolic turnover of this lipoprotein class (HDL-3 → LCAT activity→HDL-2→liver). In fact, in a previous study (Kostner *et al.*, 1985) a high hepatic lipoprotein–lipase activity (which clears HDL-2 from circulation) was found in a patient with CESD. Bile cholesterol measurements could indirectly provide a decisive answer. The intracellular accumulation of cholesteryl esters would cause arterial endothelial damage by stimulating vascular wall cells to form and secrete growth factors involved in promoting early proliferative lesions. According to the hypothesis of Ross (1986), the loss of equilibrium between growth factors and antagonist substances can be promoted by other injuries too, such as increased arterial blood pressure.

In our patient both these predisposing conditions (biochemical derangement and mechanical insult) are present, leading to high risk of early arterial damage.

REFERENCES

Brown, M. S. and Goldstein, J. L. A receptor mediated pathway for cholesterol homeostasis. *Science* 232 (1986) 34–47

Freedman, D. S., Srinivasan, S. R., Shear, C. L., Franklin, F. A., Webber, L. S. and Berenson, G. S. The relation of apolipoproteins A-I and B in children to parental myocardial infarction. *N. Engl. J. Med.* 315 (1986) 721–726

Grundy, S. M., Lena Vega, G. and Bilheimer, D. W. Causes and treatment of hypercholesterolemia. In Grundy, S. M. (ed.) *Bile Acids and Atherosclerosis*, Raven Press, New York, 1986, pp. 13–39

Kostner, G. M., Hadorn, B., Roscher, A., Zechner, R. Plasma lipids and lipoproteins of a patient with cholesteryl ester storage disease. *J. Inher. Metab. Dis.* 8 (1985) 9–12

Ross, R. The pathogenesis of atherosclerosis – an update. *N. Engl. J. Med.* 314 (1986) 488–500

J. Inher. Metab. Dis. 11 Suppl. 2 (1988) 146–148

Short Communication

Cholesteryl Ester Storage Disease with Secondary Lecithin Cholesterol Acyl Transferase Deficiency

S. Van Erum[1], D. Gnat[1], C. Finne[1], D. Blum[3], C. Vanhelleput[4], E. Vamos[2] and F. Vertongen[1]

Departments of [1]Clinical Chemistry, [2]Genetics and [3]Pediatrics, Free University of Brussels, 322, rue Haute, 1000 Brussels, Belgium and [4]Department of Pediatrics Centre Hospitalier Tivoli, av. Max Duvet, 34, 7100 La Louvière, Belgium

Cholesterol ester storage disease (CESD; McKusick 21500) is a rare lipid storage disorder inherited in an autosomal recessive manner. The disease results from a marked deficiency of lysosomal acid esterase activity which gives characteristic blood lipid abnormalities (plasma lipoprotein pattern type IIa or IIb). Wolman's disease is the more severe form of acid lipase deficiency, nearly always fatal before the age of 1 year. Clinical features of CESD include hepatomegaly, splenomegaly, oesophageal varices and finally hepatic fibrosis. In many cases hepatomegaly is the only clinical abnormality and the disease may therefore not be diagnosed until adulthood (Assmann *et al.*, 1983).

We present a case of CESD with secondary deficiency of lecithin cholesterol acyl transferase (LCAT) due to hepatocellular disease (Sabesin *et al.*, 1977). This latter enzyme is responsible for cholesterol esterification.

CASE REPORT

A 4-year-old girl was referred for hepatitis and septicaemia. The physical and psychomotor development were normal. Hepatomegaly (2 cm below the right costal margin) was noted for the first time 2 years previously. The patient had a history of familial lipid abnormalities with Fredrickson type IIb plasma lipoprotein pattern. Her parents were not related and she had an 8-year-old sister.

Haemocultures were positive for serratia (of unknown origin) and treatment with cephalosporins and aminoglycosides was started.

RESULTS

Initial laboratory investigations showed disturbed liver function (Table 1), anaemia (8 g dL^{-1} haemoglobin, usual value 12.6 g dL^{-1}) and hyponatraemia (129 mmol L^{-1}, usual value 136–145 mmol L^{-1}).

Fasting blood lipid studies showed marked increase of total cholesterol (TC), unesterified cholesterol (UC), low-density lipoprotein cholesterol (LDL-C), phos-

146

Journal of Inherited Metabolic Disease. ISSN 0141–8955. Copyright © SSIEM and MTP Press Limited, Queen Square, Lancaster, UK.

Table 1 Results of laboratory investigations in patient and family members

	Patient		Sister	Father	Mother	Normal adult value
	initially	after 2 months				
Total cholesterol (mg dL^{-1})	480	206	218	194	160	130–250
Free cholesterol (mg dL^{-1})	425	59	57	48	35	(25–40% TC)
HDL-C (mg dL^{-1})	4	19	17	30	42	40–70
LDL-C (mg dL^{-1})	450	180	195	115	80	<150
Total PL (mg dL^{-1})	760	280	209	182	166	150–250
HDL-PL (mg dL^{-1})	18	5	53	60	79	70–150
LDL-PL (mg dL^{-1})	660	115	125	80	55	—
TG (mg dL^{-1})	720	134	165	154	68	20–150
Lecithin (mg dL^{-1})	667	201	205	189	133	
ApoAI (mg dL^{-1})	12	83	71	—	—	110–140
ApoB (mg dL^{-1})	158	134	125	—	—	70–100
LCAT activity (μmol L^{-1} h^{-1})	<5	104	99	109	66	33–155
ALP (UI)	—	398	405	98	82	<270
AST-ALT (UI)	1055–738	98–87	57–39	15–20	10–10	<35
γGT (UI)	83	32	16	38	8	<40
LDL CT/PL	0.87	1.42	1.53	1.45	1.44	1.42
CT/HDLC	18.8	7.3	11.6	5.1	5.8	<6.5
% TG in HDL	15.7	18.4	17.4	14.1	11.7	8.0

pholipids (PL), triglycerides (TG), lecithin and apoprotein B (ApoB), whereas high-density lipoprotein cholesterol (HDL-C) and apoprotein AI (ApoAI) were decreased (Table 1). α-Lipoproteins were undetectable and β- and pre-β-lipoproteins were fused on agarose gel electrophoresis.

The analysis of plasma lipoproteins after separation by ultracentrifugation and selective precipitation of LDL and LDL-PL (Alcindor *et al.*, 1983) revealed a decreased LDL TC/PL ratio (Table 1). This result, as well as the high levels of UC and lecithin and the low levels of α-lipoproteins, led to the measurement of the plasma LCAT activity (Glomset *et al.*, 1983). This was nearly absent. Finally, CESD was demonstrated by complete absence of cholesteryl esterase and triolein lipase in skin fibroblasts and peripheral leucocytes.

After 2 months a clinical improvement was noted with normalization of LCAT activity, LDL TC/PL ratio and TC/UC ratio. Liver function was still slightly disturbed. TC, LDL−C, PL, LDL-PL and ApoB were still too high or at the upper limit for age. HDL−C, α-lipoproteins and apoAI remained low (Table 1).

The patient's sister had the same lipid profile (Table 1) and an acid lipase deficiency was also demonstrated in skin fibroblasts and peripheral leucocytes.

The only serum lipid abnormality found for the parents was an increased proportion of TG in HDL. A slightly decreased HDL-C was noted for the father (Table 1).

DISCUSSION

The most important biochemical features of CESD are marked elevation of plasma cholesterol (reported in all patients with CESD) and triglycerides, reduction of plasma HDL and increased LDL (Assmann *et al.*, 1983). The precise origins of these features are not yet understood (Kostner *et al.*, 1985).

Hepatomegaly and fatty infiltration of liver cause disturbed liver function and eventually hepatitis (Assmann *et al.*, 1983) in patients with CESD (Hoeg *et al.*, 1984). The defective cholesterol esterification in our patient could be attributed to impaired LCAT synthesis due to hepatocellular disease. A CESD with secondary LCAT deficiency was therefore likely.

REFERENCES

Alcindor, L. G., Aalam, H. and Piot, M. C. Intérêt pratique de la précipitation sélective des LDL, LDL-phospholipides et rapport molaire cholestérol sur phospholipides des LDL (LDL-RMCP) dans les déficiencies en lecithin–cholestérol acyl-transférase. *Ann. Clin.* 41 (1983) 311–314

Assmann, G. and Fredrickson, D. S. Acid lipase deficiency: Wolman's disease and cholesteryl ester storage disease. In Stanbury, J. B., Wyngaarden, J. B., Fredrickson, D. S., Goldstein, J. L., Brown, M. S. (eds), *The Metabolic Basis of Inherited Disease*. McGraw-Hill Book Company, New York, 1983, pp. 803-830

Glomset, J. A., Norum, K. R. and Gjone, E. Familial lecithin: cholesterol acyltransferase deficiency. In Stanbury, J. B., Wyngaarden, J. B., Fredrickson, D. S., Goldstein, J. L., Brown, H. S. (eds). *The Metabolic Basis of Inherited Disease*. McGraw-Hill Book Company, New York, 1983, pp. 643–654

Hoeg, J. M., Demosky, S. J. (Jr), Pescovitz, O. H. and Brewer, B. H. (Jr). Cholesteryl ester storage disease and Wolman disease: Phenotypic variants of lysosomal acid cholesteryl ester hydrolase deficiency. *J. Hum. Genet.* 36 (1984) 1190–1203

Kostner, G. M., Hadorn, B., Roscher, A. and Zechner. Plasma lipids and lipoproteins of a patient with cholesteryl ester storage disease. *J. Inher. Metab. Dis.* 8 (1985) 9–12

Sabesin, S. M., Hawkins, H. L., Kruken, L. and Ragland, J. B. Abnormal plasma lipoproteins and lecithin–cholesterol acyltransferase deficiency in alcoholic liver disease. *Gastroenterology* 72 (1977) 510–518

J. Inher. Metab. Dis. 11 Suppl. 2 (1988) 149–152

Short Communication

A Treatable Familial Neuromyopathy with Vitamin E Deficiency, Normal Absorption, and Evidence of Increased Consumption of Vitamin E

A. Kohlschütter, C. Hübner, W. Jansen and S. G. Lindner
Department of Pediatrics, University of Hamburg, D-2000 Hamburg 20, Federal Republic of Germany

A 19-year-old boy who was born of consanguineous parents suffered from a familial progressive neuromuscular disease. He was first seen at the age of 12 years with ataxia, sensory neuropathy, lipopigment deposition in muscle and nerve, and constantly subnormal serum vitamin E levels ($<1\,\mathrm{mg\,L^{-1}}$; normal range 3–14) (Burck *et al.*, 1978).

The patient was followed up for 7 years. His neurological findings remained typical of chronic vitamin E deficiency. All common explanations for such a deficiency were excluded. He ate normal food, had no evidence of maldigestion, malabsorption, haemolytic anaemia, or a storage disorder, and his serum lipoproteins (responsible for transport of vitamin E) were normal. Reports on similar patients have since been published (Laplante *et al.*, 1984; Harding *et al.*, 1985).

INVESTIGATIONS

Absorption of vitamin E

Absorption from the intestinal tract was studied by an oral vitamin E tolerance test according to recommendations by Dr Sokol, University of Colorado, Denver. The patient and two controls were given 100 IU of dl-α-tocopherol per kg body weight with a glass of milk, and serum vitamin E levels were followed for 3 days. High serum levels ($26\,\mathrm{mg\,L^{-1}}$) were reached after 6 h, indicating a perfectly normal absorption of vitamin E from the intestinal tract. Serum levels in the patient, however, fell more rapidly than in the controls and reached subnormal levels after 48 h.

Effects of treatment with vitamin E

'Conventional' oral supplements of vitamin E did not lead to normalization of the low vitamin E serum levels in the patient. Beginning at the age of 15, 'high' oral supplements ($1.8\,\mathrm{g}$ α-tocopherol acetate in gelatine capsules per day in three doses) were given over a period of 4 years. Serum vitamin E levels were controlled regularly and were constantly elevated (20–$40\,\mathrm{mg\,L^{-1}}$). The treatment was well

Journal of Inherited Metabolic Disease. ISSN 0141-8955. Copyright © SSIEM and MTP Press Limited, Queen Square, Lancaster, UK.

tolerated. Now, at the age of 19 years, the general health of the patient and his muscular strength are good. His neurological symptoms did not progress, but ataxia is still striking. The good clinical condition of the patient is in contrast to the course of two of his afflicted relatives, who were not treated with vitamin E, were wheelchair-bound, and died as adolescents.

Failure of tissues to load with vitamin E

Although it was possible to maintain high serum levels of vitamin E over several years, no 'loading' of the patient's tissues with vitamin E was observed. Withdrawal experiments (5 days without vitamin E supplements) were performed after 1 year, 3 years, and 4 years of continuous treatment with high vitamin E supplements. Even after these prolonged periods of treatment, serum vitamin E levels dropped to subnormal levels ($<1\,mg\,L^{-1}$) in a few days.

Relevance of the low serum vitamin E levels for biological membranes

The vulnerability of erythrocyte membranes to free oxygen radicals was measured by the hydrogen peroxide–haemolysis test (Horwitt *et al.*, 1956). This test was repeatedly positive in the patient during periods of low vitamin E serum levels. The abnormal test result could be normalized within one day by high oral doses of vitamin E.

Evidence for increased peroxidation of lipids

Substances which react with thiobarbituric acid to form a red compound are called TBA-reactive substances (TBARS). TBARS consist mainly of lipoperoxides. The concentration of TBARS in tissues and body fluids is thought to depend on the pressure of free radicals (Plaa and Witschi, 1976). We measured serum TBARS in the patient and in controls by a modification of the colorimetric procedure used by Hunter and Mohamed (1986). For the results see Figure 1. Irrespective of high or low serum levels of vitamin E, TBARS were always elevated.

CONCLUSIONS

(1) All neurological and biopsy findings in this disease are typical of chronic vitamin E deficiency.

(2) There is no common explanation of the vitamin E deficiency in this patient; absorption of vitamin E from the gut is normal.

(3) The deficiency is characterized by an abnormally rapid clearance of vitamin E from the blood (even after prolonged treatment with high supplements), by the increased vulnerability of red cell membranes to hydrogen peroxide, and by elevated lipoperoxides in serum.

(4) These findings could best be explained by a mechanism involving an increased consumption of vitamin E. An increase of free oxygen radical pressure would

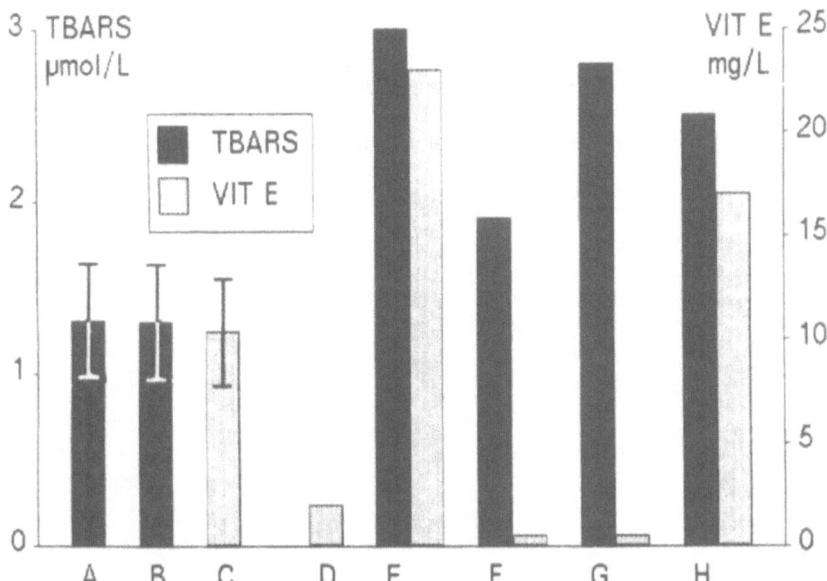

Figure 1 Serum levels of thiobarbituric acid reactive substances (TBARS) and vitamin E in controls (A–C) and patient (D–H). Controls: A, young adult males, fasted ($n = 9$; mean±SD); B, young adult males, non-fasted ($n = 9$; mean±SD); C, vitamin E in adults (Bieri *et al.*, 1964).
Patient: D, after 1 year's treatment with high vitamin E supplements and 5 days treatment withdrawal (peroxide test positive); E, 1 day after D after 1.8 g vitamin E orally (peroxide test normal); F, after 3 years high supplements and 5 days of withdrawal; G, 1 day after F, no supplement (peroxide test positive); H, 1 day after G, 1 h after 0.6 g vitamin E

lead to an increased metabolism of vitamin E (Scarpa *et al.*, 1984). The basic metabolic error should therefore be sought in tissue redox systems.

(5) The disease could be treated with constant high oral supplements of vitamin E.

REFERENCES

Bieri, J. G., Teets, L., Belavady, B. and Andrews, E. L. Serum vitamin E levels in a normal adult population in the Washington, D.C., Area. *Proc. Soc. Exp. Biol. Med. (N.Y.)* 117 (1964) 131–135

Burck, U., Goebel, H. H., Kuhlendahl, H. D., Meier, C. and Goebel, K. M. Neuromyopathy and vitamin E deficiency in man. *Neuropediatrics* 12 (1978) 267–278

Harding, A. E., Matthews, S., Jones, S., Ellis, C. J. K., Booth, I. W. and Muller, D. P. R. Spinocerebellar degeneration associated with a selective defect of vitamin E absorption. *N. Engl. J. Med.* 313 (1985) 32–35

Horwitt, M. K., Harvey, C. C., Duncan, G. D. and Wilson, W. C. Effects of limited tocopherol intake in man with relationships to erythrocyte hemolysis and lipid peroxidation. *Am. J. Clin. Nutr.* 4 (1956) 408–419

Hunter, M. I. and Mohamed, J. B. Plasma antioxidants and lipid peroxidation products in Duchenne muscular dystrophy. *Clin. Chim. Acta* 155 (1986) 123–131

Laplante, P., Vanasse, M., Michaud, J., Geoffroy, G. and Brochu, P. A progressive neurological syndrome associated with an isolated vitamin E deficiency. *Can. J. Neurol. Sci.* 11 (1984) 561–564

Plaa, G. L., and Witschi, H. Chemicals, drugs, and lipid peroxidation. *Annu. Rev. Pharmacol. Toxicol.* 16 (1976) 125–141

Scarpa, M., Rigo, A., Maiorino, M., Ursini, F., and Gregolin, C. Formation of α-tocopherol radical and recycling of α-tocopherol by ascorbate during peroxidation of phosphatidylcholine liposomes. *Biochim. Biophys. Acta* 801 (1984) 215–219

Short Communication

Fat Malabsorption, Vitamin E Deficiency, Scoliosis and Cataracts

R. D. Griffiths[1], C. J. Taylor[2], D. M. Isherwood[3] and M. J. Jackson[1]
[1]*Department of Medicine, University of Liverpool, L69 3BX, UK,* [2]*Department of Paediatrics, University of Sheffield, S10 2TH, UK,* [3]*Department of Chemical Pathology, Royal Liverpool Childrens Hospital, Alder Hey, L12 2AP, UK*

The neurological sequelae of vitamin E deficiency occurring as a result of steatorrhoea, specific defects in fat absorption (abetalipoproteinaemia), and in a selective defect of vitamin E absorption (Harding *et al.*, 1985) are now well described (Muller *et al.*, 1983). The importance of assessing vitamin E status is illustrated in this report on the presentation, treatment and progress of an atypical case of hypolipoproteinaemia associated with cataracts and scoliosis.

CASE REPORT

A male infant presented from 6 weeks of age a history of poor weight gain and loose offensive stools. He was found to be hypotonic with muscle wasting and fine sutural cataracts. When investigated at 7 months a diagnosis of coeliac disease was made, although abnormalities in the jejunal biopsy were atypical. A sweat test excluded cystic fibrosis; while biliary, pancreatic and intestinal secretions were normal as were all other investigations for malabsorption. Lipid analysis demonstrated a slightly reduced serum cholesterol at $2.3 \, mmol \, L^{-1}$ (normal 2.5–4.9) and normal serum triglycerides.

The response to a gluten-free diet, initially poor, temporarily improved when an MCT-based formula milk was added. Subsequently growth was maintained at or below the third centile, and gastrointestinal symptoms and loose stools persisted.

With hypotonia and hyporeflexia he walked at 3 years and developed a long thoraco-lumbar scoliosis by 4 years of age. He had a clumsy lordotic ataxic gait with some distal muscle weakness. A mild sensory neuropathy, not evident clinically, was demonstrated by nerve conduction studies. Plasma creatine kinase activity was elevated at $320 \, IU \, L^{-1}$. From 6 to 11 years of age there was only slow deterioration in performance. Of normal intelligence, he managed well at school until visual deterioration required removal of bilateral cataracts. There was no evidence of retinal pigmentation.

Although his vision had improved after cataract surgery, when re-investigated at $11\frac{1}{2}$ years he had worsening ataxia, a tendency to fall frequently and poor handwriting. He could not hop nor run, held onto furniture when walking, and climbed stairs only with great difficulty. Muscle power was reduced, reflexes were undetectable, with vibratory sense loss and marked truncal ataxia. Plasma creatine kinase

Journal of Inherited Metabolic Disease. ISSN 0141–8955. Copyright © SSIEM and MTP Press Limited, Queen Square, Lancaster, UK.

was $580 \, IU \, L^{-1}$. Nerve condition studies showed a predominantly sensory neuropathy most apparent in the sural and median nerves. Median nerve motor conduction was also abnormal. Ulnar motor and sensory conduction was normal. Predominantly neuropathic features were evident on muscle biopsy.

Oral challenge with a variety of fats showed complete fat malabsorption with a low serum total cholesterol ($1.73 \, mmol \, L^{-1}$) and low high-density lipoprotein cholesterol ($0.5 \, mmol \, L^{-1}$) with a normal triglyceride ($0.43 \, mmol \, L^{-1}$). Apolipoprotein B ($0.26 \, g \, L^{-1}$ (normal 0.63–1.2)) and apolipoprotein A $0.7 \, g \, L^{-1}$ (normal>1.2)) were reduced. Jejunal biopsy after gluten challenge excluded coeliac disease but marked vacuolation of jejunal epithelial cells was present. Plasma tocopherol levels (1.9 and $0.6 \, mg \, L^{-1}$) were low (normal $3{-}15 \, mg \, L^{-1}$). Muscle vitamin E content ($8.7 \, mg \, kg^{-1}$) was low normal. These findings in the absence of other causes of fat malabsorption support a diagnosis of hypobetalipoproteinaemia.

Rapid relief of all intestinal symptoms followed a low fat diet ($<20 \, g \, day^{-1}$). Treatment with 800 mg daily of vitamin E maintained low normal plasma vitamin E concentrations ($5.0 \pm 0.77 \, mg \, L^{-1}$ (normal 3–15)). After 2 years of treatment steady neurological improvement has become evident. Ataxia is less and he can run and hop five steps. Handwriting and other manual tasks have also improved.

DISCUSSION

Although scoliosis is associated with other disorders of spinocerebellar degeneration, its presence and also that of cataracts have not been reported in the vitamin E deficiency disorders. The disturbances of peripheral nerve function are now clearly recognised to be associated with a tissue deficiency of the vitamin (Traber *et al.*, 1987). Although plasma creatine kinase activity has been persistently elevated (an indication of muscle damage in the absence of any cardiac disease) the muscle vitamin E content was reasonably preserved. The skeletal muscle abnormalities are presumed secondary to the neuropathy. The striking neurological improvement witnessed with relatively simple treatment in this case emphasises the importance of assessing vitamin E status.

ACKNOWLEDGEMENT

We gratefully acknowledge the support of the Muscular Dystrophy Group of Great Britain and N. Ireland.

REFERENCES

Harding, A. E., Matthews, S., Jones, S., Ellis, C. J. K., Booth, I. W. and Muller, D. P. R. Spinocerebellar degeneration associated with a selective defect of vitamin E absorption. *N. Engl. J. Med.* 313 (1985) 32–35

Muller, D. P. R., Lloyd, J. K. and Wolff, O. H. Vitamin E and neurological function. *Lancet* 1 (1983) 225–228

Traber, M. G., Sokol, R. J., Ringel, S. P., Neville, H. E., Thellman, C. A., and Kayden, H. J. Lack of tocopherol in peripheral nerves of vitamin E-deficient patients with peripheral neuropathy. *N. Engl. J. Med.* 317 (1987) 262–265

J. Inher. Metab. Dis. 11 Suppl. 2 (1988) 155–157

Short Communication

Familial High-Density Lipoprotein Deficiency (Tangier Disease): The Third Italian Case

G. Bracco[1], G. Dotti[1], F. Levis[1]†, E. David[2], G. Saracco[3], M. Rizzetto[3] and G. Verme[3]

[1]*Laboratorio Analisi chimico-cliniche, Ospedale Infantile Regina Margherita, Torino;* [2]*Dipartimento di Scienze Biomediche e Oncologia Umana, Università di Torino;* [3]*Divisione di Gastroenterologia, Ospedale San Giovanni Battista, Torino, Italy*

Tangier disease (McKusick 20540) is caused by a severe deficiency of normal high density lipoprotein (HDL). Its clinical and pathological features depend on storage of cholesteryl esters in liver, spleen, lymph nodes, thymus, intestinal mucosa and tonsils. Diagnosis is usually suspected because of enlarged tonsils (one third of patients), peripheral neuropathy (one third), and spleen enlargement or hypercholesterolaemia or familial screening (one third) (Herbert *et al.*, 1983). Liver enlargement is present in only one third of patients and is usually not severe. Diarrhoea and abdominal pain are common. Peripheral neuropathy has been found in a recent review (Pietrini *et al.*, 1985) in 19 out of 33 patients.

Serum cholesterol is low, apolipoproteins AI and AII are abnormally low or absent, apo-C is about half the normal level. No other lipoprotein abnormality has been described (Herbert *et al.*, 1983; Dumon *et al.*, 1986; Visvikis *et al.*, 1987; Heinen *et al.*, 1978).

CASE REPORT

G. M. was born on 17 July, 1961. Both his parents were born in a town of 15 000 inhabitants in the Region of Puglia. They are not related, and they are both in good health.

The patient was healthy until 1982. Since that time he has been referred many times to hospital with many different diagnoses. In 1982 gum hypertrophy and tonsillar monolateral abscess were diagnosed. In 1983 jaundice, asthenia and hyperchromic urine brought him to hospital. He was found to be HBsAg-positive and also positive for anti-HBc IgM. Spleen enlargement and thrombocytopenia were also present. After 2 months the patient was HBsAg-negative and antibody-positive. Because of persistence of spleen and liver enlargement and of thrombocytopenia, a bone-marrow needle biopsy was performed. Histology showed increase of foam reticular cells, which was not interpreted as really abnormal. In 1985 the patient was admitted to another hospital because of the same problems. Laparoscopy and liver biopsy were performed. Macroscopically the liver appeared scarred

Journal of Inherited Metabolic Disease. ISSN 0141–8955. Copyright © SSIEM and MTP Press Limited, Queen Square, Lancaster, UK.

and the spleen was enlarged. Periportal fibrosis was described on histological observation.

In 1987 the patient was referred to our division of gastroenterology. Physical examination revealed liver and spleen enlargement and moderate tonsil hypertrophy. Admission laboratory tests were normal (i.e. ESR, blood glucose, BUN, creatinine, serum electrophoresis, immunoglobulins, AST, ÁLT, LD, AlPh, cholinesterase, amylase, prothrombin time, Na$^+$, K$^+$, faeces and urine). Platelets were low (71 000 per mm^3), serum cholesterol (46 mg dL^{-1}) and triglycerides (244 mg dL^{-1}) and lipoprotein electrophoretic pattern were abnormal, with absence of HDL-cholesterol and lipoproteins of α-motility.

Echosonography revealed liver enlargement without any specific pattern. Liver needle biopsy showed periportal fibrosis and the presence of many histiocytes with foamy cytoplasm.

A storage disease was then suspected. More investigations were undertaken on the patient's blood in order to assess his lipid state: apolipoproteins were measured by immunological methods. Their concentrations are reported in Table 1.

Table 1 Apolipoproteins (immunologically determined) (mg dL^{-1}) of the Tangier patient compared to reference values

	G.M.	Reference interval
Apo-AI	5.0	90–130
Apo-AII	absent	30–50
Apo-B	101.0	80–100
Apo-CII	0.6	3–8
Apo-CIII	2.0	8–15
Apo-E	absent	3–6

DISCUSSION

The patient was healthy until the age of 21 years. A documented hepatitis B virus infection in 1983 diverted attention as the pathological cause of liver enlargement. Tonsil hypertrophy, spleen enlargement, and the presence of foam cells in bone marrow and liver needle biopsy had been overlooked for years.

The finding of low total cholesterol, absence of HDL-cholesterol, and apolipoprotein determination eventually led to the diagnosis of Tangier disease.

Absent or nearly absent apo-AI and apo-AII and half normal apo-CII and apo-CIII are common findings in Tangier disease. The low level of apo-C in these patients may be related to the absence of HDL, as HDL and very low density lipoprotein (VLDL) each contain about half of the total apo-C (Heinen *et al.*, 1978). We are not aware of any previous description of apo-E absence: apo-E absence is more difficult to explain in Tangier disease, since apo-E is mainly contained in VLDL.

J. Inher. Metab. Dis. 11 Suppl. 2 (1988)

More investigations are being undertaken in order to assess the patient's neurological and vascular involvement, and lipoprotein abnormalities of the patient and his relatives.

REFERENCES

Dumon, M. F., Visvikis, S., Manabe, T. and Clerc, M. Immunochemical study of the plasma low and high density lipoproteins in Tangier disease. *FEBS Lett.* 201 (1986) 163–167

Heinen, R. J., Herbert, P. N., Fredrickson, D. S., Forte, T. and Lindgren, F. T. Properties of the plasma very low and low density lipoproteins in Tangier disease. *J. Clin. Invest.* 61 (1978) 120–132

Herbert, P. N., Assmann, G., Gotto, A. M., Jr. and Fredrickson, D. S. Familial lipoprotein deficiency: Abetalipoproteinemia, hypobetalipoproteinemia, and Tangier disease. In Stanbury, J. B., Wyngaarden, J. B., Fredrickson, D. S., Goldstein, J. L., Brown, M. S. (eds.) *The Metabolic Basis of Inherited Disease*, McGraw-Hill, New York, 1983, pp. 589–621

Visvikis, S., Dumon, M. F., Steinmetz, J., Manabe, T., Gaiteau, M. M., Clerc, M. and Siest, G. Plasma apolipoproteins in Tangier disease, as studied with two-dimensional electrophoresis. *Clin. Chem.* 33 (1987) 120–122

Pietrini, V., Rizzuto, N., Vergani, C., Zen, F. and Ferro Milone, F. Neuropathy in Tangier disease: A clinicopathologic study and a review of the literature. *Acta Neurol. Scand.* 72 (1985) 495–505

J. Inher. Metab. Dis. 11 Suppl. 2 (1988) 158–160

Short Communication

Failure of Taurine to Improve Fat Absorption in Cystic Fibrosis

G. N. THOMPSON

Department of Chemical Pathology, Adelaide Children's Hospital, North Adelaide, South Australia

Fat absorption in cystic fibrosis (CF) often remains suboptimal despite seemingly adequate pancreatic enzyme replacement and the use of agents which reduce duodenal acidity (Nassif *et al.*, 1981). Children with CF commonly have reduced taurine bile acid conjugates relative to glycine conjugates (i.e. elevated G/T ratio) (Robb *et al.*, 1985) and bile acids conjugated with taurine have theoretically superior qualities to glycine-conjugated bile acids in relation to fat absorption (Darling *et al.*, 1985). These findings have led to two trials of taurine supplementation in CF, results of which are conflicting. While Darling *et al.* (1985) showed significantly increased fat absorption with taurine supplementation, Thompson *et al.* (1987) were unable to reproduce these results in a similar group of patients. In neither study was the duodenal juice G/T ratio, which reflects most accurately the adequacy of the taurine dose, routinely measured.

To determine whether differences in taurine dose or in correction of the G/T ratio could explain the discrepancies between the results of the above trials, this study examined the effect of various taurine doses on the G/T ratio and on fat malabsorption in CF.

METHODS

Six CF subjects (five males, one female, mean age 11.3 years, range 8–14 years) were studied. All had normal liver function and five of the six had grossly impaired pancreatic function and were taking pancreatic enzyme supplements. The subjects were supplemented for 10 days with oral taurine, doses ranging from 30 to $300\,\text{mg}\,\text{kg}^{-1}(24\,\text{h})^{-1}$ (Table 1) in three divided doses with meals. Fat absorption studies and duodenal intubation for measurement of the duodenal juice G/T ratio (Robb *et al.*, 1985) were undertaken immediately prior to supplementation (baseline) and during the final 3 days of supplementation. The fat absorption studies were performed by careful collection of a 3-day duplicate diet and corresponding faecal collection as defined by carmine red markers. Samples were assayed for fat using the van de Kamer (1949) method and a coefficient of fat absorption was calculated. The study was carried out with the approval of the Adelaide Children's Hospital Research Ethics Committee and all subjects and their parents participated with informed consent.

Journal of Inherited Metabolic Disease. ISSN 0141–8955. Copyright © SSIEM and MTP Press Limited, Queen Square, Lancaster, UK.

RESULTS

G/T ratio and fat absorption measurements before and during taurine supplementation are presented in Table 1. Four of the six subjects had an elevated G/T ratio prior to supplementation. Taurine reduced the ratio in all subjects; higher taurine

Table 1 Effect of taurine on duodenal juice G/T ratio and fat absorption in individual patients

	Taurine dose $(\text{mg kg}^{-1}(24\,\text{h})^{-1})$					
	300	300	215	90	30	30
G/T ratio						
Baseline	12.4	7.4	2.4	5.4	2.2	>12*
Taurine	0.2	0.1	<0.1*	0.3	0.7	<0.1*
Fat absorption (%)						
Baseline	69	85	81	97	71	63
Taurine	71	88	73	93	61	82

* Low bile acid levels prevented accurate determination of G/T ratio. G/T ratio normal range 1.0–2.5

doses tended to result in greater reduction in the ratio. Despite this clearcut reversal of bile acid taurine deficiency only one subject showed improved fat absorption with taurine supplementation. The mean ±SD percentage fat absorption prior to supplementation (77.7±12.4%) was not significantly different to that during supplementation (78.0±11.9%).

DISCUSSION

Fat malabsorption in CF appears to be a function of lipase deficiency due to pancreatic insufficiency, and of decreased solubility of bile-acid micelles in the unstirred water layer. Taurine-conjugated bile acids have a lower pKa (1.8 vs 3.8–4.3) and therefore greater solubility at low pH (Regan et al., 1979) than their glycine-conjugated counterparts and are less susceptible to bacterial degradation (Hepner et al., 1973). Thus taurine deficiency in bile acids is a plausible explanation for the bile acid related component of fat malabsorption in CF. Against these theoretical advantages of taurine-conjugated bile acids is the finding that taurine supplementation does not change lipid solubilisation in patients with ileal resection (Fitzpatrick et al., 1986).

The current study has shown that the G/T ratio is remarkably sensitive to even relatively small doses of taurine; high taurine doses grossly overcorrect the ratio. Fat absorption was improved significantly in one subject. This patient had a grossly elevated G/T ratio prior to supplementation. It is possible that the heterogeneity of fat malabsorption in CF could implicate taurine deficiency as a causative factor in occasional cases. However, the facile correction of the G/T ratio by taurine suggests that most CF children would have responded to the supplements were taurine deficiency significant in CF. Thus, the study supports the conclusion of

Thompson *et al.* (1987) that taurine supplementation is not indicated in the nutritional management of CF.

It remains possible that overcorrection of the G/T ratio may not have favoured improved fat absorption. However, the taurine dose range included the doses used by Darling *et al.* ($30\,\mathrm{mg\,kg^{-1}}(24\,\mathrm{h})^{-1}$) and Thompson *et al.* ($30{-}45\,\mathrm{mg\,kg^{-1}}(24\,\mathrm{h})^{-1}$). Thus, differences in taurine dose or in the correction of the G/T ratio do not appear to be an adequate explanation for the discrepancies between the results of previous trials of taurine in CF.

REFERENCES

Darling, P. B., Lepage, G., Leroy, C., Masson, P. and Roy, C. C. Effect of taurine supplements on fat absorption in cystic fibrosis. *Pediatr. Res.* 19 (1985) 578–582

Fitzpatrick, W. J. F., Zentler-Munto, P. L. and Northfield, T. C. Ileal resection: effect of cimetidine and taurine on intrajejunal bile acid precipitation and lipid solubilisation. *Gut* 27 (1986) 66–72

Hepner, G. W., Sturman, J. A., Hofmann, A. F. and Thomas, P. J. Metabolism of steroid and amino acid moieties of conjugated bile acids in man. III. Cholyltaurine (taurocholic acid). *J. Clin. Invest.* 52 (1973) 433–440

Nassif, E. G., Younoszai, M. K., Weinberger, M. M., Nassif, C. M. Comparative effects of antacids, enteric coating, and bile salts on the efficacy of oral pancreatic enzyme therapy in cystic fibrosis. *J. Pediatr.* 98 (1981) 320–323

Regan, P. T., Malagelada, J. R., Dimagno, E. P. and Go, V. L. M. Reduced intraluminal bile acid concentrations and fat maldigestion in pancreatic insufficiency: correction by treatment. *Gastroenterology* 77 (1979) 285–289

Robb, T. A., Davidson, G. P. and Kirubakaran, C. Conjugated bile acids in serum and secretions in response to cholecystokinin/secretin stimulation in children with cystic fibrosis. *Gut* 26 (1985) 1246–1256

Thompson, G. N., Robb, T. A. and Davidson, G. P. Taurine supplementation, fat absorption and growth in cystic fibrosis. *J. Pediatr.* 111 (1987) 501–506

Van de Kamer, J. H., ten Bokkell Huinink, H. and Weyers, H. A. Rapid method for the determination of fat in feces. *J. Biol. Chem.* 177 (1949) 347–355

J. Inher. Metab. Dis. 11 Suppl. 2 (1988) 161–164

Short Communication

Peroxisomes and Peroxisomal Functions in Hyperpipecolic Acidaemia

R. J. A. WANDERS[1], C. W. T. van ROERMUND[1], M. J. A. van WIJLAND[1], R. B. H. SCHUTGENS[1], J. M. TAGER[2], H. van den BOSCH[3] and G. H. THOMAS[4]

[1]*Department of Pediatrics, University Hospital Amsterdam, Meibergdreef 9, 1105 AZ Amsterdam, The Netherlands;* [2]*Laboratory of Biochemistry, University of Amsterdam, Meibergdreef 15, 1105 AZ Amsterdam, The Netherlands;* [3]*Laboratory of Biochemistry, State University Utrecht, Padualaan 8, 3584 CH Utrecht, The Netherlands;* [4]*Department of Pediatrics, The Johns Hopkins School of Medicine, Baltimore, Maryland, USA*

Hyperpipecolic acidaemia (McKusick 23940) has so far been reported in only four patients (Gatfield *et al.*, 1968; Thomas *et al.*, 1975; Burton *et al.*, 1981). The hallmarks of this disorder are delayed development, hepatomegaly, hypotonia, retinopathy and progressive neurological deterioration with death occurring before 2–2½ years of age. Loading tests performed in three of the patients suggested a block in pipecolic acid degradation.

In 1975, Danks and co-workers reported that pipecolic acid levels are also elevated in body fluids from patients with the cerebro-hepato-renal syndrome of Zellweger. In recent years it has become clear that abnormalities in Zellweger's syndrome are not restricted to the accumulation of pipecolic acid, but also include elevated levels of very-long-chain fatty acids, di- and trihydroxy-coprostanoic acid and phytanic acid in body fluids from affected patients. The strong deficiency of peroxisomes in these patients has generally been held responsible for the multitude of biochemical abnormalities in these patients (see Schutgens *et al.*, 1986; Moser, 1987) including the accumulation of L-pipecolic acid which we have now found to result from a deficient activity of the newly discovered peroxisomal enzyme L-pipecolic acid oxidase as measured in livers from Zellweger patients (Wanders *et al.*, in preparation).

Based upon the apparent similarity in clinical and biochemical abnormalities in hyperpipecolic acidaemia and Zellweger's syndrome we decided to study peroxisomal functions in cultured skin fibroblasts available from one of the four hyperpipecolic acidaemia patients (Thomas *et al.*, 1975). Complementation studies were also carried out. The results are described in this paper.

MATERIALS AND METHODS

Enzyme activity measurements: De novo plasmalogen biosynthesis, the activity of dihydroxyacetone phosphate acyltransferase, the amount of particle-bound cata-

Journal of Inherited Metabolic Disease. ISSN 0141–8955. Copyright © SSIEM and MTP Press Limited, Queen Square, Lancaster, UK.

lase, very-long-chain fatty acid oxidation and C26/C22 ratios were determined as described in detail before (see Wanders *et al.*, 1987a,b).

Hyperpipecolic acidaemia patient: The hyperpipecolic acidaemia patient studied in this paper has been described by Thomas *et al.* (1975). The patient's fibroblasts were obtained from The Human Genetic Mutant Cell Repository, Institute for Medical Research, Copeland and Davis Streets, Camden, New Jersey 08103, USA (code GM 3605).

RESULTS

In order to find out whether biochemical abnormalities in hyperpipecolic acidaemia are restricted to the accumulation of pipecolic acid in body fluids, we studied peroxisomes and peroxisomal functions in fibroblasts from one of the four hyper-

Table 1   Peroxisomal functions in fibroblasts from control subjects and patients with Zellweger syndrome and hyperpipecolic acidaemia

Parameter measured	Control subjects	Zellweger patients	Hyperpipecolic acidaemia patient
De novo plasmalogen biosynthesis			
(% dpm in PE)	57.3±9.9(11)	14.8±3.0(15)	17.30
(% pPE in PE)	91.4±4.1(11)	48.9±15.8(15)	63.60
(% dpm in PC)	31.9±4.9(11)	63.1±4.4(15)	56.30
(% pPC in PC)	20.3±7.1(11)	1.4±0.9(15)	1.90
Dihydroxyacetonephosphate acyltransferase activity*	7.8±2.0(59)	0.66±0.59(9)	1.53
C26/C22 ratio	0.067±0.036(39)	0.57±0.23(17)	0.40
C26:0 β-oxidation activity**	2.23±0.41(10)	0.18±0.10(6)	0.12
Amount of particle-bound catalase (% of total)	65±8(9)	≤5(6)	≤5

PE = phosphatidyl serine; PC = phosphatidyl choline. Activities are in *nmol $(2h)^{-1}$ (mg protein)$^{-1}$, and **pmol min^{-1} (mg protein)$^{-1}$

pipecolic acidaemia cases described in literature. The results are described in Table 1. The data indicate that there is an impairment in *de novo* plasmalogen biosynthesis and very-long-chain fatty acid (C26:0) β-oxidation in the patient's fibroblasts. Furthermore, dihydroxyacetone phosphate acyltransferase and particle-bound catalase are deficient and the C26/C22 very-long-chain fatty acids ratio elevated. The same set of biochemical abnormalities has previously been found in fibroblasts from patients with Zellweger syndrome, neonatal adrenoleukodystrophy and infantile Refsum disease (see Moser, 1987).

DISCUSSION

The results described in this paper indicate that there is a multiple loss of peroxisomal functions in hyperpipecolic acidaemia as in Zellweger syndrome, neonatal adrenoleukodystrophy and infantile Refsum disease. The finding that catalase is not contained within subcellular particles (peroxisomes) but localized in the soluble cytoplasm in these fibroblasts suggests that the deficiency of these enzyme activities is the direct consequence of the strong deficiency of peroxisomes in the patient's fibroblasts as in Zellweger syndrome. The accumulation of pipecolic acid in hyperpipecolic acidaemia can thus be considered as only one of a series of biochemical abnormalities resulting from the deficiency of peroxisomes in hyperpipecolic acidaemia as well as in Zellweger patients.

In order to study the genetic relationship between hyperpipecolic acidaemia and the three other peroxisomal disorders in which peroxisomes are strongly deficient (Zellweger syndrome, neonatal adrenoleukodystrophy and infantile Refsum disease) we have carried out complementation studies after somatic cell fusion (Wanders *et al.*, 1986; Tager *et al.*, 1987; Brul *et al.*, 1988). So far we have identified four different complementation groups (apart from rhizomelic chondrodysplasia punctata): Group I, Zellweger cell line 1; Group II, Zellweger cell line 2, infantile Refsum disease and hyperpipecolic acidaemia; Group III, Zellweger cell line 3; Group IV, neonatal adrenoleukodystrophy. Based on these results it can be concluded that hyperpipecolic acidaemia, infantile Refsum disease and Zellweger's syndrome (cell line 2) are caused by allelic mutations or represent phenotypic variations of the same mutation although lack of complementation resulting from the absence of pre-exisiting peroxisomes cannot be ruled out (see Tager *et al.*, 1987).

Taken together the data described in this paper show that hyperpipecolic acidaemia, Zellweger syndrome, neonatal adrenoleukodystrophy and infantile Refsum disease are all related disorders caused by mutations leading to the inability to form functional peroxisomes.

REFERENCES

Brul, S., Westerveld, A., Strijland, A., Wanders, R. J. A., Schram, A. W., Heymans, H. S. A., Schutgens, R. B. H., van den Bosch, H. and Tager, J. M. Genetic heterogeneity in the cerebro-hepato-renal (Zellweger) syndrome and other inherited disorders with a generalized impairment of peroxisomal functions: A study using complementation analysis. *J. Clin. Invest.* (1988) (in press).

Burton, B. K., Reed, S. P. and Remy, W. T. Hyperpipecolic acidemia: clinical and biochemical observations in two male siblings. *J. Pediatr.* 99 (1981) 729–734

Danks, D. M., Tippett, P., Adams, C. and Campbell, P. Cerebro-hepato-renal syndrome of Zellweger. A report of eight cases with comments upon the incidence, the liver lesion, and a fault in pipecolic acid metabolism. *J. Pediatr.* 86 (1975) 382–387

Gatfield, P. D., Taller, E., Hinton, G. G., Wallace, A. C., Abdelnour, G. M. and Haust, M. D. Hyperpipecolic acidemia: a new metabolic disorder associated with neuropathy and hepatomegaly. *Can. Med. Assoc. J.* 99 (1986) 1215–1233

Moser, H. W. New approaches in peroxisomal disorders. *Dev. Neurosci.* 9 (1987) 1–18

Schutgens, R. B. H., Heymans, H. S. A., Wanders, R. J. A., van den Bosch, H. and Tager, J. M. Peroxisomal disorders: a newly recognized group of genetic diseases. *Eur. J. Pediatr.* 144 (1986) 430–440

Thomas, G. H., Haslam, R. H. A., Batshaw, M. L., Capute, A. J., Niedengard, L. and Ransom, J. L. Hyperpipecolic acidaemia associated with hepatomegaly, mental retardation, optic nerve dysplasia and progressive neurological disease. *Clin. Genet.* 8 (1975) 376–382

Tager, J. M., Westerveld, A., Strijland, A., Schram, A. W., Schutgens, R. B. H., van den Bosch, H. and Wanders, R. J. A. Complementation analysis of peroxisomal disorders by somatic cell fusion. In Fahimi, H. D. and Sies, H. (eds.), *Peroxisomes in Biology and Medicine*, Springer-Verlag, Berlin, Heidelberg, 1987, pp. 353–357

Wanders, R. J. A., Saelman, D., Heymans, H. S. A., Schutgens, R. B. H., Westerveld, A., Poll-Thé, B. T., Saudubray, J. M., van den Bosch, H., Strijland, A., Schram, A. W. and Tager, J. M. Genetic relationship between the Zellweger syndrome, infantile Refsum's disease and rhizomelic chondrodysplasia punctata. *N. Engl. J. Med.* 314 (1986) 787–788

Wanders, R. J. A., Schutgens, R. B. H., Schrakamp, G., Tager, J. M., van den Bosch, H., Moser, A. B. and Moser, H. W. Neonatal adrenoleukodystrophy. *J. Neurol. Sci.* 77 (1987a) 331–340

Wanders, R. J. A., van Wijland, M. J. A., van Roermund, C. W. T., Schutgens, R. B. H., van den Bosch, H., Tager, J. M., Nijenhuis, A. and Tromp, A. Prenatal diagnosis of Zellweger syndrome by measurement of very long chain fatty acid (C26:0) β-oxidation in cultured chorionic villous fibroblasts: implications for early diagnosis of other peroxisomal disorders. *Clin. Chim. Acta* 165 (1987b) 303–310

NOTE ADDED IN PROOF

Recently similar studies have been carried out by Moser and co-workers (Baltimore, USA) in fibroblasts from the hyperpipecolic acidaemia patients described by Gatfield *et al.* (1968) and Burton *et al.* (1981). Also in these fibroblasts a generalized loss of peroxisomal functions was found resulting from a strong deficiency of peroxisomes (H. W. Moser and A. E. Moser, personal communication).

J. Inher. Metab. Dis. 11 Suppl. 2 (1988) 165–168

Short Communication

Bile Acid Analyses in "Pseudo-Zellweger" Syndrome; Clues to the Defect in Peroxisomal β-Oxidation

P. T. Clayton[1], B. D. Lake[1], M. Hjelm[1], J. B. P. Stephenson[2],
G. T. N. Besley[3], R. J. A. Wanders[4], A. W. Schram[4], J. M. Tager[4],
R. B. H. Schutgens[4] and A. M. Lawson[5]
[1]*Institute of Child Health, London WC1, UK;* [2]*Royal Hospital for Sick
Children, Yorkhill, Glasgow, UK;* [3]*Royal Hospital for Sick Children,
Edinburgh, UK;* [4]*University of Amsterdam, Amsterdam, The Netherlands; and*
[5]*Clinical Research Centre, Harrow, UK*

In Zellweger syndrome, β-oxidation of very long chain fatty acids (VLCFA) and C_{27} bile acids is impaired because peroxisomes are absent (Schutgens *et al.*, 1986). In 1986, Goldfischer *et al.* described an infant with 'pseudo-Zellweger' syndrome in whom peroxisomes were present but oxidation of VLCFA and C_{27} bile acids impaired, due to a deficiency of peroxisomal 3-oxoacyl-CoA thiolase (Schram *et al.*, 1987). This paper describes three siblings with deficient peroxisomal β-oxidation but structurally normal peroxisomes. Immunoreactive acyl CoA-oxidase, bifunctional protein and thiolase were all present in the liver but analysis of the C_{27} bile acids suggested a functional deficiency of peroxisomal thiolase (or possibly of the bifunctional protein).

PATIENTS AND METHODS

Three children out of six born to first cousin Asian parents presented as neonates with severe nuchal hypotonia, seizures and a large fontanelle and metopic suture. The affected girl (N.A.) was born at 34 weeks gestation weighing 1.92 kg. She became hypoglycaemic at 6 h and continued to have episodic hypoglycaemia for 3 weeks. She developed a haemorrhagic diathesis at 24 h with hypoprothrombinaemia and a poor response to vitamin K. Her two brothers (N.B. and I.B.) were born at term weighing 2.76 and 3.12 kg. All three had persistent nuchal hypotonia, severe developmental delay and refractory seizures. The electroretinogram was almost absent and the brainstem auditory evoked response considerably reduced. Radiology showed delayed osseous maturation. Hepatomegaly became apparent at $1\frac{1}{2}$–3 months but did not persist and liver biopsies (N.A. and N.B.) were normal on light microscopy. N.A. died at 5 months, N.B. at 9 months and I.B. at $8\frac{1}{2}$ months. A limited autopsy (N.B.) showed adrenal atrophy and renal cortical cysts.

Tests of peroxisomal structure and function were performed as described previously (Lawson *et al.*, 1986; Schutgens *et al.*, 1986, Clayton *et al.*, 1987). Detection

165

Journal of Inherited Metabolic Disease. ISSN 0141-8955. Copyright © SSIEM and MTP Press Limited, Queen Square, Lancaster, UK.

of peroxisomal β-oxidation proteins by immunoblotting was performed on liver biopsy material (N.B.) and post mortem liver tissue (N.A.). Identification of C_{27} bile acids in bile and plasma by gas chromatography-mass spectrometry was facilitated by purification of the methyl ester of (24Z)-3α,7α,12α-trihydroxy-5β-cholest-24-en-26-oic acid (Δ^{24}-THCA) from the bile of *Varanus salvator* (Ali *et al.*, 1982; Clayton *et al.*, 1987). On alumina chromatography it eluted just before the two isomers of methyl varanate (methyl-3α,7α,12α,24-tetrahydroxy-5β-cholestanoate). The relative retention time on gas chromatography of the trimethylsilyl ether was identical with that of synthetic methyl Δ^{24}-THCA (Une *et al.*, 1983). The mass spectrum (Lawson and Setchell, 1987) was compatible with this structure. Hydrogenation (H_2/PtO_2) yielded the methyl ester of THCA (3α,7α,12α-trihydroxy-5β-cholestanoic acid).

RESULTS

Electron microscopy and histochemistry of the liver biopsy from N.B. showed abundant structurally normal peroxisomes containing catalase. His fibroblasts showed normal catalase latency, dihydroxyacetone phosphate acyl transferase activity and *de novo* plasmalogen synthesis. Immunoblotting experiments indicated the presence of all three peroxisomal β-oxidation proteins: acyl-CoA oxidase, the bifunctional protein and 3-oxoacyl-CoA thiolase. Plasma VLCFA in N.B. were: C_{26} 3.4 μg mL^{-1} (normal 0.27±0.11); C_{24}/C_{22} ratio 1.68 (0.83±0.13); C_{26}/C_{22} ratio 0.26 (0.020±0.010). Oxidation of [1-^{14}C]hexacosanoic acid by his fibroblasts was 0.01 pmol min^{-1}(mg protein)$^{-1}$ (2.23±0.41).

A chromatogram of duodenal bile acids is shown in Figure 1. The largest C_{27} bile acid peaks were produced by varanic acid (two isomers) and Δ^{24}-THCA. In bile from Zellweger patients each of these accounted for <5% of the total and only traces of Δ^{24}-THCA were detected. Peak 5 was tentatively identified as 3α,7α,12α-trihydroxy-24-oxo-5β-cholestanoic acid (24-oxo-THCA). The mass spectrum of the methyl trimethylsilyl derivative contained the following ions: 694(M), 604, 514, 424, 281, 253. When bile from N.B. was analysed by fast atom bombardment mass spectrometry the major ion was m/z 465, the quasimolecular ion for unconjugated varanic acid. Analysis of plasma from N.B. and I.B. also yielded chromatograms which were different from those of Zellweger patients (Clayton *et al.*, 1987). Varanic acid (one isomer) was a major component, Δ^{24}-THCA was readily detectable and there was no C_{29}-dicarboxylic acid. The major bile acids in urine (analysed following alkaline hydrolysis) were THCA, its 1β- and 6α-hydroxylated derivatives (whose taurine conjugates produced an ion of m/z 572 on fast atom bombardment mass spectroscopy), and two Δ^{24}-C^{27} bile acids with four nuclear hydroxyl groups (producing ions of m/z 463 [unconjugated] and 570 [taurine conjugate] on fast atom bombardment mass spectroscopy).

DISCUSSION

Our three patients had some features of classical Zellweger syndrome (hypotonia, seizures, impaired vision and hearing, developmental delay, large fontanelle) but

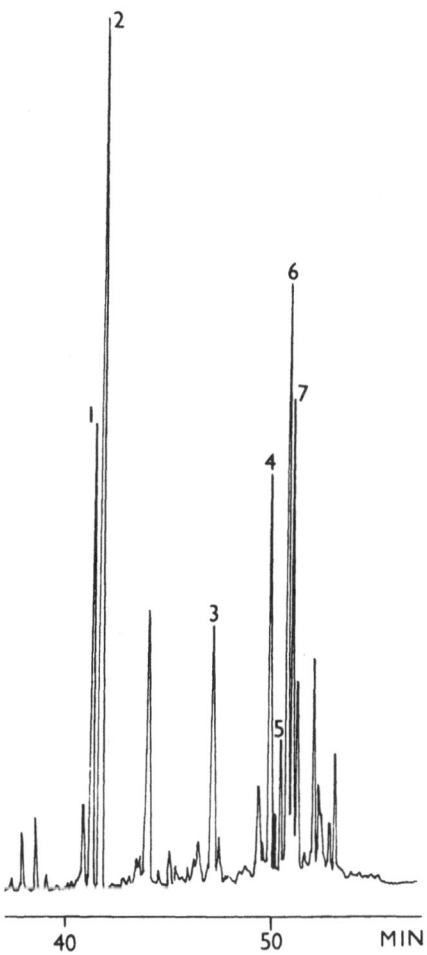

Figure 1 Capillary gas chromatogram of methyl ester trimethylsilyl ethers of bile acids from the duodenal juice of N.B. analysed following deconjugation with alkali. Identities indicated by mass spectrometry: 1, chenodeoxycholic acid; 2, cholic acid; 3 THCA; 4, Δ^{24}-THCA; 5, 24-oxo-THCA (tentative identification); 6, varanic acid (isomer 1); 7, varanic acid (isomer 2)

lacked the gross dysmorphic features and extensive hepatic fibrosis (cf. patient of Goldfischer *et al.*, 1986). Peroxisomes were present but oxidation of VLCFA and C_{27} bile acids was impaired. The three peroxisomal β-oxidation proteins were present in a form which reacted with the antibody used for immunoblotting. It was concluded that one of them may be inactive. The C_{27} bile acid pattern differed from that of patients with absent peroxisomes. THCA was present, but in smaller amounts than two intermediates in the conversion of THCA to cholic acid: Δ^{24}-THCA and varanic acid. This suggested that the oxidase which converts THCA to Δ^{24}-THCA and the hydratase which converts Δ^{24}-THCA to varanic acid were active and that the defect was in the 3-hydroxyacyl-CoA dehydrogenase component of the bifunctional protein or the thiolase. If the presence of 24-oxo-THCA can be confirmed, this would favour thiolase deficiency.

REFERENCES

Ali, S. S., Stephenson, E. and Elliott, W. H. Bile acids, LXVII. The major bile acids of *Varanus monitor. J. Lipid Res.* 23 (1982) 947–954

Clayton, P. T., Lake, B. D., Hall, N. A., Shortland, D. B., Carruthers, R. A. and Lawson, A. M. Plasma bile acids in patients with peroxisomal dysfunction syndromes: analysis by capillary gas chromatography–mass spectrometry. *Eur. J. Pediatr.* 146 (1987) 166–173

Goldfischer, S., Collins, J., Rapin, I., Neumann, P., Neglia, W., Spiro, A., Ishil, T., Roels, F., Vamecq, J. and Van Hoof, F. Pseudo-Zellweger syndrome: Deficiencies in several peroxisomal oxidative activities. *J. Pediatr.* 108 (1986) 25–32

Lawson, A. M., Madigan, M. J., Shortland, D. and Clayton, P. T. Rapid diagnosis of Zellweger syndrome and infantile Refsum's disease by fast atom bombardment-mass spectrometry of urine bile salts. *Clin. Chim. Acta* 161 (1986) 221–231

Lawson, A. M and Setchell, K. D. R. Chapter 5. Mass spectrometry of bile acids. In *The Bile Acids*. Volume 4 (1987) Plenum Press, New York, London

Schram, A. W., Goldfischer, S., van Roermund, C. W. T., Brouwer-Kelder, E. M., Collins, J., Hashimoto, T., Heymans, H. S. A., Bosch, H. v.d., Schutgens, R. B. H., Tager, J. M. and Wanders, R. J. A. Human peroxisomal 3-oxoacyl-coenzyme A thiolase deficiency. *Proc. Natl. Acad. Sci. USA* 84 (1987) 2494–2496

Schutgens, R. B. H., Heymans, H. S. A., Wanders, R. J. A., Bosch, H. v.d. and Tager, J. M. Peroxisomal disorders: A newly recognised group of genetic diseases. *Eur. J. Pediatr.* 144 (1986) 430–440

Une, M., Nagel, F., Kihira, K., Kuramoto, T. and Hoshita, H. Synthesis of four diastereoisomers at carbons 24 and 25 of $3\alpha,7\alpha,12\alpha,24$-tetrahydroxy-$5\beta$-cholestan-26-oic acid, intermediates of bile acid biosynthesis. *J. Lipid Res.* 24 (1983) 924–929

J. Inher. Metab. Dis. 11 Suppl. 2 (1988) 169–172

Short Communication

Adrenomyeloneurodystrophy with Late Cerebral Involvement and Evidence of a Multiple Autoimmune Disorder

A. Federico[1], M. T. Dotti[1], P. Annunziata[1], U. Bonuccelli[2], G. Fenzi[2], G. Ciacci[1], A. Malandrini[1], G. Meucci[2] and G. C. Guazzi[1]

[1]*Istituto di Scienze Neurologiche e Centro per lo Studio delle Encefalo-Neuro-Miopatie Genetiche, Università di Siena;* [2]*Clinica Neurologica, Università di Pisa, Italy*

The classical form of adrenoleukodystrophy (ALD; McKusick 20237) has infantile onset. Adrenomyeloneurodystrophy (AMN), having genetic, biochemical and ultrastructural features closely correlated to those of ALD, has been described in adults (Griffin *et al.*, 1977). AMN is clinically characterized by spastic paraparesis, adrenal insufficiency, peripheral neuropathy and variable degree of hypogonadism.

We report the case of a 53-year-old man with central white matter demyelination at CT scan and evidence of multisystemic immunological disorder including autoimmune thyroiditis, anti-gastric mucosa and anti-smooth muscle antibodies and CSF abnormalities. Beside the congenital alteration of the long-chain fatty acid metabolism, these findings suggest that an immunological mechanism may underlie the pathogenesis of this disorder (Moser *et al.*, 1984).

CASE REPORT

The patient was a 53-year-old man whose family history was negative. At 18 years he had epileptic seizures. At 47 years he began to complain of difficulty in walking which progressively worsened. Two years later loss of libido and sexual activity were reported. Clinical examination at that time revealed widespread melanodermia and sparse body and scalp hair. Clear pyramidal tract signs were evident in the lower limbs. Sensitivity was normal. There was slight intellectual deterioration. Routine blood chemistry parameters were normal. EEG, EMG and auditory evoked potentials were substantially normal. Nerve conduction velocities were clearly reduced (motor conduction velocity of median nerve, $45.3\,m\,s^{-1}$; sciatic nerve, $33.7\,m\,s^{-1}$; sensory conduction velocity of median nerve, $29.2\,m\,s^{-1}$; ulnar nerve, $35.4\,m\,s^{-1}$; sural nerve, $38.9\,m\,s^{-1}$; normal value $>41\,m\,s^{-1}$) with increased latency. Parieto-occipital bilateral hypodensity and marked atrophy were evident by CT scan (Figure 1).

INVESTIGATIONS

Endocrinological and immunological study: Investigation of adrenocortical function showed a reduction to the lower limits of basal serum cortisol and its circadian

Journal of Inherited Metabolic Disease. ISSN 0141–8955. Copyright © SSIEM and MTP Press Limited, Queen Square, Lancaster, UK.

rhythm (h 8 = 94 ng mL^{-1}, normal value 161–76; h 16 = 52 ng mL^{-1}, normal value 73–52; h 24 = 25 ng mL^{-1}, normal value 64–54) without significant increase after ACTH stimulation test.

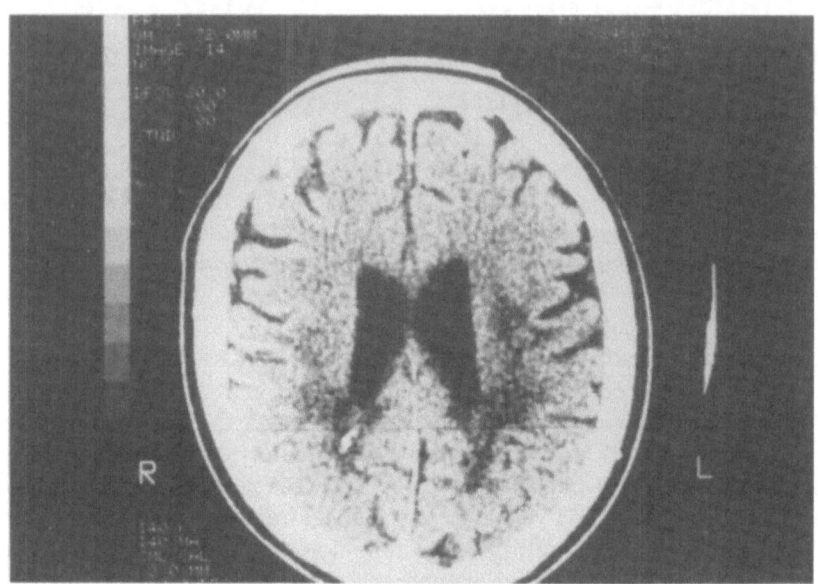

Figure 1 Brain CT scan showing bilateral demyelination and marked atrophy

Primitive hypogonadism with low testosterone values (3.5 ng mL^{-1}, normal value 3.6–9.9) and high levels of luteinizing hormone (LH) (33 U L^{-1}, normal value 3–14) were present. Thyroid function was found to be normal but high levels of anti-thyroid microsome antibodies indicated autoimmune reaction in progress (titre 1:102 400). Anti-gastric mucosa and anti-smooth muscle antibodies were also present. Examination for IgG antibodies against the adrenal gland was negative. Cerebral spinal fluid (CSF) examination showed an increase in IgG Index (0.84, normal value = 0.033) and several oligoclonal bands with an alkaline isoelectric point.

Very-long-chain fatty acids analysis: Plasma levels of very-long-chain fatty acids were analyzed by high performance liquid chromatography (HPLC) according to Alberghina *et al.* (1984) and gave significantly elevated values of the C26:0 fraction (1.67 nmol mL^{-1}, controls 0.47±0.14) and of the ratios C26:0/C22:0 (0.093, controls 0.072±0.001) and C24:0/C22:0 (1.53, controls 1.033±0.28). The subject's sister had normal values.

Neuropathological findings: A biopsy of the superficial peroneal nerve was performed. Light microscopy revealed a significant reduction in the total number of myelin fibres, especially those of large diameter (5016 fibres per μm^2, normal value 7000–10000), and an increase in collagen tissue of the endoneurium. Axons having a reduced myelin sheath and surrounded by agglomerations of Schwann cell

POLYMORPHISM OF Lp(a) LIPOPROTEIN

In 1963 Berg demonstrated that a rabbit antibody against human betalipoproteins was able to distinguish two genetic lipoprotein phenotypes that he designated Lp(a+) and Lp(a−). Later it was shown by several groups that the inherited antigenic property resides in a distinct lipoprotein subpopulation called the Lp(a) lipoprotein that has a density of about 1.05–1.1 g/ml (Wiegandt *et al.*, 1968; Ehnholm *et al.*, 1972). The concentration of this lipoprotein varies widely between individuals (from virtually non-detectable to 200 mg/dl) but is very constant in a given individual.

In caucasians the distribution of Lp(a) concentrations is highly skewed with most subjects having low Lp(a) concentrations (Albers and Hazzard, 1974). Several groups have studied the inheritance of the quantitative Lp(a) trait but the genetics remained largely unclear. A major dominant gene for high Lp(a) levels with polygenic background has been postulated (Sing *et al.*, 1974). Recently Hasstedt and Williams (1986) proposed a model where three alleles Lp^A, Lp^a, and Lp^0 determine high, intermediate, and low concentrations. Research on the Lp(a) lipoprotein was greatly stimulated by reports that suggested an implication of Lp(a) lipoprotein in the aetiology of atherosclerosis (Renninger *et al.*, 1965; Berg *et al.*, 1974; Kostner *et al.*, 1981; Rhoads *et al.*, 1986).

Biochemical characterization of the isolated Lp(a) lipoprotein revealed that it is a complex molecule composed of a particle resembling LDL in protein and lipid composition and of a high MW glycoprotein that is unique for Lp(a) lipoprotein. This glycoprotein is bonded to apoB100 in the LDL by a disulphide bridge (Gaubatz *et al.*, 1983; Fless *et al.*, 1984; Utermann *et al.*, 1987). We have recently developed a method of demonstrating the Lp(a) glycoprotein, or apo(a) as it is called by others, directly in human plasma or sera by SDS-gel electrophoresis under reducing conditions followed by immunoblotting with poly- and monoclonal anti-Lp(a)

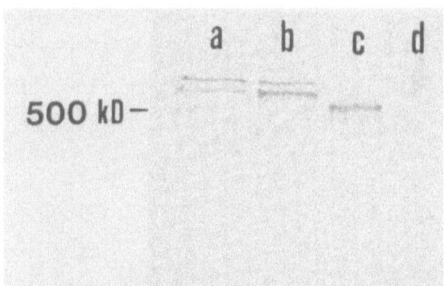

Figure 6 Western blot demonstrating the principal Lp(a) glycoprotein phenotypes. Total delipidated sera were subjected to SDS-polyacrylamide-gel electrophoresis under reducing conditions. Proteins were blotted onto nitrocellulose and Lp(a) glycoprotein was demonstrated with an affinity purified rabbit-anti-Lp(a) antibody followed by a gold-labelled sheep anti-rabbit IgG. Lanes a and b: double-band phenotypes (S1/S2); lane c: single band phenotype (B); lane d: 0-phenotype

antibodies (Utermann *et al.*, 1987, 1988a; Figure 6). Application of this technique in population and family studies has yielded the following results: at least six

REFERENCES

Alberghina, M., Fiumara, A., Pavone, L. and Giuffrida, A. M. Determination of C20–C30 fatty acids by reversed-phase chromatographic techniques: an efficient method to quantitate minor fatty acids in serum of patients with adrenoleukodystrophy. *Neurochem. Res.* 9 (1984) 1719–1727

Bernheimer, H., Budka, H. and Muller, P. Brain tissue immunoglobulins in adrenoleukodystrophy: a comparison with multiple sclerosis and systemic lupus erythematosus. *Acta Neuropathol.* 59 (1983) 95–102

Bouteille, M., Guazzi, G. C., Martin, J. J., Masselin, S., Houdart, R. and Delarue, J. Un cas d'éncéphalite périaxile diffuse de Schilder. I. Etude anatomo-clinique. *Ann. Anat. Pathol.* 13 (1968) 43–54

Griffin, J. W., Goren, E., Schaumburg, H., Engel, W. K. and Loriaux, L. Adrenomyeloneuropathy: a probable variant of adrenoleukodystrophy. I. Clinical and endocrinological aspects. *Neurology* 27 (1977) 1107–1113

Martin, J. J., Dompas, B., Ceuterick, C. and Jacobs, K. Adrenomyeloneuropathy and adrenoleukodystrophy in two brothers. *Eur. Neurol.* 19 (1980) 281–287

Moser, H. W., Moser, A. E., Singh, A. E. and O'Neill, B. P. Adrenoleukodystrophy: survey of 303 cases: biochemistry, diagnosis and therapy. *Ann. Neurol.* 16 (1984) 628–641

Schaumburg, H. H., Powers, J. M., Raine, C. S., Spencer, P. S., Griffin, J. W., Prineas, J. W. and Boheme, D. M. Adrenomyeloneuropathy: a probable variant of adrenoleukodystrophy. II. General pathologic, neuropathologic and biochemical aspects. *Neurology* 27 (1977) 1114–1119

J. Inher. Metab. Dis. 11 Suppl. 2 (1988) 173–177

Short Communication – SSIEM Award

X-Linked Adrenoleukodystrophy: Identification of the Primary Defect at the Level of a Deficient Peroxisomal Very Long Chain Fatty Acyl-CoA Synthetase Using a Newly Developed Method for the Isolation of Peroxisomes from Skin Fibroblasts

R. J. A. WANDERS[1], C. W. T. VAN ROERMUND[1], M. J. A. VAN WIJLAND[1],
R. B. H. SCHUTGENS[1], A. W. SCHRAM[2], J. M. TAGER[2],
H. VAN DEN BOSCH[3], and C. SCHALKWIJK[3]

[1]*Department of Pediatrics, University Hospital Amsterdam, Meibergdreef 9, 1105
AZ Amsterdam;* [2]*Department of Biochemistry, University of Amsterdam,
Meibergdreef 15, 1105 AZ Amsterdam;* [3]*Laboratory of Biochemistry, State
University Utrecht, Padualaan 8, 3584 CH Utrecht, The Netherlands*

Several types of adrenoleukodystrophy (ALD) differing in age of onset, mode of inheritance and clinical presentation have been described in the literature. The X-linked form of adrenoleukodystrophy, which is the most common, is characterized by demyelination, adrenal sufficiency and the accumulation of very long chain fatty acids, particularly hexacosanoic acid ($C_{26:0}$), in tissues and body fluids (Moser *et al.*, 1984). It is now generally accepted that the accumulation of very long chain fatty acids in X-linked ALD is caused by their impaired degradation via the peroxisomal β-oxidation system as first shown by Singh and co-workers in 1981 (see also Rizzo *et al.*, 1984; Singh *et al.*, 1984; Tsuji *et al.*, 1985). Recent studies have shown that the deficient peroxisomal oxidation of very long chain fatty acids in X-linked ALD is not caused by a deficiency of one of the peroxisomal β-oxidation enzymes (acyl-CoA oxidase, bifunctional protein with enoyl-CoA hydratase and 3-hydroxyacyl-CoA dehydrogenase activities and 3-oxoacyl-CoA thiolase) since all three enzyme proteins were not only found to be immunologically present as shown by immunoblotting (Wanders *et al.*, 1987) but also functionally active (Hashmi *et al.*, 1986; Wanders *et al.*, 1987). The latter finding led Hashmi and co-workers to suggest that the defect in X-linked ALD is at the level of a deficient activation of very long chain fatty acids to their CoA-esters. However, measurement of this activity in X-linked ALD fibroblasts using tetracosanoic acid ($C_{24:0}$) and hexacosanoic acid ($C_{26:0}$) as substrates revealed only a partial (about 25%) very long chain acyl-CoA synthetase deficiency (Wanders *et al.*, 1987a). The finding of a dual subcellular localization of very long chain acyl-CoA synthetase in peroxisomes and microsomes from rat liver (Wanders *et al.*, 1987c,d) together with the partial

173

Journal of Inherited Metabolic Disease. ISSN 0141–8955. Copyright © SSIEM and MTP Press Limited, Queen Square, Lancaster, UK.

deficiency of this enzyme in X-linked ALD fibroblasts, led us to suggest that in X-linked ALD there is only a deficiency of the *peroxisomal* very long chain fatty acyl-CoA synthetase, whereas the *microsomal* activity of this enzyme is present in normal amounts (Wanders *et al.*, 1987c).

Using a newly developed method to isolate peroxisomes from cultured skin fibroblasts we have now obtained definitive evidence that the defect in X-linked ALD is, indeed, at the level of a deficient peroxisomal activation of very long chain fatty acids, the microsomal activity of this enzyme being present in normal amounts. The results of this study are described here.

MATERIALS AND METHODS

Measurement of fatty acid oxidation and activation: Peroxisomal fatty acid β-oxidation was measured exactly as described before (Wanders *et al.*, 1987d) using $[1-{}^{14}C]$hexacosanoic acid as substrate. The activity of hexacosanoyl-CoA synthetase (cerotoyl-CoA synthetase) was measured as described for rat liver homogenates (Wanders *et al.*, 1987d) and fibroblast postnuclear supernates (Wanders *et al.*, 1987c).

Preparation of postnuclear supernatants: Postnuclear supernatants were prepared as previously described (Wanders *et al.*, 1987a).

Preparation of peroxisomal fractions from cultured skin fibroblasts: Cultured human skin fibroblasts were grown and harvested according to standard procedures (Wanders *et al.*, 1987c). After harvesting, the cells were washed several times, taken up in a medium containing $250 \, \text{mmol} \, L^{-1}$ sucrose, $2 \, \text{mmol} \, L^{-1}$ morpholino-propanesulphonic acid (MOPS) and $2 \, \text{mmol} \, L^{-1}$ EDTA (final pH 7.4; SME buffer) put on ice and subjected to homogenization using a motor-driven homogenizer. The homogenate was subsequently centrifuged at 500 rpm for 10 min in a Sorvall RC-58 Refrigerated Superspeed Centrifuge using a SS-34 rotor at 4°C. The resulting supernatant was then centrifuged at 13 000 rpm in the same centrifuge to remove most of the mitochondria. Subsequently aliquots of 0.9 ml were carefully pipetted into tubes containing 2.7 mL of a solution containing 22% (w/v) metrizamide, $2 \, \text{mmol} \, L^{-1}$ MOPS–NaOH plus $2 \, \text{mmol} \, L^{-1}$ EDTA (final pH 7.40) layered on top of 0.9 mL of a solution containing 50% (w/v) metrizamide, $2 \, \text{mmol} \, L^{-1}$ MOPS-NaOH and 2 mM EDTA (pH 7.40). Centrifugation was subsequently carried out at 4°C for 2.5 h at 32 000 rpm in a Beckmann L8-70 Ultracentrifuge using a SW Ti rotor. The gradient was carefully unloaded and the lower fractions containing the peroxisomes were pooled, diluted with 9 volumes of SME buffer and centrifuged to remove the metrizamide (32 000 rpm, 30 min). The final pellet was taken up in a buffer containing 0.025% (w/v) Triton X-100, $2 \, \text{mmol} \, L^{-1}$ MOPS-NaOH and $2 \, \text{mmol} \, L^{-1}$ EDTA (final pH 7.4). This peroxisomal fraction was used directly to measure the activity of very long chain fatty acyl-CoA synthetase.

RESULTS

We have previously shown that in cultured skin fibroblasts from X-linked adreno-leukodystrophy patients tetracosanoic acid ($C_{24:0}$) oxidation is impaired and lignoceroyl-CoA synthetase is partially deficient (Wanders *et al.*, 1987a). Since the accumulation of hexacosanoic acid ($C_{26:0}$) rather than tetracosanoic acid ($C_{24:0}$) is the most marked abnormality in X-linked adrenoleukodystrophy, studies were also done with hexacosanoic acid as substrate. Table 1 shows that oxidation of this very long chain fatty acid is impaired in X-linked ALD fibroblasts although the deficiency is not so marked as in Zellweger fibroblasts. Table 1 further shows that there is a partial deficiency of hexacosanoyl-CoA ($C_{26:0}$-CoA) synthetase in both X-linked ALD and Zellweger fibroblasts as found previously with tetracosanoic acid as substrate (Wanders *et al.*, 1987a). These data would agree with the previous suggestion of a deficient peroxisomal activation of very long chain fatty acids to their CoA esters in X-linked ALD.

Of course, a rigorous demonstration that it is the peroxisomal form of the very long chain fatty acyl-CoA synthetase that is deficient in X-linked ALD, requires the isolation of peroxisomes from fibroblasts. Although various techniques have been described for the isolation of peroxisomes especially from rat tissues, no methods are available for the isolation of micro-peroxisomes from, for instance, fibroblasts. Using differential and density gradient centrifugation to remove mito-chondria and microsomes, respectively (see Materials and Methods for details), we have now succeeded in obtaining a peroxisomal fraction from cultured skin fibro-blasts which is essentially free from mitochondria but still contains some contami-nating microsomes (see Discussion).

As shown in Table 1 the activity of very long chain fatty acyl-CoA synthetase was found to be markedly deficient in peroxisomal fractions isolated from X-linked ALD fibroblasts.

DISCUSSION

The results of Table 1, showing that there is a marked deficiency of very long chain fatty acyl-CoA synthetase in peroxisomal fractions isolated from X-linked ALD fibroblasts, provide the first direct evidence for the suggestion that X-linked ALD is due to a peroxisomal very long chain fatty acyl-CoA synthetase deficiency, the microsomal activity of this enzyme being present in normal amounts. Remarkably, residual very long chain fatty acyl-CoA synthetase activities were found to be similar in peroxisomal fractions isolated from X-linked ALD and Zellweger fibroblasts.

Clearly, since peroxisomes are grossly deficient in Zellweger fibroblasts, *perox-isomal* very long chain fatty acyl-CoA synthetase must be deficient in these cells. Hence, the residual very long chain fatty acyl-CoA synthetase activity found in these peroxisomal fractions must reflect microsomal very long chain fatty acyl-CoA synthetase activity. Indeed, peroxisomal fractions isolated from fibroblasts were found to be contaminated by microsomes as shown by the presence of the microso-mal enzyme NADPH cytochrome *c* reductase. Taken together the data of Table 1 would suggest that there is a virtually complete deficiency of peroxisomal very long chain fatty acyl-CoA synthetase in X-linked ALD. Future research will be devoted to isolate peroxisomes free from microsomes to show this more rigorously.

The finding that the impaired oxidation of very chain fatty acids in X-linked ALD

Table 1 Hexacosanoic acid β-oxidation and the activity of hexacosanoyl-CoA synthetase in cultured skin fibroblasts from control subjects, Zellweger patients and X-linked ALD patients and the activity of hexacosanoyl-CoA synthetase in peroxisomal fractions isolated from these fibroblasts

Parameter	Controls	X-linked ALD	Zellweger syndrome
Hexacosanoic acid β-oxidation activity	2.23±0.41(10)	0.67±0.21(8)	0.18±0.10(6)
Hexacosanoyl-CoA synthetase activity in:			
fibroblasts	29.7±2.2(13)	23.1±1.3(15)	22.3±1.6(20)
isolated peroxisomal fractions	63.9(3)	16.2(3)	18.8(3)

Activities are expressed as $nmol\,min^{-1}(mg\,protein)^{-1}$; means±SD; n in parentheses

is due to a deficiency of peroxisomal very long chain fatty acyl-CoA synthetase, suggests that there is a tight coupling between the peroxisomal activation of very long chain fatty acids and their β-oxidation in the interior of the peroxisome, since the microsomal activating enzyme is apparently not able to allow peroxisomal very long chain fatty acid β-oxidation to proceed normally. This would suggest that metabolism of very long chain fatty acids (β-oxidation in the peroxisome and incorporation into phospholipids and cholesterol esters in the endoplasmic reticulum) is functionally compartmentalized. This intriguing aspect is currently under investigation.

ACKNOWLEDGEMENTS

This work was supported by grants from the Princess Beatrix Fund (The Hague, The Netherlands) and the Netherlands Organisation for Pure Scientific Research (ZWO) under auspices of the Netherlands Foundation for Medical and Health Research (MEDIGON). The expert technical assistance of Ellen Meyboom, Annie Vandenput, Paul Bentlage, Anneke Strijland and Stanley Brul is gratefully acknowledged. We are grateful to Truus Klebach and Paula Zwaal for preparation of the manuscript.

REFERENCES

Hashmi, M., Stanley, W. and Singh, I. Lignoceroyl-CoASH ligase: enzyme defect in fatty acid β-oxidation in X-linked adrenoleukodystrophy. *FEBS Lett.* 86 (1986) 247–250

Moser, H. W., Moser, A. B., Singh, I. and O'Neill, B. P. Adrenoleukodystrophy: survey of 303 cases: biochemistry, diagnosis and therapy. *Ann. Neurol.* 16 (1984) 628–641

Rizzo, W. B., Avigan, J., Chemke, J. and Schulman, J. D. Adrenoleukodystrophy: very long chain fatty acid metabolism in fibroblasts. *Neurology* 34 (1984) 163–169

Singh, I., Moser, A. E., Goldfischer, S. and Moser, H. W. Lignoceric acid is oxidized in the peroxisome: implications for the Zellweger cerebro-hepato-renal syndrome and adrenoleukodystrophy. *Proc. Natl. Acad. Sci. USA* 81 (1984) 4203–4207

Singh, I., Moser, H. W., Moser, A. E. and Kishimoto, Y. Adrenoleukodystrophy: impaired oxidation of long chain fatty acids in cultured skin fibroblasts and adrenal cortex. *Biochem. Biophys. Res. Commun.* 102 (1981) 1223–1229

Tsuji, S., Sano-Kawamuru, T., Ariga, T. and Miyatake, T. Metabolism of $[17,18\text{-}^3H_2]$ hexacosanoic acid and $[15,16\text{-}^3H_2]$ lignoceric acid in cultured skin fibroblasts from patients with adrenoleukodystrophy (ALD) and adrenomyeloneuropathy (AMN). *J. Neurol. Sci.* 71 (1985) 359–367

Wanders, R. J. A., van Roermund, C. W. T., van Wijland, M. J. A., Heikoop, J., van den Put, A., Bentlage, P., Meyboom, E., Tager, J. M., Schram, A. W., van den Bosch, H. and Schutgens, R. B. H. Peroxisomal fatty acid β-oxidation in human skin fibroblasts: X-linked adrenoleukodystrophy, a peroxisomal very long chain fatty acyl-CoA synthetase deficiency? *J. Inher. Metab. Dis.* 10 Suppl. 2 (1987a), 220–224

Wanders, R. J. A., van Roermund, C. W. T., van Wijland, M. J. A., Nijenhuis, A. A., Tromp, A., Schutgens, R. B. H., Brouwer-Kelder, E. M., Schram, A. W., Tager, J. M., van den Bosch, H. and Schalkwijk, C. X-linked adrenoleukodystrophy: defective peroxisomal oxidation of very long chain fatty acids but not of very long chain fatty acyl-CoA esters. *Clin. Chim. Acta* 165 (1987b) 321–329

Wanders, R. J. A., van Roermund, C. W. T., van Wijland, M. J. A., Schutgens, R. B. H., Heikoop, J., van den Bosch, H., Schram, A. W. and Tager, J. M. Peroxisomal fatty acid β-oxidation in relation to the accumulation of very long chain fatty acids in cultured skin fibroblasts from patients with Zellweger syndrome and other peroxisomal disorders. *J. Clin. Invest.* (1987c) (in press)

Wanders, R. J. A., van Roermund, C. W. T., van Wijland, M. J. A., Schutgens, R. B. H., Schram, A. W., van den Bosch, H. and Tager, J. M. Studies on the peroxisomal oxidation of palmitate and lignocerate in rat liver. *Biochim. Biophys. Acta* 919 (1987d) 21–25

J. Inher. Metab. Dis. 11 Suppl. 2 (1988) 178–182

Short Communication

Infanto-Juvenile Encephaloneuropathy and Pigmentary Retinopathy in a Girl Associated with Congenital Adrenal Insufficiency and Altered Plasma Medium-Chain Fatty Acid Levels

A. Federico[1], G. Baracchini[4], M. T. Dotti[1], L. Ibba[2], A. Malandrini[1], G. Ciacci[1], M. Meloni[1], S. Palmeri[1], A. Pompella[3] and G. C. Guazzi[1]

Centro per lo studio delle Encefalo-Neuro-Miopatie Genetiche della Università di Siena, [1]Istituto di Scienze Neurologiche, Viale Bracci 2, 53100 Siena, [2]Istituto di Istologia ed Embriologia Generale, Ospedale di Pisa, [3]Istituto di Patologia Generale, Università di Siena; and [4]Divisione di Neuropsichiatria Infantile, Ospedale di Pisa, Italy

The combination of adrenocortical insufficiency, pigmentary retinopathy, seizures, peripheral neuropathy and hepatomegaly was first reported by Dyck *et al.* (1981) in two male subjects and was associated with decreased hepatic arachidonic acid. In 1978, Allgrove *et al.* reported four cases, one of them female, from two different families, with adrenocortical deficiency, achalasia, decreased lachrymation and autonomic dysfunction.

Here we report the genetic, clinical, ultrastructural, endocrinological and biochemical study of a complex case having certain similarities to the previously reported patients.

CASE REPORT

B.J., 13 years of age, is the first of two daughters of consanguineous parents (first cousins). The family history was negative for similar disorders. Delivery was normal as was early psychomotor development, except for the absence of tear production.

At 13 months, severe diarrhoea with foamy, liquid, malodourous stools first appeared. On admission to a paediatric hospital, EEG showed diffuse irritative electric activity. Routine blood chemistry parameters were normal and no parasites were found in the faeces. The xylose test gave an activity of 15.7% (control values 20–36%). A rectal biopsy, performed at 4 years of age, was normal as were lactose and sucrose loading tests. Intestinal problems continued until the age of 6 years, when they disappeared spontaneously.

At 6 years of age, seasonal urticaria and allergic conjunctivitis appeared and a slight IgA deficiency, later confirmed, was found.

Journal of Inherited Metabolic Disease. ISSN 0141-8955. Copyright © SSIEM and MTP Press Limited, Queen Square, Lancaster, UK.

At 7 years, sight and hearing decrease and seizures appeared. The EEG showed spindles and spike-wave complexes. Visual evoked potentials and electroretinograms were abnormal. Audiometry confirmed auditory nerve conduction deafness. Lysosomal enzymes, skin and muscle biopsy and cranial computed tomography (CT) scan were normal. Phenobarbital and carbamazepine therapy were given.

At 12 years of age, there was a bout of pyrexia leaving the subject in a confused state for 2 weeks. Following admission to a neurologic clinic, a tentative diagnosis of acute encephalitis was proposed. CSF examination was normal.

The patient was examined by us at 12 years and 6 months of age: she was easily tired, had walking difficulties, poor sight and deafness. Other findings included absence of tears, diffuse melanodermia, slight hepatomegaly, optic subatrophy, slight ataxia, generalized hypotonia, depressed upper limb and increased lower limb reflexes. Psychological examination showed a moderate oligophrenia (IQ75). Routine blood chemistry parameters, including serum lipoproteins, sulphatides, vitamin E, urinary mucopolysaccharides and oligosaccharides were normal. Severe adrenocortical deficiency was found with reduced plasma levels of cortisol ($16 \, ng \, mL^{-1}$; normal range 50–220) and elevated ACTH ($400–1200 \, pg \, mL^{-1}$; normal, $80 \, pg \, mL^{-1}$). Stomach and oesophagus X-ray examination was normal. EEG showed diffuse irritative subcontinuous abnormalities. Visual evoked potentials gave a normal trace with increased P100 wave latency to stimuli of 5/10 angular magnitude. Electroretinography confirmed the presence of the photopic and scotopic components, that were hypovolted. CT scan was normal. The electromyogram was normal. Normal motor nerve conduction velocity ($46 \, m \, s^{-1}$; normal >41.5) with an increased distal motor latency ($7.2 \, m \, s^{-1} (9 \, cm)^{-1}$; normal <5.8) was present in deep peroneal nerve. Sensory potentials of sural nerve were decreased in velocity ($36 \, m \, s^{-1}$; normal > 40.8) as well in amplitude ($6.8 \, \mu V$; normal >11.4). All these data suggested a distal sensory and motor neuropathy in the lower limbs affecting both myelin and axons.

Biopsy of the superficial peroneal nerve showed a decrease in fibre numbers ($5908 \, per \, \mu m^2$; normal 7000–10 000) involving more severely the large diameter myelinated fibres. Axonal rarefaction, Renaut bodies in the endoneurium and cytoplasmic inclusions resembling Reich bodies in some Schwann cells were also evident.

ANALYSIS OF SERUM FATTY ACIDS

Gas chromatographic analysis of fatty acids (Yao and Dyck, 1978) showed (Table 1) increased concentrations of C18 in the fractions of triglycerides, phospholipids and total lipids, lower values of C22:6 in all fractions (it was undetectable in triglyceride, cholesterol esters and total lipid fractions) and of C18:2 in the triglyceride fraction, where the C20:4/C18:2 ratio was increased. The C20:4/C18 ratio was decreased in the phospholipids, total lipids and cholesterol esters fractions and slightly increased in the triglyceride fraction. The amount of serum very-long-chain fatty acids (C26:0/C22:0 ratio, 0.068, normal 0.072±0.01; C24:0/C22:0, 1.02, normal 1.033±0.2) was normal; phytanic acid was absent in serum and urine.

Table 1 Serum lipid fatty acids

	C_{16}	$C_{16:1}$	C_{18}	$C_{18:1}$	$C_{18:2}$	$C_{20:3}$	$C_{20:4}$	$C_{22:6}$	$C_{20:4}/C_{18:2}$	$C_{20:4}/C_{18}$
Phospholipids										
Patient	28.75	1.24		12.59	22.4	2.39	13.17	2.98	0.592	0.796
Controls	32±2.2	1.4±0.2	13.7±0.9	12.9±0.8	20.8±1.1	2.6±0.2	13.1±0.6	3.5±0.3	0.629	0.956
Triglycerides										
Patient	29.59	4.12		44.92	*10.94*	n.d.	3.62	n.d.	0.330	0.572
Controls	27.4±1.6	3.9±0.06	3.9±0.8	42.4±1.2	20.0±2.1	0.5±0.05	1.8±0.2	0.2±0.1	0.09	0.461
Cholesterol esters										
Patient	14.82	3.695	*1.64*		52.91	0.6	7.17	n.d.	0.135	4.37
Controls	13.3±0.9	3.2±0.3	1.3±0.1	21.11±1.3	52.0±3.2	0.8±0.2	8.3±0.5	0.3±0.1	0.160	6.38
Total lipids										
Patient	26.19		*9.585*	21.72	28.72	*1.26*	8.46	n.d.	0.294	0.882
Controls	24.90±1.5	3.30±0.2	7.85±1.1	26.85±1.8	24.40±2.1	2.35±0.3	8.85±0.6	1.55±0.1	0.362	1.127

n.d. = undetectable

Values are expressed as % of total, the more significant changes are in italic

DISCUSSION

The present case of seemingly autosomic recessive transmission with involvement of many systems and organs (the retina, central and peripheral nervous system, liver and adrenal cortex) and medium-chain fatty acid abnormalities suggests a genetic disease of lipid metabolism. Compared to the other diseases of this group so far identified, our case shows some clinical and biochemical peculiarities. Refsum's disease (Refsum, 1946) and adrenoleukodystrophy (Moser *et al.*, 1981) were excluded because of clinical and genetic differences (adrenoleukodystrophy is an X-linked disorder, although a few cases have been reported in females) and also by the absence of their biochemical characteristics.

In 1978 Allgrove *et al.* reported four cases, one female, with adrenocortical deficiency associated with achalasia, convulsions and changes in tear secretion. Although many of the clinical characteristics of our case are analogous, there is no achalasia of the oesophagus.

Family dysautonomy or Riley–Day syndrome was excluded for ethnic reasons, for the normality of fungiform papillae of the tongue and for the absence of neurovegetative changes other than deficient tear production.

Dyck *et al.* (1981) described a multi-system neuronal degeneration with epilepsy, peripheral neuropathy, adrenocortical deficiency, retinitis pigmentosa and hepatomegaly associated with lipid metabolism defect and reduced tissue concentrations of arachidonic and other polyunsaturated fatty acids in two brothers. According to the authors, these changes are compatible with a genetically determined (X-linked or autosomal recessive) metabolic abnormality of the polyenoic 6-ω fatty acids. Later Yao *et al.* (1982) identified prostanate and phytanate in the plasma of the two patients and put forward the hypothesis of a phytanic acid metabolic defect, different from that of Refsum's disease, involving δ-5-desaturase. An attempt to correct this anomaly by diet confirmed the hypothesis (Yao *et al.*, 1983).

Similarly, a defect in the oxidation of short-chain fatty acid was reported by Bennett *et al.* (1985) in a 15-month-old girl with recurrent episodes of gastroenteritis, delayed growth, repeated loss of consciousness episodes, and death occurring at the age of 20 months. No adrenal insufficiency was described.

Our case is similar to those reported by Dyck *et al.* (1981) except for the deficient tear production. The multi-system involvement including the severe adrenal insufficiency and the changes of serum fatty acid concentrations suggest a disorder in lipid metabolism. However, the reported changes of linoleate and fatty acid patterns of serum lipids in other neurological disorders (Yao *et al.*, 1978) suggest prudence in the interpretation of these results. *In vitro* studies on cultured fibroblasts of this patient will better clarify the biochemical pathogenesis of the syndrome.

ACKNOWLEDGEMENT

This research has been in part supported by a grant from CNR, Rome (Progetto Finalizzato Medicina Preventiva e Riabilitativa).

REFERENCES

Allgrove, J., Clayden, G. S., Grant, D. B. and Macaulay, J. C. Familial glucocorticoid deficiency with achalasia of the cardia and deficient tear production. *Lancet* 1 (1978) 1284–1286

Bennett, M. J., Gray, R. G. F., Isherwood, D. M., Murphy, N. and Pollitt, R. J. The diagnosis and biochemical investigation of a patient with a short-chain fatty acid oxidation defect. *J. Inher Metab. Dis.* 8 Suppl. 2 (1985) 135–136

Dyck, P. J., Yao, J. K., Knickerbocker, D. E., Holman, R. T., Gomez, M. R., Hayles, A. B. and Lambert, E. H. Multisystem neuronal degeneration, hepatosplenomegaly and adrenocortical deficiency associated with reduced tissue arachidonic acid. *Neurology* 31 (1981) 925–934

Moser, H. W., Moser, A. B., Frayer, K. K., Chen, W., Schulman, J. D., O'Neill, B. P. and Kishimoto, Y. Adrenoleukodystrophy: increased plasma content of saturated very-long-chain fatty acids. *Neurology* 31 (1981) 1241–1249

Refsum, S. Heredopathia atactica polyneuritiformis: a familial syndrome not hitherto described. A contribution to the clinical study of hereditary diseases of the nervous system. *Acta Psychiatr. Scand.* Suppl. 38 (1946) 1–303

Yao, J. K. and Dyck, P. J. Lipid abnormalities in hereditary neuropathy, Part 2 (serum phospholipids). *J. Neurol. Sci.* 36 (1978) 225–236

Yao, J. K., Jardine, I. and Dyck, P. J. Presence of plasma branched-chain fatty acids in multineuronal degeneration, hepatosplenomegaly and adrenocortical insufficiency. *J. Neurol. Sci* 55 (1982) 185–195

Yao, J. K., Cannon, P. C., Holman, R. T. and Dyck, P. J. Effects of polyunsaturated fatty acid diets on plasma lipids of patients with adrenomultineuronal degeneration, hepatosplenomegaly and fatty acid derangement. *J. Neurol. Sci.* 62 (1983) 67–76

J. Inher. Metab. Dis. 11 Suppl. 2 (1988) 183–185

Short Communication

Familial Hypoketotic Hypoglycaemia Associated with Peripheral Neuropathy, Pigmentary Retinopathy and C_6–C_{14} Hydroxydicarboxylic Aciduria. A New Defect in Fatty Acid Oxidation?

B. T. Poll-The[1], J. P. Bonnefont[1], H. Ogier[1], C. Charpentier[1], A. Pelet[1], J. M. Le Fur[2], C. Jakobs[3], R. M. Kok[3], M. Duran[4], P. Divry[5], J. Scotto[6] and J. M. Saudubray[1]

[1]*Hôpital Enfants-Malades, Clinique Génétique Médicale, 149 rue de Sèvres, 75015 Paris, France;* [2]*CHU Augustin-Morvan, Brest, France;* [3]*Free University of Amsterdam, Amsterdam, The Netherlands;* [4]*University Children's Hospital "Het Wilhelmina Kinderziekenhuis", Utrecht, The Netherlands;* [5]*Hôpital Debrousse, Lyon, France;* [6]*Hôpital Bicêtre et INSERM U 56, Paris, France*

Hypoketotic hypoglycaemia is a frequent feature of defective mitochondrial β-oxidation (Gregersen, 1985), whereas pigmentary retinopathy and peripheral neuropathy are frequent symptoms in peroxisomal disorders (Schutgens *et al.*, 1986). Two siblings presented a combination of these features associated with hydroxydicarboxylic aciduria, pointing to a defect of the β-oxidation at the level of 3-hydroxyacyl-CoA dehydrogenase or 3-ketothiolase. In addition, accumulation of di- and trihydroxycoprostanoic acids in plasma and abnormal morphology of hepatic peroxisomes suggest a combination of impaired mitochondrial and peroxisomal β-oxidation.

CASE REPORTS

Two sibs, a girl (patient 1) and a boy (patient 2), born to healthy unrelated parents developed a similar clinical picture. Patient 1 has presented, at the ages of 9 and 11 months, attacks of hypoketotic hypoglycaemic coma with metabolic acidosis preceded by a febrile illness and poor feeding during several days. Patient 2, the younger brother, has had a similar episode at 3 years of age. Both patients improved clinically with infusion of glucose. Transient hepatomegaly has been noted following these episodes. They developed signs of peripheral neuropathy of the lower extremities, which have been confirmed by decreased conduction velocities and abnormal nerve biopsy. A muscle biopsy of patient 1 showed lipidosis. Subsequently, pigmentary retinopathy was diagnosed by fundoscopy and electroretinogram. Their mental development and hearing remained normal.

Plasma values of vitamins A and E and phytanic acid were normal, and plasma

183

Journal of Inherited Metabolic Disease. ISSN 0141–8955. Copyright © SSIEM and MTP Press Limited, Queen Square, Lancaster, UK.

free carnitine concentrations were mildly decreased. In both patients a 24-h fasting test showed hypoglycaemia (1.5–2 mmol L^{-1}), and hypoketonaemia (<1 mmol L^{-1}) contrasting to the increase of free fatty acids (>2.5 mmol L^{-1}). Medium chain triglyceride (MCT) loading tests (1.5 g per kg b.wt), after an overnight fast or after a 19-h fast, revealed an increase of blood ketone bodies: patient 1 (age 1 year) from 0.66 to 2.02 mmol L^{-1}, and patient 2 (age 2 years) from 0.75 to 3.0 mmol L^{-1}.

RESULTS AND DISCUSSION

The oxidation of [1-^{14}C]palmitic acid, with or without addition of carnitine to the cell culture of skin fibroblasts, was decreased in both patients (30% of controls). However, measurements in fibroblasts of all three chain length specific acyl-CoA dehydrogenases (DH), palmitoyl-CoA synthetase, palmitoyl carnitine transferase 1, glutaryl-CoA DH (ETF, ETF DH) and pyruvate DH were all normal.

Normal peroxisomal functions (patient 1) included very long chain fatty acids ($\geq C_{22}$), plasmalogens and pipecolic acid. However, di- and trihydroxycoprostanoic acids were found in plasma (0.11 and 1.14 μmol L^{-1}, respectively), pointing to a defective conversion of these intermediates to bile acids, which is known to be catalysed in the peroxisomes (Kase *et al.*, 1983).

Urine organic acid analysis by gas chromatography–mass spectroscopy revealed C_6–C_{14} hydroxydicarboxylic acids (Table 1) during an attack of hypoketotic hypo-

Table 1   Urinary organic acids

	Patient 1 (basal condition)	Patient 2 (attack)
3-Hydroxybutyric acid	254	768
3-Ketobutyric acid		453
5-Hydroxyhexanoic acid	40	138
Adipic acid	314	3732
Hexenedioic acid		31
Hydroxyadipic acid lactone		1247
Hydroxyadipic acid	311	148
Pimelic acid	61	147
Suberic acid	109	969
Octenedioic acid	44	216
3-hydroxysuberic acid	present	present
Sebacic acid	present	375
Decenedioic acid I		707
Decenedioic acid II		118
Hydroxydecenedioic acid	60	154
Hydroxysebacic acid	144	1373
Hydroxydodecenedioic acid I		249
Hydroxydodecenedioic acid II		169
Hydroxydodecanedioic acid	present	958
Hydroxytetradecenedioic acid		200
Hydroxytetradecanedioic acid		220

Values are μmol (g creatinine)$^{-1}$

glycaemic coma (patient 2) and less important amounts when there was no clinical decompensation (patient 1). The profile of hydroxydicarboxylic acids with the corresponding unsaturated derivatives is very similar to the profile found in a patient described by Riudor *et al.* (1986). Hydroxydicarboxylic aciduria may be related to an impaired fatty acid β-oxidation, e.g. at the level of the 3-hydroxyacyl-CoA DH or 3-ketothiolase.

In our sibs, the increased concentrations of blood ketone bodies following MCT-loading and the impaired oxidation of palmitic acid in fibroblasts suggest a specific defect of long-chain fatty acid oxidation, which has been shown not to be due to a long-chain acyl-CoA DH deficit. Accumulation of medium-chain hydroxydicarboxylic acids may be a consequence of a decreased activity of a specific 3-hydroxyacyl-CoA DH rather than a deficit of thiolase which is probably an ubiquitous enzyme. In addition, the presence of pigmentary retinopathy and peripheral neuropathy, the accumulation of bile acid precursors and the morphologic abnormalities of liver peroxisomes are features which would make us think of the possibility of a defect concerning not only mitochondrial but also peroxisomal β-oxidation. In this view an impaired activity of the peroxisomal 3-hydroxyacyl-CoA epimerase may also be considered.

REFERENCES

Gregersen, N. The acyl-CoA dehydrogenation deficiences. *Scand. J. Clin. Lab. Invest.* 45 Suppl. 174 (1985)

Kase, F., Björkhem, I. and Pedersen, J. I. Formation of cholic acid from 3α,7α,12α-trihydroxy-5β-cholestanoic acid by rat liver peroxisomes. *J. Lipid Res.* 24 (1983) 1560–1567

Riudor, E., Ribes, A., Boronat, M., Sabado, C., Dominguez, C. and Ballabriga, A. A new case of C_6–C_{14} dicarboxylic aciduria with favourable evolution. *J. Inher. Metab. Dis.* 9 Suppl. 2 (1986) 297–299

Schutgens, R. B. H., Heymans, H. S. A., Wanders, R. J. A., v.d. Bosch, H. and Tager, J. M. Peroxisomal disorders: a newly recognized group of genetic diseases. *Eur. J. Pediatr.* 144 (1986) 430–440

J. Inher. Metab. Dis. 11 Suppl. 2 (1988) 186–188

Short Communication

A New Type of Mitochondrial Encephalomyopathy with Stroke-like Episodes due to Cytochrome Oxidase Deficiency

P. Maertens[1], R. Richardson[1], F. Bastian[2], J. P. Williams[3] and F. Hommes[4]

Departments of Neurology[1], Pathology[2] and Radiology[3], University of South Alabama, Mobile, Alabama, USA; and [4]Department of Cell and Molecular Biology, Medical College of Georgia, Augusta, Georgia, USA

Several clinical syndromes have been associated with cytochrome oxidase (EC 1.9.3.1) deficiency. Mitochondrial encephalomyopathy, lactic acidosis and stroke-like episodes (MELAS) might be one such syndrome (Pavlakis *et al.*, 1984). Here we present a patient with a progressive encephalomyopathy, stroke-like episodes and cytochrome oxidase deficiency in the biopsied skeletal muscle.

CASE REPORT

J.M., a black female born at term after an uncomplicated pregnancy, was the first child of a second marriage between unrelated and healthy parents. Her family history included an aunt on the father's side and a first cousin on the mother's side who died at 2 and 28 years, respectively. The aunt had seizures, microcephaly and progressive developmental delay, and the first cousin had seizures and severe mental retardation.

The patient had a normal development up to the age of 5 months when she had her first seizure. Recurrence of seizures was followed by a progressive psychomotor deterioration. Microcephaly and intracranial calcifications were first noted at 6 months of age. She was first admitted to our division at 23 months of age after a partial simple seizure followed by a transient right hemiparesis. Microcephaly, staturoponderal and psychomotor delay were severe. Fundoscopy did not reveal abnormalities. Laboratory investigation was unremarkable except for an elevated CSF lactate (2.6 mmol L^{-1}). Echocardiography studies were unremarkable. Electromyography and nerve conduction studies were normal. Computed tomography (CT) scan of the head showed diffuse subcortical calcifications, prominent in the frontal region, another calcification in the pons, effacement of the cortical sulci on the left, and normal sized ventricles.

Three months later she was readmitted after a partial simple seizure followed by a transient left hemiparesis. Venous lactate was normal (1.8 mmol L^{-1}; normal

186

Journal of Inherited Metabolic Disease. ISSN 0141–8955. Copyright © SSIEM and MTP Press Limited, Queen Square, Lancaster, UK.

value $0.5-2.2\,\text{mmol}\,L^{-1}$) while CSF lactate was elevated ($2.3\,\text{mmol}\,L^{-1}$; normal value $0.5-2.0\,\text{mmol}\,L^{-1}$). Mild aminoaciduria was documented by quantitative analysis. Repeated CT scan of the head showed changes consistent with cortical atrophy on the left and cortical swelling on the right. The quadriceps muscle biopsy showed normal histochemical staining except for a virtual absence of cytochrome oxidase. Ultrastructural examination revealed focal subsarcolemmal accumulation of large mitochondria and glycogen associated with multiple lipid droplets between the myofibrils. The vasculature showed endothelial swelling and accumulation of mitochondria.

MATERIAL AND METHODS

For biochemical analysis a sample of fresh muscle (250 mg) was immediately placed on ice and homogenized at 0°C with a Potter-Elvejhem homogenizer in $250\,\text{mmol}\,L^{-1}$ sucrose. After the centrifugation at 1000 g for 5 min, supernatant was collected and centrifuged at 10 000 g for 10 min. The mitochondria-containing pellet was stored at $-196°C$. The mitochondrial fraction was solubilized in $0.1\,\text{mol}\,L^{-1}$ sodium hydroxide prior to determination of the protein content by the method of Lowry and associates. An Aminco-DW2c UV/VIS spectrophotometer was used to measure the activity of the various components of the respiratory chain at 30°C. NADH dehydrogenase (King and Howard, 1967), NADH-CoQ reductase (EC 1.6.99.2) (Sanadi *et al.*, 1967), NADH-cytochrome c reductase (EC 1.6.99.3) (Hatefi and Rieske, 1967), succinate dehydrogenase (King, 1967), succinate-CoQ reductase (Ziegler and Rieske, 1967), succinate-cytochrome c reductase (EC 1.3.99.1) (Tisdale, 1967), and cytochrome c oxidase (Wharton and Tzagoloff, 1967) were assayed according to the published methods.

RESULTS AND DISCUSSION

In the patient's mitochondria, the rate of oxidation of reduced cytochrome c is less than 1% of control values. The values for complexes I, II and III are well within normal range (Table 1).

Table 1 Enzyme activities of respiratory chain in muscle mitochondria

	Patient	Controls	
NADH dehydrogenase	1619.	1513	(471)
NADH-CoQ reductase	91	98	(88)
NADH-cytochrome *c* reductase	51	91	(69)
Succinate dehydrogenase	62	41	(15)
Succinate-CoQ reductase	7.6	9.6	(3.1)
Succinate-cytochrome *c* reductase	28	14.0	(5.5)
Cytochrome oxidase	4	4687	(3129)

Values are $\text{nmol}\,\text{min}^{-1}$ (mg mitochondrial protein)$^{-1}$; SD in parentheses for control values, $n = 6$

The diagnosis of mitochondrial encephalomyopathy, lactic acidosis and stroke-like episodes (MELAS) is suggested in our patient by the presence of a mitochondrial myopathy and recurrent episodes of stroke associated with seizures (Pavlakis *et al.*, 1984). Peripheral chronic lactic acidaemia, commonly found in this syndrome, was absent in our patient although the CSF lactate was increased. The CT scan of the head in MELAS syndrome shows focal low density areas of the cortex, calcification of the basal ganglia and ventricular dilatation (Hasuo *et al.*, 1987). Our patient displays a unique pattern of intracranial calcifications and no ventriculomegaly. This case may represent a new form of cytochrome oxidase deficiency. It furthermore suggests that the CSF lactate level is more diagnostic than the blood lactate level.

REFERENCES

Hatefi, Y. and Rieske, J. S. The preparation and properties of DPNH-cytochrome *c* reductase (Complex I-III of the respiratory chain). *Methods Enzymol.* 10 (1967) 225-231

Hasuo, K., Tamura, S., Yasumori, K., Uchino, A., Goda, S., Ishimoto, S., Kamikaseda, K., Wakuta, Y., Kishi, M. and Masuda, K. Computed tomography and angiography in MELAS (mitochondrial myopathy, encephalopathy, lactic acidosis and stroke-like episodes); report of 3 cases, *Neuroradiology* 29 (1987) 393-397

King, T. E. Preparation of succinate dehydrogenase and reconstitution of succinate oxidase. *Methods Enzymol.* 10 (1967) 322-331

King, T. E. and Howard, R. L. Preparations and properties of soluble NADH dehydrogenases from cardiac muscle. *Methods Enzymol.* 10 (1967) 275-294

Pavlakis, S. G., Phillips, P. C., Di Mauro, S., De Vivo, D. C. and Rowland, L. P. Mitochondrial myopathy, encephalopathy, lactic acidosis and stroke-like episodes: a distinctive clinical syndrome. *Ann. Neurol.* 16 (1984) 481-488

Sanadi, D. R., Pharo, R. L. and Sordahl, L. A. NADH-CoQ reductase – Assay and purification. *Methods Enzymol.* 10 (1967) 297-302

Tisdale, H. D. Preparation and properties of succinic-cytochrome *c* reductase (Complex II-III). *Methods Enzymol.* 10 (1967) 213-225

Wharton, D. C. and Tzagoloff, A. Cytochrome oxidase from beef heart mitochondria. *Methods Enzymol.* 10 (1967) 245-250

Ziegler, D. and Rieske, J. S. Preparation and properties of succinate dehydrogenase-coenzyme Q reductase (Complex II). *Methods Enzymol.* 10 (1967) 231-234

J. Inher. Metab. Dis. 11 Suppl. 2 (1988) 189–192

Short Communication

Cytochrome c Oxidase Deficiency in Three Patients with Leigh's Disease

M. Di Rocco[1], E. Veneselli[2], M. O. Ciccone[1], A. Taccone[3],
M. Stroppiano[1] and F. Cottafava[4]

[1]*III Divisione Pediatria,* [2]*Divisione di Neuropsichiatria Infantile,* [3]*Servizio di Radiologia, and* [4]*Clinica Pediatrica I, Istituto "G. Gaslini", Via V Maggio 39, 16148 Genova, Italy*

Subacute necrotizing encephalomyopathy, first reported by Leigh more than 30 years ago, is an autosomal recessive disorder characterized by multiple symmetrical foci of incomplete necrosis (spongy degeneration) in brainstem, spinal cord, basal ganglia and cerebellum. The clinical picture is heterogeneous, depending on the variability of lesions, but usually it consists of disorders of respiratory rhythm, nuclear and supranuclear oculomotor paralysis, other signs of cranial nerve dysfunction, abnormal movements, ataxia and optic atrophy.

Different biochemical causes have been proposed: defects of pyruvate carboxylase, pyruvate dehydrogenase complex and the respiratory chain are reported in patients affected with Leigh's disease. We report here three new patients, in whose fibroblasts we found partial defects of cytochrome oxidase. We emphasize that the association of typical neuroradiological lesions with one of the proposed biochemical markers can lead to premortem diagnosis in living patients and to early genetic counselling.

CASE REPORT

Case 1: R. Cosimo is the first child of healthy, related parents (first cousins). He had normal psychomotor development during the first 2 years of age; after which time he began to show failure to thrive, vomiting, arm and head tremor and jerky eye movements. At the age of 3 years he was first admitted to our department. Clinical examination revealed hypotonia, paralysis of vertical gaze, horizontal nystagmus, cerebellar intentional tremor, inco-ordination and hyporeflexia; mental abilities were quite normal.

Lactic acidaemia ranged between 40 and 30.3 mg dL^{-1} (normal values 5–20) and pyruvic acidaemia between 0.56 and 0.38 mg dL^{-1} (normal values 0.36–0.59). CSF proteins were 43 mg dL^{-1} with normal electrophoretic pattern. The boy continued to deteriorate and at the age of 4 years he was readmitted to hospital. At this time the clinical picture consisted of cerebellar ataxia, optic atrophy and mental deterioration; nerve conduction velocity was decreased. CT scan showed typical basal ganglia lesions (Figure 1). The child is still alive at the age of 6 years and

Journal of Inherited Metabolic Disease. ISSN 0141-8955. Copyright © SSIEM and MTP Press Limited, Queen Square, Lancaster, UK.

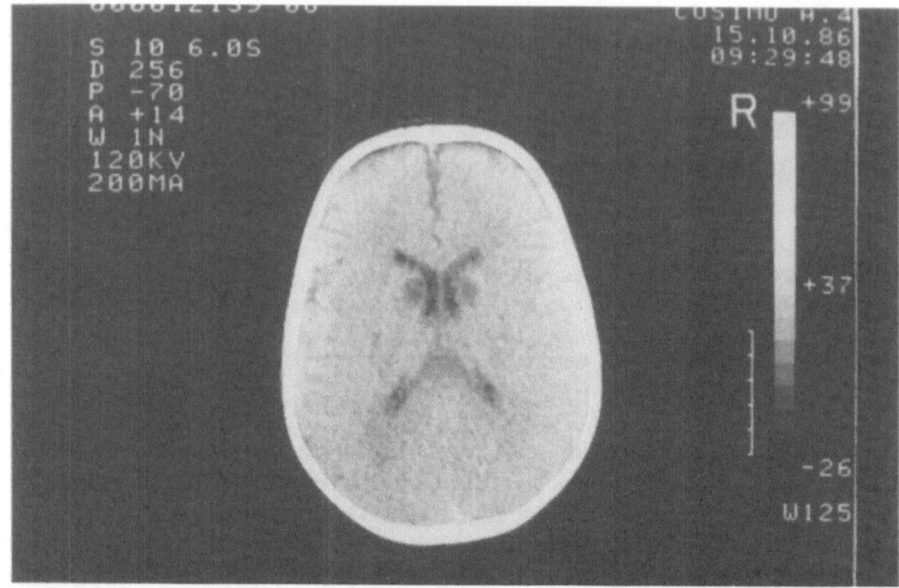

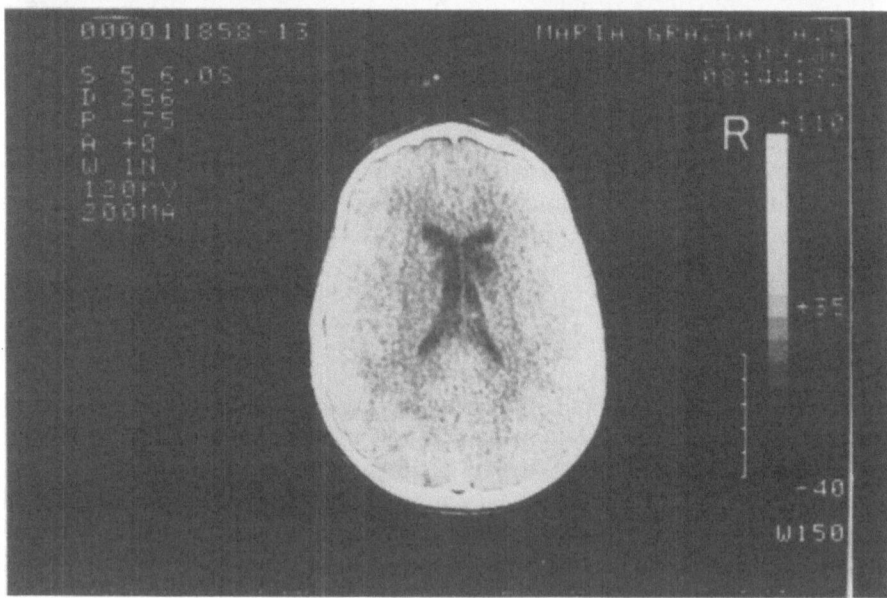

Figure 1 CT scans of patient 1 (upper) and patient 3 (lower) showing hypodense lesions of basal ganglia

now he also shows respiratory disorders (irregular breath rhythm with apnoea and hyperpnoea).

Cytochrome oxidase was assayed spectrophotometrically on skin fibroblasts by the decrease in absorbance at $550\,\mu m$ of reduced cytochrome c (Wharton and

Tzagoloff, 1967). We found reduced activity of this enzyme $(0.0028 \mu mol\,min^{-1}$ (mg protein)$^{-1}$; controls, $n = 10$, 0.0148 ± 0.0028), while rotenone-sensitive NADH-cytochrome c reductase and citrate synthase were normal.

Case 2: R. Guiseppe is Cosimo's brother. His parents did not notice anything until the age of 8 months, when Cosimo was diagnosed as having Leigh's disease. For this reason they asked for a paediatric control for Giuseppe. At the age of 8 months he could sit alone and he had good psychic development; mild hypotonia and hyporeflexia were the only neurological signs. In view of the diagnosis of his brother a CT scan was performed, showing basal ganglia hypodense lesions. The activity of cytochrome oxidase was $0.0019 \mu mol\,min^{-1}$ (mg protein)$^{-1}$. He is still alive at the age of 15 months and clear psychomotor regression is now evident.

Case 3: B. Maria Grazia is the first child of healthy unrelated parents. She has a normal brother and familial anamnesis is negative. The girl showed normal psychomotor development during the first two years of life; after this age her parents noticed loss of motor skills. She was first admitted to hospital at the age of 4 years and clinical examination demonstrated failure to thrive, oculomotor paralysis, cerebellar ataxia, optic atrophy, muscular hypotrophy, hyporeflexia and dysarthria. Despite this suggestive clinical picture a diagnosis of Leigh's disease was not suspected before the finding of typical CT lesions of basal ganglia (Figure 1). The girl is still alive at the age of 6 years.

The activity of cytochrome oxidase on fibroblasts was $0.0025 \mu mol\,min^{-1}$ (mg protein)$^{-1}$, with normal rotenone-sensitive NADH cytochrome c reductase and citrate synthase.

DISCUSSION

The three cases here reported have clinical and neuroradiological findings suggestive of Leigh's disease and a defect of cytochrome c oxidase has been demonstrated on skin fibroblasts.

The attempt to link the proposed biochemical defects to Leigh's disease is not easy. Pincus *et al.* (1974) suggested an abnormality in thiamine metabolism and proposed a urine inhibitor test as specific for Leigh's disease, but this specificity was later disclaimed. Hommes *et al.* (1968) reported pyruvate carboxylase deficiency in a patient affected with this disease. A defect of *in vitro* activation of pyruvate dehydrogenase (De Vivo *et al.*, 1979) and deficiency of pyruvate decarboxylase (Farmer *et al.*, 1973) are described. However studies of pyruvate metabolism on fibroblasts of patients who had been shown neuropathologically to be affected with Leigh's disease, failed to show defects of pyruvate carboxylase, pyruvate decarboxylase and lipoamide dehydrogenase, in subacute necrotizing encephalomyelopathy (Hansen *et al.*, 1982).

In 1977 Willems and co-workers first reported a defect of cytochrome c oxidase in muscle and heart of a girl with Leigh's disease; at this moment seven other patients are known to have cytochrome oxidase deficiency (DiMauro *et al.*, in press). This defect is generalized, being detectable also in skin fibroblasts and it is partial, residual cytochrome oxidase activity varying from tissue to tissue. Further observations need to establish that this is the primary defect (or one of the primary defects) of Leigh's disease, but in our opinion the detection of this biochemical abnormality in live patients with characteristic clinical symptoms and typical CT scan hypodense lesions of basal ganglia could permit premortem diagnosis and early genetic counselling.

REFERENCES

De Vivo, D. C., Haymond, M. W., Obert, K. A., Nelson, J. S. and Pagliara, A. S. Defective activation of pyruvate dehydrogenase complex in subacute necrotizing encephalomyelopathy (Leigh's disease). *Ann. Neurol.* 6 (1979) 483–494

Di Mauro, S., Servidei, S., Zeviani, M., Di Rocco, M., De Vivo, D. C., Di Donato, S., Uziel, G., Berry, K., Hoganson, G., Johnsen, S. D. and Johnson, P. C. Cytochrome *c* oxidase deficiency in Leigh's syndrome. *Ann. Neurol.* 22 (1987) 438–506

Farmer, T. W., Veath, L., Miller, A. L., O'Brien, J. S. and Rosenburg, R. N. Pyruvate decarboxylase deficiency in a patient with subacute necrotizing encephalomyelopathy. *Neurology* 23 (1973) 429

Hansen, T. L., Christensen, E. and Brandt, N. J. Studies on pyruvate carboxylase, pyruvate decarboxylase and lipoamide dehydrogenase in subacute necrotizing encephalomyelopathy. *Acta Paediatr. Scand.* 71 (1982) 263–267

Hommes, F. A., Polman, H. A. and Reerink, J. D. Leigh's encephalomyelopathy: an inborn error of gluconeogenesis. *Arch. Dis. Child.* 43 (1968) 423–426

Pincus, J. H., Cooper, H. R., Piros, K. and Turner, V. Specificity of the urine inhibitor test for Leigh's disease. *Neurology* 24 (1974) 855–890

Wharton, D. C. and Tzagoloff, A. Cytochrome oxidase *b* from beef heart mitochondria. *Methods Enzymol.* 10 (1967) 245–250

Willems, J. L., Monnens, L. A. H., Trijbels, J. M. F., Veerkamp, J. H., Meyer, A. F., Van Dam, K. and Van Haelst, U. Leigh's encephalomyelopathy in a patient with cytochrome *c* oxidase deficiency in muscle tissue. *Pediatrics* 60 (1977) 850–857

J. Inher. Metab. Dis. 11 Suppl. 2 (1988) 193–197

Short Communication

Histochemical, Ultrastructural and Biochemical Study of Muscle Mitochondria in Leber's Hereditary Optic Atrophy

A. Federico[1], L. Manneschi[1], M. Meloni[1], C. Alessandrini[2], A. M. Bardelli[3], M. T. Dotti[1] and P. Sabatelli[1]

[1]*Istituto di Scienze Neurologiche e Centro per lo studio delle Encefalo-Neuro-Miopatie Genetiche, Università di Siena;* [2]*Istituto di Istologia ed Embriologia Generale e Centro per lo studio delle Encefalo-Neuro-Miopatie Genetiche, Università di Siena;* [3]*Istituto di Scienze Oftalmologiche, Università di Siena, Italy*

Leber's hereditary optic atrophy (McKusick 30890) is characterized by severe abiotrophy of the pregeniculate optic pathway with acute onset in young adults, often without any other neurological symptoms. Most of the affected patients are male, but the disease is usually transmitted by the female. The mechanism of this genetic non-mendelian maternal hereditary transmission is not yet known, but the most probable hypothesis is mitochondrial inheritance, as mitochondria and their genetic heritage are of maternal origin.

It has been hypothesized that there is an impairment of cyanide detoxication in the biochemical pathogenesis of the disease and rhodanese, a cyanide-related enzyme, has been studied (Nikoskelainen *et al.*, 1984; Cagianut *et al.*, 1984; Poole and Kind, 1986; Pallini *et al.*, 1987).

Nikoskelainen *et al.* (1984) reported ultrastructural abnormalities in muscle mitochondria from Leber's hereditary optic atrophy (Leber's disease) patients. In order to verify the mitochondrial hypothesis of the pathogenesis of this disorder, we performed histological, histochemical, ultrastructural and biochemical studies of muscle biopsy specimens of three patients with Leber's disease. Attention was focused on possible changes in the number and disposition of mitochondria and, biochemically, in the activity of the respiratory enzymes in the isolated muscle mitochondria.

MATERIALS AND METHODS

In three typical Leber's disease patients (according to Lunsgaard's criteria) a biopsy specimen was taken of the sural quadriceps muscle under local anesthesia. The samples were immediately divided into two fractions, one for histochemical and ultrastructural studies and the other for biochemical investigations. Routine histological and histochemical techniques were used, including Gomori's trichromic, succinic dehydrogenase, α-glycerophosphate dehydrogenase menadione-linked and ATPase stains. For ultrastructural studies samples were fixed in 2.5% glutaral-

Journal of Inherited Metabolic Disease. ISSN 0141–8955. Copyright © SSIEM and MTP Press Limited, Queen Square, Lancaster, UK.

dehyde in $0.1 \, mol \, L^{-1}$ cacodylate buffer, pH 7.2, postfixed in 1% OsO_4, dehydrated in ethanol and embedded in araldite. Ultrathin sections stained with uranyl acetate and lead citrate, were observed in a Philips 400 electron microscope. Muscle mitochondria were prepared immediately after biopsy. In isolated mitochondria, respiratory chain enzymes for complex I, II and IV were analysed; for details of the methods, see Federico *et al.* (1987).

RESULTS

Histological, histochemical and ultrastructural findings: In all three cases there were variations in fibre size and images of splitting. No atrophic fibres were evident. ATPase staining was normal. Histochemical staining for oxidative enzymes showed peripheral thickening of the staining (Figure 1a) and the presence of many moth-eaten fibres. α-Glycerophosphate dehydrogenase menadione-linked staining showed type I fibre predominance without grouping.

Ultrastructural examination showed subsarcolemmal collection of enlarged mito-chondria in all the examined cases (Figure 1b) some with paracristalline inclusions.

Biochemical studies: We studied three respiratory enzymes and citrate synthase. Citrate synthase activity was normal (Patient 1, 349; Patient 2, 425; Patient 3, $230 \, nmol \, min^{-1} (mg \, mitochondrial \, protein)^{-1}$; controls 331 ± 91.4). Normal values were also found for cytochrome *c* oxidase (Patient 1, 268; Patient 2, 291; Patient 3, $180.5 \, nmol \, min^{-1} (mg \, mitochondrial \, protein)^{-1}$; controls 238 ± 62) and NADH cytochrome *c* reductase (Patient 1, 291; Patient 2, 386; Patient 3, 287; controls 263 ± 68). Succinate cytochrome *c* reductase activity tended to be high in all cases (Patient 1, 96.7; Patient 2, 94.7; Patient 3, 111; controls 58 ± 19). When mean enzyme activity is expressed as a ratio to citrate synthase, a common mitochondrial matrix marker, a two times increased succinate cytochrome *c* reductase activity is evident whereas all the other enzymes studied gave normal results (cytochrome *c* oxidase: patients, 0.74 ± 4; controls, 0.72 ± 6; NADH cytochrome *c* reductase: patients, 0.99 ± 15; controls, 0.79 ± 25; succinate cytochrome *c* reductase: patients, 0.33 ± 6; controls, 0.17 ± 4).

DISCUSSION

Morphological abnormalities in mitochondrial encephaloneuromyopathies include subsarcolemmal collection and enlargement of mitochondria, proliferation of cry-stae and presence of paracrystalline inclusions. The first two morphological findings appear in the muscle of the three Leber's disease patients examined, as previously

---

Figure 1 Histochemical and electron microscopic findings in muscle biopsy from Leber's disease patients: (a) succinic dehydrogenase staining (×25) showing increased peripheral oxidative staining; and (b) electron microscopic aspect of increased subsarcolemmal mito-chondrial number (magnification × 16 500) with presence of some mitochondria with paracry-stalline inclusions (magnification × 45 000)

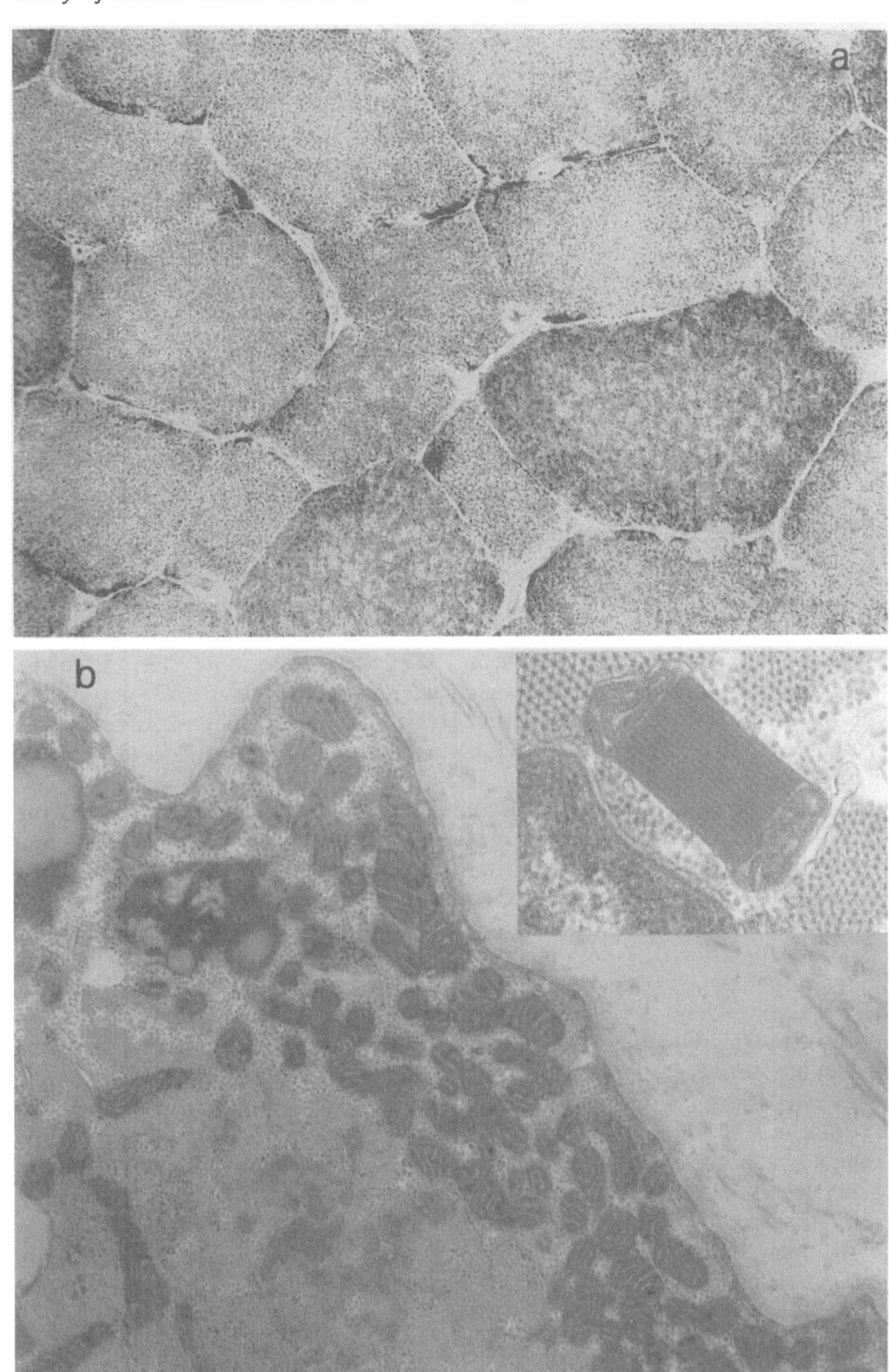

reported by Nikoskelainen *et al.* (1984) who also described the presence of paracrystalline inclusions in one out of six patients. Histochemical staining indicated abnormalities of oxidative enzymes and revealed moth-eaten fibres.

Few biochemical studies on muscle from Leber's disease patients are reported. Nikoskelainen *et al.* (1984) found normal rhodanese (thiosulphate sulphur transferase) enzyme activity in muscle. However this enzyme has been found to be decreased in liver (Cagianut *et al.*, 1984) and in rectal mucosa (Poole and Kind, 1968) and normal in peripheral leukocytes (Pallini *et al.*, 1987). The decreased liver rhodanese enzyme suggests that an impairment in the cyanide detoxication mechanism could be responsible for the pathogenesis of the disease, as hypothesized many years ago.

As interaction of rhodanese and many mitochondrial enzymes has been reported (Bonomi *et al.*, 1977), the study of respiratory chain enzymes in the muscle of Leber's disease patients appears justifiable. Obba and Kagoshima, at the Symposium of Paediatric Metabolic Ophthalmology (Parma, May 1986) reported results similar to ours in two cases with the same morphological and biochemical evidence (increased succinate cytochrome reductase activity).

Our results confirm the involvement of muscle in Leber's disease, showing morphological changes and a tendency to increased activity in one mitochondrial enzyme, notably succinate cytochrome *c* reductase. The mitochondrial pathogenesis of Leber's disease is still uncertain, supported by the typical maternal inheritance and the multisystem involvement.

Many authors have reported that in Leber's disease the first signs are in the retinal vessels and neuropathy appears later, suggesting that optic nerve fibre atrophy is secondary to vasculopathy. Heart involvement (Nikoskelainen *et al.*, 1984) has even been reported in presymptomatic cases as well as high serum pyruvate and lactate concentrations.

Vascular disorders in mitochondrial encephaloneuromyopathy characterize the so-called MELAS-syndrome (mitochondrial encephalopathy, lactic acidosis and stroke-like episodes). Some cases with this syndrome show minor histological and ultrastructural evidences of mitochondrial changes.

The multisystem involvement of Leber's disease is confirmed by several articles describing its association with other more generalized neurological disorders such as spastic paraplegia and peripheral neuropathy (Pages and Pages, 1983) or dystonia with clinical findings of infantile striatal bilateral necrosis or subacute necrotizing encephalopathy (Novotny *et al.*, 1986), a disease in which a cytochrome *c* oxidase defect has been reported in muscle, brain, heart and cultured fibroblasts.

In conclusion, our results indicate muscle involvement in Leber's disease and are one more argument in favour of multisystem participation in this disorder. We cannot however exclude the possibility that these changes are secondary to a more generalized, as yet unidentified, metabolic disorder due to the presence of a substance that becomes toxic for the cells in high amounts (in particular at the level of the optic nerve). We are currently studying fibroblast cultures in order to clarify this problem.

ACKNOWLEDGEMENT

This research was partially financed by a grant from the Regione Toscana.

REFERENCES

Bonomi, F., Pagani, S., Cerletti, P. and Cannella, C. Rhodanese-mediated sulphur transfer to succinate dehydrogenase. *Eur. J. Biochem.* 72 (1977) 17–24

Cagianut, B., Schnebli, H. P., Rhyner, K. and Furrer, J. Decreased thiosulphate sulphur transferase (rhodanese) in Leber's hereditary optic atrophy. *Klin. Wochenschr.* 62 (1984) 850–854

Federico, A., Manneschi, L. and Paolini, E. Biochemical difference between intermyofibrillar and subsarcolemmal mitochondria from human muscle. *J. Inher. Metab. Dis.* 10 Suppl. 2 (1987) 242–246

Nikoskelainen, E., Hassimem, I. E., Paljarvi, L., Lang, H. and Kalimo, K. Leber's hereditary optic neuroretinopathy; a mitochondrial disease? *Lancet* 2 (1984) 1474

Novotny, E. J., Sing, G., Wallace, D. C., Dorfman, L. J., Louis, A., Sogg, R. L. and Steiman, L. Leber's disease and dystonia: a mitochondrial disease. *Neurology* 36 (1986) 1053–1060

Pages, M. and Pages, A. M. Leber's disease with spastic paraplegia and peripheral neuropathy. Case report with nerve biopsy study. *Eur. Neurol.* 22 (1983) 181–185

Pallini, R., Martelli, P., Bardelli, A. M., Guazzi, G. C. and Federico, A. Thiosulphate: cyanide sulphurtransferase in leucocytes from Leber's patients. Enzyme characterization and activity levels. *Neurology* 37 (1987) 1878–1882

Poole, C. J. M. and Kind, P. R. M. Deficiency of thiosulphate sulphurtransferase (rhodanese) in Leber's hereditary optic neuropathy. *Br. Med. J.* 292 (1986) 1229–1230

J. Inher. Metab. Dis. 11 Suppl. 2 (1988) 198–201

Short Communication

Morphometric and Biochemical Study of Muscle Mitochondria in Adult Chronic Progressive External Ophthalmoplegia

A. Federico[1], L. Manneschi[1], P. Sabatelli[1], M. T. Dotti[1],
G. Ciacci[1], L. Ibba[2] and R. Gerli[2]
Centro per lo studio delle Encefalo-Neuro-Miopatie Genetiche dell'Università di Siena, [1]Istituto di Scienze Neurologiche, [2]Istituto di Istologia ed Embriologia Generale, Viale Bracci, 53100 Siena, Italy

Adult chronic progressive external ophthalmoplegia (ACPEO) is characterized by slowly progressive paralysis of the extraocular muscles, with or without other neurological manifestations. The presence of ragged red fibres (Olsen *et al.*, 1972) and the fact that abnormalities in the mitochondrial compartment are often the most striking ultrastructural findings in the syndrome raises the possibility of a failure of oxidative energy metabolism in the pathogenesis of the disease. This supposition is supported by the finding of excessive lactic acidosis after exercise in several cases, suggesting a defective utilization of pyruvate (Scarlato *et al.*, 1980).

Few biochemical and morphometric studies have until now been reported. Mitsumoto *et al.* (1983) found an increased number of structurally abnormal mitochondria in patients with ragged red fibres and an increased number of structurally normal mitochondria in patients without ragged red fibres. In the latter group a decrease in cytochrome *b* and *cc1* was found. More recently Byrne *et al.* (1985) reported a partial cytochrome oxidase deficiency in two patients.

Here we report a morphometric and biochemical analysis of muscle mitochondria in four cases of ACPEO.

MATERIALS AND METHODS

Diagnosis of ACPEO: The four patients showed the characteristic clinical signs of ACPEO. Other causes of ophthalmoplegia were excluded. There were no heart abnormalities. They presented ophthalmoplegia and moderate signs of systemic myopathy, confirmed by EMG. A moderate increase in serum lactate and pyruvate concentrations was found (Dotti *et al.*, 1987).

Muscle histological, histochemical and ultrastructural findings: Muscle biopsy from sural quadriceps was performed under local anaesthesia. Immediately after biopsy, samples were divided into two fractions, one for histochemical and ultrastructural studies and another for biochemical investigations.

Histological examination showed unmodified fascicular architecture and increased variability in size and shape of fibres, some of which were hypertrophic.

198

Journal of Inherited Metabolic Disease. ISSN 0141-8955. Copyright © SSIEM and MTP Press Limited, Queen Square, Lancaster, UK.

Atrophic and/or angular fibres were also evident. With the modified Gomori's trichrome stain, approximately 5% of the fibres were 'ragged red'. Histoenzymatic analysis showed a slight predominance of type I fibres. Ultrastructural observations showed an increased number of subsarcolemmal (SSM) and intermyofibrillar (IMF) mitochondria, some with the typical morphological alterations of mitochondrial myopathies, i.e. paracrystalline inclusions, increased cristae, large mitochondria, absence of cristae, ring cristae, dark inclusions, etc.

Morphometric analysis: The electron microscope images of the different cases were analysed by Images Analyzer Ibas I-Kontron. Area and maximal diameter of IMF and SSM mitochondria were calculated according to Jerusalem *et al.* (1975). Normal values have been obtained by examining muscle samples of subjects undergoing orthopaedic interventions, showing no histological, histochemical nor ultrastructural changes.

Biochemical analysis: Muscle mitochondria were prepared immediately after biopsy and analysed biochemically for respiratory chain enzymes. For methods see Federico *et al.* (1987).

RESULTS

Morphometric studies: Table 1 shows the results of the morphometric study of SSM and IMF mitochondria. Under normal conditions the area of SSM is 64%

Table 1 **Morphometric data of subsarcolemmal (SSM) and intermyofibrillar (IMF) mitochondria in normal subjects and in adult chronic progressive external ophthalmoplegia (ACPEO)**

	Normal	ACPEO	% Difference
SSM area (μm^2)	0.297 ±0.227	0.240 ±0.175*	−19.08
Mode	0.165	0.132	
SSM maximum diameter (μm)	0.899 ±0.545	0.767 ±0.423*	−14.73
Mode	0.673	0.504	
IMF area (μm^2)	0.190 ±0.208	0.123 ±0.121*	−35.05
Mode	0.0599	0.0718	
IMF maximum diameter (μm)	0.765 ±0.793	0.491 ±0.375*	−35.8
Mode	0.259	0.353	
SSM/IMF area	1.56	1.948*	+19.92
SSM/IMF diameter	1.17	1.56*	+13.3

*In comparison with normal subjects $p < 0.001$ (Student's *t* test)

larger than that of IMF mitochondria. The SSM mitochondrial diameter is 8.5% larger than that of IMF. The SSM area/IMF area ratio is 1.56. In ACPEO, electron microscope examination showed an increased number of both SSM and IMF mitochondria, and a decrease in the maximal diameter and area in SSM (−19% and −14.7%, respectively) and more markedly in IMF mitochondria (−35.1% and −38.8%) was observed. The area of ACPEO SSM is 51% larger than that of IMF and the diameter of SSM is 6.4% larger than that of IMF mitochondria. The ratio SSM/IMF mitochondria diameter is 1.17 in controls and 1.56 in ACPEO, with an increase of 13.3%.

Biochemical studies: Citrate synthase activity in total ACPEO mitochondrial fraction was slightly decreased (ACPEO, 221.5±38; controls 331±91). The activities of NADH cytochrome c reductase (ACPEO, 273±31; controls, 263±68) and succinate cytochrome c reductase (ACPEO, 70.25±25; controls, 58±19) were normal or slightly increased. Cytochrome c oxidase was depressed (ACPEO, 92.38±42; controls, 238±62). The values are expressed as $nmol\,min^{-1}(mg\,mito-$ chondrial protein$)^{-1}$. When mean values were expressed as a ratio to a common mitochondrial marker, i.e. citrate synthase, NADH cytochrome c reductase (ACPEO, 1.23; controls, 0.794) and succinate cytochrome c reductase (ACPEO, 0.317; controls, 0.175) enzyme activities were increased whereas a 58% decrease was found for cytochrome c oxidase (ACPEO, 0.418; controls, 0.719).

DISCUSSION

Federico *et al.* (1987) have reported biochemical differences in SSM and IMF mitochondria in normal human muscle suggesting that these two classes of mitochondria may play a different role in muscle function.

The different participation in pathological processes of the two classes of mitochondria is evident in our study in which IMF mitochondria appear more heavily involved in the pathology of ACPEO.

Morphological alterations in ACPEO have been described in detail and include paracrystalline inclusions, overabundance or absence of cristae and large mitochondria. These changes are also reflected in alterations in the diameter and area of mitochondria and, of course, in their biochemical functions.

Although morphometric studies of mitochondria under normal conditions (Jerusalem *et al.*, 1975), during ageing (Poggi *et al.*, 1987) and in chronic progressive external ophthalmoplegia (Mitsumoto *et al.*, 1983) have been previously reported, no data have been published on the morphometric differentiation between SSM and IMF mitochondria under normal conditions or in ACPEO. The area of SSM is normally 64% higher than that of IMF mitochondria and the diameter is only 8.5% higher. In ACPEO the area is 51% higher and the diameter is about 64% higher. In ACPEO a more severe involvement of IMF mitochondria seems evident. They are smaller with a decreased area and diameter of 35%, but in higher numbers than normal.

Our results confirm the likelihood that SSM and IMF mitochondria may play different roles in normal and pathological conditions. The decreased activity of cytochrome c oxidase agrees with the reports of Mitsumoto *et al.* (1983) and Byrne *et al.* (1985). It is impossible to establish whether the decrease in size of SSM and more markedly of IMF mitochondria could, differently, be responsible for the decreased activity of cytochrome c oxidase, until micromethods for the preparation of SSM and IMF mitochondria become available.

ACKNOWLEDGEMENT

This research has been in part supported by grants from The Regione Toscana and the University of Siena.

REFERENCES

Byrne, E., Dennet, X., Troumce, I. and Henderson, R. Partial cytochrome oxidase (*aa3*) deficiency in chronic progressive external ophthalmoplegia – Histochemical and biochemical studies. *J. Neurol. Sci.* 71 (1985) 257–271

Dotti, M. T., Federico, A., Polito, E. and Guazzi, G. C. Oftalmoplegie croniche progressive a patogenesi mitocondriale. *Atti III Congr. Naz. Oftalmol. Pediatr.* Tip. Litotip, Roma, 1987, pp. 223–241

Federico, A., Manneschi, L. and Paolini, E. Biochemical differences between intermyofibrillar and subsarcolemmal mitochondria from human muscle. *J. Inher. Metab. Dis.* 10 Suppl. 2, (1987) 242–246

Jerusalem, F., Engel, A. G. and Peterson, H. A. Human muscle fiber fine structure: morphometric data on controls. *Neurology* 25 (1975) 127–134

Mitsumoto, H., Aprille, J. R., Wray, S. H., Nemni, R. and Bradley, W. G. Chronic progressive external ophthalmologia (CPEO): clinical, morphologic and biochemical studies. *Neurology* 33 (1983) 452–461

Olsen, E., Engel, G. O., Walsh, G. O. and Einaugher, R. Oculo-craniosomatic neuromuscular disease with ragged red fibres. *Arch. Neurol.* 26 (1972) 193–211

Poggi, P., Marchetti, C. and Scelsi, R. Automatic morphometric analysis of skeletal muscle fibres with aging man. *Acta Anat.* (1987) (in press)

Scarlato, G., Pellegrini, G. and Moggio, M. Ophthalmoplegia plus: metabolic studies and therapeutical trials. In Angelini, C., Daniel, G. A., and Fontanari, D. (eds.) *Muscular Dystrophy Research: Advances and New Trends*, Excerpta Medica, Amsterdam, 1980, pp. 231–241

J. Inher. Metab. Dis. 11 Suppl. 2 (1988) 202–204

Short Communication

Cytochrome *c* Oxidase: Organ-Specific Isoenzymes and Deficiencies

K. M. C. Sinjorgo, T. B. M. Hakvoort, A. O. Muijsers,
A. W. Schram and J. M. Tager
*Laboratory of Biochemistry, University of Amsterdam, P.O. Box 20151, 1000
HD Amsterdam, The Netherlands*

Ever since the first report of Luft *et al.* (1962), considerable information has become available on inherited metabolic diseases originating from defects in mitochondria. Apart from defects in the import of substrates, in the carboxylic acid cycle, and in fatty acid oxidation, many dysfunctions have been described in the mitochondrial respiratory chain and ATP synthase (Morgan-Hughes, 1980). In the present study, we focused on defects of cytochrome *c* oxidase (E.C. 1.9.3.1), the terminal component of the mitochondrial respiratory chain.

Mammalian cytochrome *c* oxidase consists of 13 subunits, the three largest of which are encoded by mitochondrial DNA. The 10 other subunits are nuclearly encoded and imported into the mitochondria. Considering the complexity of the biogenesis and processing of cytochrome *c* oxidase, numerous causes for dysfunction of the enzyme are feasible. A clue as to the origin of a certain class of cytochrome *c* oxidase defects may, however, be given by the following observations. Firstly, dysfunctions in components of the respiratory chain are often restricted to a limited number of tissues (DiMauro *et al.*, 1986). Secondly, organ-specific isoforms of many of the nuclearly encoded subunits of mammalian cytochrome *c* oxidase have been reported (Kadenbach *et al.*, 1986). Thus it can be speculated that organ-specific defects of cytochrome *c* oxidase originate from defects in biogenesis or processing of an organ-specific isoform of one of the subunits.

Due to the fact that human cytochrome *c* oxidase could not be isolated in a native state, very little is known about human cytochrome *c* oxidase isoenzymes. Therefore, we developed an isolation procedure and studied human cytochrome *c* oxidase isoforms. Furthermore, immunological techniques were made operative for the study of the subunit patterns of cytochrome *c* oxidase in small tissue samples and cultured cells.

MATERIALS AND METHODS

For the isolation of cytochrome *c* oxidase from human heart and skeletal muscle, submitochondrial particles were prepared. Cytochrome *c* oxidase was selectively solubilized by consecutive extractions with 2% and 3% laurylmaltoside, respectively, in the presence of $1 \, mol \, L^{-1}$ KCl, followed by centrifugation ($100\,000 \, g$ for

Journal of Inherited Metabolic Disease. ISSN 0141–8955. Copyright © SSIEM and MTP Press Limited, Queen Square, Lancaster, UK.

15 min). Final purification was performed on a high performance liquid chromatography (HPLC) size-exclusion column (Dupont GF-2500) as described in Sinjorgo *et al.* (1987). The cytochrome *c* oxidase peak position was determined by examining absorbance spectra of the fractions.

For studies of cytochrome *c* oxidase subunit patterns in human muscle biopts, fibroblasts and myoblasts, crude cell extracts were made by sonification followed by addition of 3% laurylmaltoside, $1 \, mol \, L^{-1}$ KCl. The holoenzyme was immunoprecipitated by addition of protein A-Sepharose CL-4B (Pharmacia) equilibrated with polyclonal antibodies directed against the bovine heart enzyme. After gel electrophoresis, the subunit patterns were visualised by immunoblotting using the same antibody. Gel electrophoresis was performed by the method of Kadenbach (1983).

Cytochrome *c* oxidase activity was determined spectrophotometrically by following the rate of ferrocytochrome *c* oxidation at a wavelength of 550 nm in $50 \, mmol \, L^{-1}$ Tris sulphate (pH 7.4), $1 \, mmol \, L^{-1}$ EDTA, 0.05 % laurylmaltoside.

RESULTS

Using high performance liquid chromatography, cytochrome *c* oxidase was purified from human heart and skeletal muscle. Both preparations showed steady-state activity and absorbance spectra that were comparable to what is generally found for bovine heart cytochrome *c* oxidase. This indicated that the method yields a native enzyme. Polyacrylamide gel electrophoresis showed differences between the subunit VI regions of cytochrome *c* oxidase, indicating that the two tissues investigated contain different isoforms of cytochrome *c* oxidase.

The immunological techniques described in Materials and Methods proved to be adequate to study the subunit pattern of cytochrome *c* oxidase in low amounts of human material (0.5–5 mg protein from muscle tissue, fibroblasts or myoblasts). In crude extracts of cultured cells (made as described in Materials and Methods), the specific activity of cytochrome *c* oxidase could readily be determined. The values found were reproducible and three times as low in fibroblasts as in myoblasts. The methods described will be applied to tissues and cultured cells of patients that show decreased cytochrome *c* oxidase activity.

DISCUSSION

We demonstrated that the cytochrome *c* oxidases from human heart and skeletal muscle contain isoforms of some subunits. The genes coding for the small subunits of mammalian cytochrome *c* oxidase have not been identified; there is no evidence that the nuclear genome contains more than one gene for any of the cytochrome *c* oxidase subunits. Thus an important question remains – whether or not the differences in apparent molecular weight observed for corresponding cytochrome *c* oxidase subunits from different tissues originate from organ-specific gene expression or from organ-specific processing of identical gene products. Genetic investigations or amino acid sequencing of purified subunits may eventually solve this problem.

Furthermore, investigation of organ-specific defects in the biogenesis of cytochrome *c* oxidase could reveal whether the enzyme dysfunctions are caused by a defect in the (as yet unknown) organ-specific or cell-specific step in subunit biosynthesis.

ACKNOWLEDGEMENTS

This work was supported by grants from the Netherlands Organization for the Advancement of Pure Research (ZWO) and the Princess Beatrix Fund.

REFERENCES

DiMauro, S., Zeviani, M., Bonilla, E., Bresolin, N., Nakagawa, M., Miranda, A. F. and Moggio, M. Cytochrome *c* oxidase deficiency. *Biochem. Soc. Trans.* 13 (1986) 651–653

Kadenbach, B., Jarausch, J., Hartmann, R. and Merle, P. Separation of mammalian cytochrome *c* oxidase into 13 polypeptides by a sodium dodecyl sulfate-gel electrophoretic procedure. *Anal. Biochem.* 129 (1983) 517–521

Kadenbach, B., Stroh, A., Ungibauer, M., Kuhn-Nentwig, L., Büge, U. and Jarausch, J. Isoenzymes of cytochrome *c* oxidase. Characterization and isolation from different tissues. *Methods Enzymol.* 126 (1986) 31–45

Luft, R., Ikkos, D., Palmieri, G., Ernster, L. and Afzelius, B. A case of severe hypermetabolism of non-thyroid origin with a defect in the maintenance of respiratory control: a correlated clinical, biochemical and morphological study. *J. Clin. Invest.* 41 (1962) 1776–1804

Morgan-Hughes, J. C. Mitochondrial diseases. *Trends Neurol. Sci.* 9 (1980) 15–19

Sinjorgo, K. M. C., Hakvoort, T. B. M., Durak, I., Draijer, J. W., Post, J. K. P. and Muijsers, A. O. Human cytochrome *c* oxidase isoenzymes from heart and skeletal muscle; purification and properties. *Biochim. Biophys. Acta* 850 (1987) 144–150

J. Inher. Metab. Dis. 11 Suppl. 2 (1988) 205–207

Short Communication

Enzymatic Heterogeneity in Primary Hyperoxaluria Type 1 (Hepatic Peroxisomal Alanine: Glyoxylate Aminotransferase Deficiency)

C. J. Danpure and P. R. Jennings
Division of Inherited Metabolic Diseases, Clinical Research Centre, Harrow, HA1 3UJ, UK

Primary hyperoxaluria type 1 (PH1; McKusick 25990) is characterized by abnormal glyoxylate metabolism, which leads to excessive synthesis and excretion of oxalate and glycolate. The disease is caused by a deficiency of peroxisomal alanine: glyoxylate aminotransferase (AGT; EC 2.6.1.44) in the liver (Danpure and Jennings, 1986). The presentation of PH1 is heterogeneous, not only with respect to the severity of the clinical symptoms and responsiveness to pyridoxine (Williams and Smith, 1983), but also in terms of residual AGT enzyme activity (Danpure *et al.*, 1987) and the presence or absence of immunologically cross-reacting AGT protein (Wise *et al.*, 1987).

In this study we have extended our earlier observations by measuring AGT activities in the livers from 13 patients with classical pyridoxine-resistant PH1 (hyperoxaluria+hyperglycolic aciduria), one patient who was a mild PH1-type variant (mild hyperoxaluria+severe hyperglycolic aciduria), one patient with primary hyperoxaluria type 2 (PH2; McKusick 26000 hyperoxaluria + hyper-L-glyceric aciduria), one patient with primary hyperoxaluria not of type 1 or 2 (PH3; hyperoxaluria only) and an individual who was heterozygous for PH1. These have been compared with 11 control livers.

METHODS

The liver samples were sonicated in 10 volumes of $100\,\text{mmol}\,\text{L}^{-1}$ potassium phosphate buffer pH 7.4, containing $100\,\mu\text{mol}\,\text{L}^{-1}$ pyridoxal phosphate, as described previously (Danpure *et al.*, 1987). The aminotransferases were measured either by the double enzyme spectrophotometric method of Rowsell *et al.* (1972) or by the radioactive micro-method of Allsop *et al.* (1987). Subcellular fractionation of the livers was performed as described previously (Danpure *et al.*, 1986).

RESULTS

Considerable quantitative heterogeneity was found in the amount of residual AGT activity in the livers of the PH1 patients (Figure 1). In 13 patients with classical

Journal of Inherited Metabolic Disease. ISSN 0141–8955. Copyright © SSIEM and MTP Press Limited, Queen Square, Lancaster, UK.

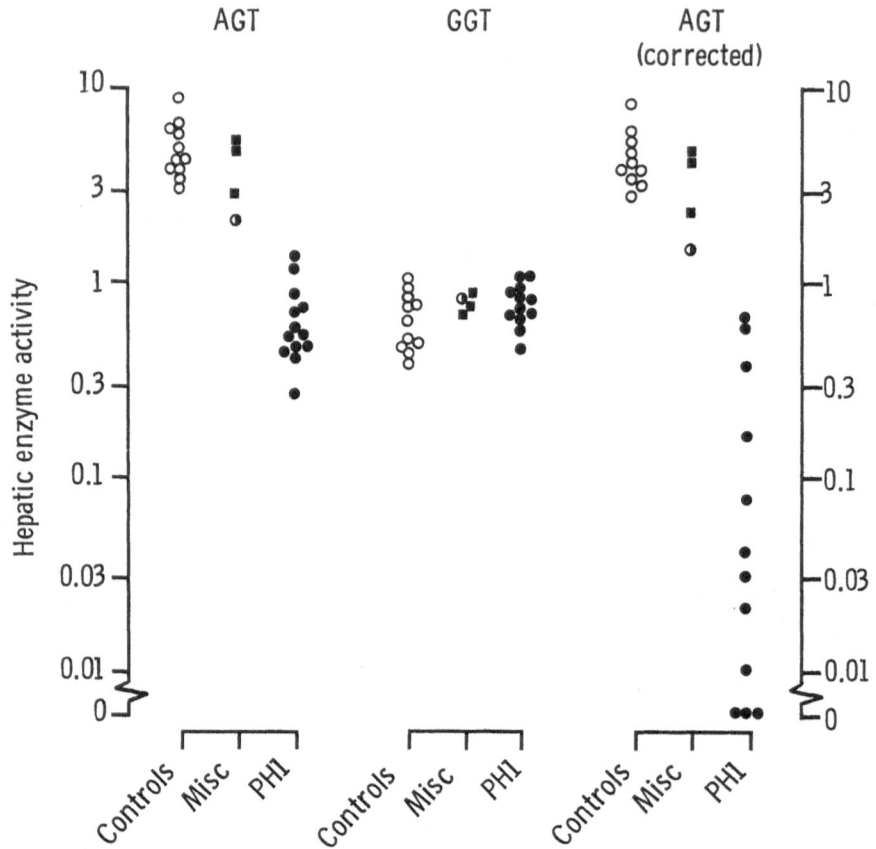

Figure 1 Scatter diagram of AGT and GGT activities in control and hyperoxaluric livers. AGT = alanine:glyoxylate aminotransferase; GGT = glutamate:glyoxalate aminotransferase; AGT (corrected) = AGT activity corrected for 66% cross-over from GGT (Thompson and Richardson, 1966). Enzyme activity is expressed as μmol h^{-1}(mg protein)$^{-1}$ on a logarithmic scale. ○ = controls; ● = classical PH1 homozygotes; ■ = hyperoxaluric variants; ◑ = PH1 heterozygote

pyridoxine-resistant PH1, the total liver AGT levels ranged from 5.4 to 26.6% of the mean control activity (0 to 14.0% when corrected for cross-over from cytosolic glutamate:glyoxylate aminotransferase (GGT; EC 2.6.1.4) (Thompson and Richardson, 1966). A single very mild PH1-type variant had 56% AGT activity (50% when corrected), while an asymptomatic PH1 heterozygote had 40% AGT activity, (32% when corrected). A patient with PH2 and one with PH3 had normal AGT activities.

In all subjects, the liver activities of the peroxisomal markers (catalase, D-amino acid oxidase and L-α-hydroxyacid oxidase) and various other aminotransferases (glutamate: glyoxylate, alanine: 2-oxoglutarate and aspartate: 2-oxoglutarate) were normal.

In all four of the PH1 patients whose livers were subcellularly fractionated (in which the residual AGT activity ranged from 8.5 to 14.1% uncorrected, or 0.2 to 1.8% corrected) there was a complete deficiency of peroxisomal AGT activity. The AGT activity in the subcellularly fractionated PH1 heterozygote was normally distributed (i.e. mainly peroxisomal).

DISCUSSION

These data demonstrate that despite considerable quantitative heterogeneity, hepatic AGT deficiency is specific and pathognomic for PH1, and is not found in other forms of primary hyperoxaluria. It also demonstrates a gene-dosage effect in the PH1 heterozygote who had AGT values intermediate between those of the homozygotes and the controls. This opens up the possibility of heterozygote detection in PH1 as well as enzymatic diagnosis of homozygotes (Danpure *et al.*, 1987). There is a good correlation between various parameters of clinical severity and residual AGT activity in PH1 homozygotes (Danpure *et al.*, 1987 and unpublished observations). However it is interesting to note that the asymptomatic PH1 heterozygote had less hepatic AGT activity than the mild atypical PH1-type variant, who had considerable hyperglycolic aciduria but only mild hyperoxaluria. Presumably this patient has a different type of mutation and should be reclassified into a new subgroup of primary hyperoxaluria.

REFERENCES

Allsop, J., Jennings, P. R. and Danpure, C. J. A new micro-assay for human liver alanine: glyoxylate aminotransferase. *Clin. Chim. Acta* 170 (1987) 187–193

Danpure, C. J. and Jennings, P. R. Peroxisomal alanine:glyoxylate aminotransferase deficiency in primary hyperoxaluria type I. *FEBS Lett.* 201 (1986) 20–24

Danpure, C. J., Purkiss, P., Jennings, P. R. and Watts, R. W. E. Mitochondrial damage and the subcellular distribution of 2-oxoglutarate:glyoxylate carboligase in normal human and rat liver and in the liver of a patient with primary hyperoxaluria type I. *Clin. Sci.* 70 (1986) 417–425

Danpure, C. J., Jennings, P. R. and Watts, R. W. E. The enzymological diagnosis of primary hyperoxaluria type I by measurement of hepatic alanine:glyoxylate aminotransferase activity. *Lancet* 1 (1987) 289–291

Rowsell, E. V., Carnie, J. A., Snell, K. and Taktak, B. Assays for glyoxylate aminotransferase activities. *Int. J. Biochem.* 3 (1972) 247–257

Thompson, J. S. and Richardson, K. E. Isolation and characterization of a glutamate: glycine transaminase from human liver. *Arch. Biochem. Biophys.* 117 (1966) 599–603

Williams, H. E. and Smith, L. H. Primary hyperoxaluria. In Stanbury, J. B., Wyngaarden, J. B., Fredrickson, D. S., Goldstein, J. L. and Brown, M. S. (eds.) *The Metabolic Basis of Inherited Disease*, McGraw-Hill, New York, 1983, pp. 204–228

Wise, P. J., Danpure, C. J. and Jennings, P. R. Immunological heterogeneity of hepatic alanine:glyoxylate aminotransferase in primary hyperoxaluria type 1. *FEBS Lett.* 222 (1987) 17–20

J. Inher. Metab. Dis. 11 Suppl. 2 (1988) 208–211

Short Communication

Diversity in Residual Alanine Glyoxylate Aminotransferase Activity in Hyperoxaluria Type I: Correlation with Pyridoxine Responsiveness

R. J. A. WANDERS[1], C. W. T. VAN ROERMUND[1], S. JURRIAANS[1], R. B. H. SCHUTGENS[1], J. M. TAGER[2], H. VAN DEN BOSCH[3], E. D. WOLFF[4], H. PRZYREMBEL[4], R. BERGER[5], F. G. SCHAAPHOK[5], W. REITSMA[5] and W. H. J. VAN LUYK[5]

[1]*Department of Pediatrics, University of Amsterdam, Meibergdreef 9, 1105 AZ Amsterdam, The Netherlands;* [2]*Laboratory of Biochemistry, University of Amsterdam, Meibergdreef 15, 1105 AZ Amsterdam, The Netherlands;* [3]*Laboratory of Biochemistry, State University Utrecht, Padualaan 8, 3584 CH Utrecht, The Netherlands;* [4]*Departments of Pediatrics and Nephrology, University Hospital Rotterdam/Sophia Children's Hospital, The Netherlands; and* [5]*Department of Pediatrics, State University Groningen, Groningen, The Netherlands*

Primary hyperoxaluria type I is a rare genetic disorder of glyoxylate metabolism in which patients usually present during the first decade of life with recurrent calcium oxalate nephrolithiasis (see Williams and Smith, 1983). Although it was long believed that hyperoxaluria type I is associated with a deficiency of the cytosolic form of 2-oxoglutarate:glyoxylate carboligase, it is now clear that the primary defect in hyperoxaluria type I is at the level of a deficient alanine: glyoxylate aminotransferase activity as first shown by Danpure and Jennings (1986). Although most hyperoxaluria type I patients die before 20 years of age from progressive renal insufficiency, several patients have been described suffering from a milder, pyridoxine responsive form of hyperoxaluria type I (see Williams and Smith, 1983). In these patients the urinary excretion of oxalate and glycolate, the two characteristic metabolites in blood and urine from hyperoxaluria type I patients, can be reduced upon administration of pyridoxine. In the present report we determined the activity of alanine glyoxylate aminotransferase (AGT) in liver needle specimens from one patient who died from pyridoxine-resistant hyperoxaluria type I and three patients with the pyridoxine-responsive form of the disease using an improved sensitive enzyme assay.

MATERIALS AND METHODS

Enzyme activity measurements: The activity of alanine glyoxylate aminotransferase (EC 2.6.1.44) was measured exactly as described before (Wanders *et al.*, 1987)

Journal of Inherited Metabolic Disease. ISSN 0141–8955. Copyright © SSIEM and MTP Press Limited, Queen Square, Lancaster, UK.

except that the concentration of L-alanine was $80\,mmol\,L^{-1}$ and incubations were carried out for 240 min. The activity of glutamate glyoxylate aminotransferase was measured as described for alanine glyoxylate aminotransferase except that $20\,mmol\,L^{-1}$ L-glutamate was used instead of $80\,mmol\,L^{-1}$ L-alanine. Further processing of the samples was as described for alanine glyoxylate aminotransferase (Wanders *et al.*, 1987). The amount of 2-oxoglutarate generated was measured according to standard procedures involving the use of glutamate dehydrogenase on a COBAS-BIO centrifugal analyser (Hoffman-La Roche, Basel, Switzerland).

Determination of oxalate and glycolate in urine: Urinary levels of oxalate and glycolate were measured by gas chromatography essentially as described before (Wanders *et al.*, 1987).

Preparation of human liver samples: Liver samples were obtained by needle biopsy or at necropsy and processed as described in detail before (Wanders *et al.*, 1987).

Case reports: A full report on patients 2, 3 and 4 will be published elsewhere (van Luyk *et al.*, in preparation). Patient 1 has been described before (Wanders *et al.*, 1987).

RESULTS

In Table 1 we present the activity of alanine glyoxylate aminotransferase in human liver specimens obtained by needle biopsy from one patient with the severe, neonatal pyridoxine-resistant form of hyperoxaluria and three patients with a pyridoxine-responsive type of hyperoxaluria. Furthermore, we measured the plasma levels and urinary excretion of oxalate and glycolate in the patients. The results of Table 1 show that administration of pyridoxine ($3 \times 50\,mg$ per day) led to a drastic reduction in the urinary excretion of glycolate and oxalate in patients

Table 1 The urinary excretion of oxalate and glycolate and the activity of alanine glyoxylate aminotransferase in four patients with hyperoxaluria type I

Subjects	Alanine glyoxylate aminotransferase activity[a]	Urinary excretion of oxalate plus glycolate[b]	
		−Vit. B6	+Vit. B6
Hyperoxaluria type I patients:			
1	0.96		
2	5.8	0.82	0.57
3	9.8	0.90	0.20
4	16.1	0.69	0.23
Controls: mean ±SD ($n = 7$)	98.8±18.8	<0.2	n.d.
range	79.8–133.7		

n.d. = not done
Values are [a] $nmol\,min^{-1}\,mg^{-1}$ and [b] $mmol\,(mmol\,creatinine)^{-1}$

2, 3 and 4. When the activity of alanine glyoxylate aminotransferase was measured in liver from the patients, the enzyme was found to be almost completely deficient in liver from the patient with the severe, neonatal pyridoxine-resistant type of hyperoxaluria type I, whereas higher residual activities were found in liver from the three pyridoxine-responsive patients amounting to 5.9, 9.9 and 16.3% of control values (Table 1). The alanine glyoxylate aminotransferase activities in Table 1 represent true AGT activities since corrections were made for crossover by glutamate glyoxylate aminotransferase as described by Danpure *et al.* (1987). Enzyme activities were measured under optimized conditions including the use of high concentrations of L-alanine and glyoxylate (80 mmol L^{-1} and 10 mmol L^{-1}, respectively) at pH 8.0. Since the specific activity of alanine glyoxylate aminotransferase is high in human liver (Table 1) and enzyme activity is linear with time up to 240 min when measured under optimized conditions, enzyme activity measurements can be performed in microgram quantities of liver.

DISCUSSION

The results described in this report indicate that alanine glyoxylate aminotransferase is indeed deficient in hyperoxaluria type I. However, in contrast to an almost complete deficiency of this enzyme in a patient who died from the pyridoxine-resistant form of hyperoxaluria type I, the deficiency of alanine glyoxylate aminotransferase was found to be less severe in three patients suffering from a pyridoxine-responsive type of hyperoxaluria (Table 1). The latter finding provides a rational basis for the effectiveness of pyridoxine in lowering the urinary excretion of oxalate and glycolate in these patients by allowing residual enzyme activity to operate optimally in the presence of surplus pyridoxal-5'-phosphate. Although it is not clear how the situation is *in vivo*, addition of pyridoxal-5'-phosphate leads to an almost twofold increase of alanine glyoxylate aminotransferase activity when measured *in vitro* (Danpure *et al.*, 1987; Wanders *et al.*, 1987). Future studies involving combined immunological (ELISA, immunoblotting, etc.) and enzymatical techniques will shed more light on the nature of the different enzyme deficiencies in the patients of Table 1.

Using the optimized assay for alanine glyoxylate aminotransferase described in Materials and Methods and a radiochemical method essentially as described by Kisaki and Tolbert (1969) we have been unable to detect alanine glyoxylate aminotransferase in tissues other than liver and kidney. No activity was found in leucocytes, platelets and cultured skin fibroblasts. Finally, no activity was found in cultured amniocytes and chorionic villi nor chorionic villous fibroblasts, which suggests that prenatal diagnosis of hyperoxaluria type I can only be done by fetal liver biopsy or DNA analysis.

REFERENCES

Danpure, C. J. and Jennings, P. R. Peroxisomal alanine glyoxylate aminotransferase deficiency in primary hyperoxaluria type I. *FEBS Lett.* 201 (1986) 20–24

Danpure, C. J., Jennings, P. R. and Watts, R. W. E. Enzymological diagnosis of primary hyperoxaluria type I by measurement of hepatic alanine:glyoxylate aminotransferase activity. *Lancet* 1 (1987) 289–291

Kisaki, T. and Tolbert, N. E. Glycolate and glyoxylate metabolism by isolated peroxisomes and chloroplasts. *Plant Physiol.* 44 (1969) 242–250

Wanders, R. J. A., van Roermund, C. W. T., Westra, R., Schutgens, R. B. H., van der Ende, M. A., Tager, J. M., Monnens, L. A. H., Baadenhuysen, H., Govaerts, L., Przyrembel, H., Wolff, H., Blom, W., Huymans, J. G. M. and van Laarhoven, F. G. M. Alanine glyoxylate aminotransferase and the urinary excretion of oxalate and glycolate in hyperoxaluria type I and the Zellweger syndrome. *Clin. Chim. Acta* 165 (1987) 311–319

Williams, H. E. and Smith, L. H. Primary hyperoxaluria. In Stanbury, J. B., Wyngaarden, J. B., Fredrickson, D. S., Goldstein, J. L. and Brown, M. S., eds. *The Metabolic Basis of Inherited Disease*, 5th Edition, McGraw-Hill, New York, 1983, pp. 204–228

J. Inher. Metab. Dis. 11 Suppl. 2 (1988) 212–214

Short Communication

The Subcellular Metabolism of Glyoxylate in Primary Hyperoxaluria Type 1: The Relationship Between Glycine Production and Oxalate Overproduction

G. N. THOMPSON, P. PURKISS and C. J. DANPURE
Division of Inherited Metabolic Diseases, Clinical Research Centre, Watford Rd., Harrow, Middlesex HA1 3UJ, UK

Primary hyperoxaluria type 1 (PH; McKusick 25990) results from deficiency of hepatic peroxisomal alanine:glyoxylate aminotransferase (AGT; EC 2.6.1.44) (Danpure and Jennings, 1986). It is proposed that this defect results in the passage of unmetabolized glyoxylate across the peroxisomal membrane into the cytosol where, in addition to being transaminated to glycine by glutamate:glyoxylate aminotransferase (GGT; EC 2.6.1.4), it is also oxidised to oxalate by lactate dehydrogenase (EC 1.1.2.3). The quantitative relationship between cytosolic glyoxylate transamination and oxidation is not, however, known. We have attempted to examine this relationship with respect to the above hypothesis by investigating the metabolism of labelled glyoxylate to glycine and oxalate in subcellular fractions of PH liver.

METHODS

Whole liver from two PH subjects (described previously as patients D and F by Danpure *et al.*, 1987) was subfractionated on continuous Percoll gradients (Neat *et al.*, 1980). Cytosolic, mitochondrial or peroxisomal fractions ($20–100\,\mu L$, approx $0.1\,mg$ protein) were then incubated at 37°C for 20 min in $400\,\mu L$ of incubation medium made up to a total volume of $500\,\mu L$ with deionised water if necessary. The incubation medium consisted of $0.25\,mol\,L^{-1}$ sucrose and $1\,mmol\,L^{-1}$ pyridoxal phosphate buffered to pH 7.4 with $16\,mmol\,L^{-1}$ phosphate buffer. Glyoxylate ($0.5\,\mu Ci\,mL^{-1}$; $9\,\mu Ci\,\mu mol^{-1}$) or glycine ($0.5\,\mu Ci\,mL^{-1}$; $113\,\mu Ci\,\mu mol^{-1}$; Amersham) were used as substrates and alanine or glutamate ($100\,\mu mol\,L^{-1}$) as cosubstrates, or no cosubstrate was added. Some liver fractions were lysed by adding $10\,\mu L$ of 10% Triton X-100 to the incubation mixture, a method which rapidly released over 90% of cytochrome oxidase (EC 1.9.3.1) and catalase (EC 1.11.1.6) activity from the fractions. No evidence of substrate limitation was noted under these conditions.

On completion of the incubation, sulphosalicylic acid ($10\,\mu L$, 50%) was added and the incubation mixture was mixed and centrifuged at $1200\,g$ for 10 min. The supernatant was removed and stored at −20°C until analysed. Radioactivity in

Journal of Inherited Metabolic Disease. ISSN 0141-8955. Copyright © SSIEM and MTP Press Limited, Queen Square, Lancaster, UK.

oxalate and glycine was separated from that in glyoxylate by high-voltage electro-phoresis on Whatman 3MM paper in formic acid (6% v/v) at 65 V cm^{-1} for 25 min using a Miles Hivolt apparatus. This method gave recoveries of 94–110% from standard solutions of glycine and oxalate. After correcting for a blank value obtained by adding sulphosalicylic acid prior to addition of the liver fraction, glycine and oxalate production rates were corrected by volume and enzyme (catalase or cytochrome oxidase) recovery and the relative quantitative contributions by each subcellular compartment to homogenate glycine and oxalate production rates from glyoxylate were expressed as nmol h^{-1} (mg liver)$^{-1}$. AGT and GGT activities were determined in each fraction by the method of Rowsell *et al.* (1972) and alanine and glutamate concentrations by amino acid analyser. AGT and GGT activities were expressed per gram of liver based on volume and enzyme recoveries.

RESULTS

Glycine and oxalate production rates from glyoxylate are presented in Table 1 and the distribution of AGT and GGT in Table 2. Glycine production in the absence of added cosubstrate (results not shown) was virtually undetectable in mitochon-drial and peroxisomal fractions. In cytosolic fractions glycine production was not

Table 1 **Glycine and oxalate production from glyoxylate in PH liver fractions**

| | Cytosol | | Mitochondria | | Peroxisomes | |
	Glycine	Oxalate	Glycine	Oxalate	Glycine	Oxalate
Subject 1						
Alanine	49(54)	42(37)	13(19)	5(18)	2(9)	0(0)
Glutamate	40(49)	42(21)	13(78)	19(21)	3(5)	4(0)
Subject 2						
Alanine	60(59)	51(40)	1(3)	11(3)	1(1)	5(4)
Glutamate	46(50)	47(35)	4(22)	11(12)	1(2)	5(4)

Values are presented in nmol h^{-1} (mg liver)$^{-1}$ for unlysed and lysed (in parentheses) fractions with alanine or glutamate added as cosubstrate

Table 2 **AGT and GGT distribution in PH liver fractions**

	Cytosol	Mitochondria	Peroxisomes
Subject 1			
AGT	35	3	1
GGT	52	6	1
Subject 2			
AGT	20	4	1
GGT	62	5	4

AGT units are μmol pyruvate h^{-1} (g liver)$^{-1}$
GGT units are μmol 2-oxoglutarate h^{-1} (g liver)$^{-1}$

altered by the addition of either cosubstrate. This may be due to the significant concentrations of glutamate (97 and 61 μmol L^{-1} in subjects 1 and 2, respectively) and alanine (11 μmol L^{-1} and 89 μmol L^{-1}) in the cytosolic fractions even before the addition of cosubstrate. Production of either glyoxylate or oxalate was virtually undetectable when glycine was used as the substrate.

DISCUSSION

The rate of metabolism of glyoxylate to glycine and oxalate by peroxisomal fractions in both subjects was much lower than that by mitochondrial and, especially, cytosolic fractions. This finding is consistent with the now established deficiency of peroxisomal AGT in PH (Danpure *et al.*, 1987). The oxidation of glyoxylate to oxalate is not thought to be impaired in PH, and the negligible production of oxalate in peroxisomal fractions in the current study suggests that, in the presence of normal AGT activity, glycine would be the principal product of glyoxylate metabolism in peroxisomes.

In contrast, metabolism of glyoxylate by PH cytosolic liver fractions resulted in oxalate production rates similar to those of glycine. Thus, when the principal site of glyoxylate metabolism changes from the peroxisome to the cytosol, as appears to occur in PH, significantly more oxalate is likely to be produced. The capacity of the hepatic cytosol to transaminate at least some glyoxylate to glycine may modulate the total body production of oxalate from glyoxylate in PH. This transamination is most likely attributable to GGT, which is known to be able to use either glutamate or alanine as the amino donor (Thompson and Richardson, 1966) suggesting that, in the absence of peroxisomal AGT, hepatic GGT activity may be a determinant of the severity of phenotypic expression in PH.

Lysis increased the metabolism of glyoxylate to glycine by over 50% in mitochondrial fractions in both subjects and in peroxisomal fractions in subject 1. The finding contrasts with similar studies in rat liver, where lysis of organelle fractions does not appear to change the rate of metabolism of glyoxylate (Thompson and Danpure, unpublished observation). The proposed pathogenesis of PH (Danpure and Jennings, 1986) requires that normal glyoxylate metabolism must be localized peroxisomally within the hepatocyte. Limited permeability of organelles may be the mechanism by which this is achieved.

REFERENCES

Danpure, C. J. and Jennings, P. R. Peroxisomal alanine: glyoxylate aminotransferase deficiency in primary hyperoxaluria type 1. *FEBS Lett.* 201 (1986) 20–24

Danpure, C. J., Jennings, P. R. and Watts, R. W. E. Enzymological diagnosis of primary hyperoxaluria type 1 by measurement of hepatic alanine: glyoxylate aminotransferase activity. *Lancet* 1 (1987) 289–291

Neat, C. E., Thomassen, M. S. and Osmundsen, H. Induction of peroxisomal β-oxidation in rat liver by high fat diets. *Biochem. J.* 186 (1980) 369–371

Rowsell, E. V., Carnie, J. A., Snell, K. and Taktak, B. Assays for glyoxylate aminotransferase activities. *Int. J. Biochem.* 3 (1972) 247–257

Thompson, J. S. and Richardson, K. E. Isolation and characterisation of glutamate: glycine transaminase from human liver. *Arch. Biochem. Biophys.* 117 (1966) 599–603

J. Inher. Metab. Dis. 11 Suppl. 2 (1988) 215–217

Short Communication

Primary Hyperoxaluria and L-Glyceric Aciduria in the Cat

W. F. Blakemore[1], M. F. Heath[1], M. J. Bennett[2], C. H. Cromby[3] and R. J. Pollitt[3]*

[1]*Department of Clinical Veterinary Medicine, University of Cambridge;* [2]*Department of Chemical Pathology, Sheffield Children's Hospital;* [3]*Neonatal Screening Laboratory, P.O. Box 134, Middlewood Hospital, Sheffield S6 1TP, UK*

L-Glyceric aciduria (primary hyperoxaluria type 2; McKusick 26000) is a rare inherited disorder characterized in man by recurrent calcium oxalate nephrolithiasis, chronic renal failure and early death from uraemia. To date only eight cases have been described in the literature (Williams and Smith, 1968; Chalmers *et al.*, 1984). A defect in D-glycerate dehydrogenase has been demonstrated in human leukocytes (Williams and Smith, 1968). Deficiency of this enzyme results in the accumulation of hydroxypyruvate which is then reduced to L-glycerate by lactate dehydrogenase. The cause of the increased urinary oxalate is unclear but may represent an alternative route of hydroxypyruvate metabolism.

We present here a number of cats with L-glyceric aciduria and hyperoxaluria who may represent the first animal model of this severe condition.

CLINICAL SIGNS AND PATHOLOGICAL FINDINGS IN AFFECTED CATS

The affected cats develop acute renal failure between 5–9 months of age. At this time they show signs of weakness, which progresses to profound weakness over a period of a few days, and muscle biopsy shows evidence of denervation atrophy. At post mortem examination such animals show prominent accumulation of neurofilaments in the proximal axons of ventral horn cells and dorsal root ganglion cells of the spinal cord. The kidneys are usually swollen and the renal tubules contain oxalate crystals.

Prior to the onset of renal failure it is difficult to demonstrate specific neurological deficits although affected cats are noticed to be weaker than their littermates. Muscle biopsy at this early stage has so far failed to show abnormalities of muscle. However, examination of one animal prior to the onset of renal failure has shown that the swelling of proximal axons is present in animals not in renal failure.

The condition has been seen in cats of both sexes and all affected cats are related; present evidence indicates that the mode of inheritance is an autosomal recessive.

---

* Corresponding Author

Journal of Inherited Metabolic Disease. ISSN 0141–8955. Copyright © SSIEM and MTP Press Limited, Queen Square, Lancaster, UK.

BIOCHEMICAL INVESTIGATIONS

Oxalate was determined using a commercial kit (Crider and Curran, 1984). Organic acids were extracted from urine by ethyl acetate and ether solvent extractions, converted to their trimethysilyl derivatives using bis(trimethylsilyl)-trifluoroacetamide containing 1% chlorotrimethylsilane, and examined by gas chromatography–mass spectrometry. Glyceric acid was quantitated by capillary gas chromatography as the trimethylsilyl derivative against a standard of DL-glyceric acid extracted under similar conditions. The configuration of the abnormal feline glyceric acid peak was determined by capillary gas chromatography of the *O*-acetylated l-menthyl esters (Kamerling *et al.*, 1977).

Examination of the urine of affected animals indicated L-glyceric aciduria and intermittent hyperoxaluria. These features are the hallmarks of the human disorder primary hyperoxaluria type 2. Data for random cystocentesis samples from affected

Table 1   Urine oxalate and glycerate concentrations in L-glyceric aciduria

	Oxalate	Glycerate
Cats:		
affected ($n = 6$)	17–523	120–11 100
related ($n = 11$)	2–66	0
controls ($n = 6$)	1–2	0
Humans[a]:		
cases ($n = 4$)	44–136	290–1350
related ($n = 10$)	14–86	0

Values are mmol (mol creatinine)$^{-1}$
[a] Data of Chalmers *et al.* (1984)

cats, related but unaffected animals, and unrelated controls are shown in Table 1; for comparison data are shown from human cases of primary hyperoxaluria type 2 and their relatives.

DISCUSSION

Peripheral neuropathy of the type present in these cats is not a feature of primary hyperoxaluria type 2 in man. Accumulations of neurofilaments in the proximal portion of motor and sensory axons of the spinal cord are a feature of β,β'-iminodipropionitrile intoxication and of a dominantly inherited condition in the dog (canine spinal muscular atrophy in Brittany spaniels). The renal lesions are more severe in the cats than those seen in primary hyperoxaluria type 2 in man. These differences indicate either that D-glycerate dehydrogenase deficiency manifests differently in the cat or that we are dealing with a different condition. Nevertheless this may be a useful animal model in which to study the mechanism of hyperoxaluria as a consequence of L-glyceric aciduria.

ACKNOWLEDGEMENT
R. J. Pollitt is a member of the external scientific staff of the Medical Research Council.

REFERENCES

Chalmers, R. A., Tracey, B. M., Mistry, J., Griffiths, K. D., Green, A. and Winterborn, M. H. L-Glyceric aciduria (primary hyperoxaluria type 2) in siblings in two unrelated families. *J. Inher Metab. Dis.* 7 Suppl. 2 (1984) 133–134

Crider, Q. E. and Curran, D. F. Simplified method for enzymic urine oxalate assay. *Clin. Biochem.* 17 (1984) 351–355

Kamerling, J. P., Gerwig, G. J., Vliegenthart, J. F. G., Duran, M., Ketting, D. and Wadman, S. K. Determination of the configurations of lactic and glyceric acids from human serum and urine by capillary gas–liquid chromatography. *J. Chromatogr.* 143 (1977) 117–123

Williams, H. E. and Smith, L. H. L-Glyceric aciduria. A new genetic variant of primary hyperoxaluria. *N. Engl. J. Med.* 278 (1968) 233–239

J. Inher. Metab. Dis. 11 Suppl. 2 (1988) 218–220

Short Communication

Clinical Effects of Serine Medication in Non-ketotic Hyperglycinaemia Due to Deficiency of P-Protein of the Glycine Cleavage Complex

F. A. Wijburg[1], C. J. de Groot[1], R. B. H. Schutgens[1], P. G. Barth[1] and K. Tada[2]

[1]*Department of Pediatrics, University Hospital of Amsterdam (AMC), 1105 AZ Amsterdam, The Netherlands;* [2]*Department of Pediatrics, Tohoku University School of Medicine, Sendai 980, Japan*

Non-ketotic hyperglycinaemia (McKusick 23830) is an inborn error of metabolism in which, due to a defect in the glycine cleavage reaction, there is accumulation of glycine in plasma and CSF.

There is exact knowledge about the site of the metabolic lesions in non-ketotic hyperglycinaemia (Hayasaka *et al.*, 1987); less is known about the pathogenesis of the symptoms. The clinical characteristics are: hypotonia, myoclonic seizures, lethargy and irritability, leading to death after a few weeks or severe psychomotor retardation. Symptoms have been attributed to the effects of glycine as an inhibitory neurotransmitter and to a shortage of one-carbon units which are provided by the glycine cleavage reaction (Tada and Hayasaka, 1987). Therapeutic trials have been based on these assumptions: strychnine and diazepam to counteract the neurotransmitter effect of glycine, and methionine and leucovorine to supply one-carbon fragments. However, these treatments have not essentially changed the clinical course of the disease. As a further attempt for treatment we tried giving L-serine (as a donor of one-carbon units) to a patient with non-ketotic hyperglycin-aemia and observed a striking and unexpected change in the clinical condition.

CASE HISTORY

The patient, a boy, was born from healthy, unrelated, Moroccan parents. After 3 days he became hypotonic and intubation and artifical ventilation were necessary due to respiratory failure. The EEG showed a characteristic burst suppression pattern. At 23 days the plasma glycine concentration was $2.0 \, \text{mmol} \, \text{L}^{-1}$ (normal: 0.1–0.6) and CSF glycine was $204 \, \mu\text{mol} \, \text{L}^{-1}$ (normal 6.3–8.0). There was no ketosis and blood pH was normal. The child started to have myoclonic jerks and tonic fits but became less hypotonic and could be extubated. The diagnosis of non-ketotic hyperglycinaemia was confirmed by a liver biopsy specimen. Enzyme analysis showed no activity of the glycine cleavage reaction due to a complete defect in the

218

Journal of Inherited Metabolic Disease. ISSN 0141–8955. Copyright © SSIEM and MTP Press Limited, Queen Square, Lancaster, UK.

Table 1 Enzyme activity of the glycine cleavage reaction in the patient

	Glycine cleavage	P-protein	H-protein	T-protein
Patient	0	0	25.8	61.5
Controls	4.4–10.8	4.5–5.7	14.6–18.3	52.1–77.9

Activity is expressed as μmol product (g protein)$^{-1}$h^{-1}

P-protein (Table 1). Restriction of protein intake (1.4 g kg^{-1}day^{-1}) together with administration of sodium benzoate (500 mg kg^{-1}day^{-1}) resulted in a drop of plasma glycine to normal values and CSF glycine to 122 μmol L^{-1}. The infant however remained hypotonic and suffered from almost continuous myoclonic jerks and seizures. There were periods of persistent crying and he was very irritable.

SERINE TRIAL

While increasing benzoate dosage to 1 g kg^{-1}day^{-1}, L-serine was administered (200 mg kg^{-1}day^{-1}, after 3 days 400 mg kg^{-1}day^{-1}). A striking change in the clinical condition was observed: the stereotypic movements, the convulsions, the myoclonic jerks and the persistent crying disappeared after 2 days and the child was remarkably quiet. He became increasingly drowsy and hypotonic. Tube feeding was necessary (while he could drink sufficiently before). The CSF glycine level was lowered (56 μmol L^{-1}) while plasma glycine had remained normal. CSF and plasma serine remained within normal values.

Serine medication was stopped and within a few days the symptoms of restlessness, crying and tonic convulsions reappeared. Gavage feeding could be stopped. Reintroduction of serine (130 mg kg ^{1}day^{-1}) again resulted in the same severe apathic state without convulsions or myoclonic jerks. In this period the child suddenly died due to aspiration.

DISCUSSION

L-Serine has been administered to patients with non-ketotic hyperglycinaemia in amino-acid loading tests for diagnostic reasons and to investigate the metabolic pathways involved (Palmer and Oberholzer, 1985; Baumgartner et al., 1969). A change in clinical condition (increased lethargy) was described and explained by the simultaneous increase in (plasma) glycine due to serine–glycine interconversion (Krieger and Hart, 1974). We observed striking (and reversible) clinical effects with administration of L-serine without a rise in CSF glycine, while during L-serine medication the benzoate dosage was increased to 1 g kg^{-1}day^{-1}. The increased somnolence and lethargy, with disappearance of the myoclonic jerks, can therefore not be attributed to the neurotoxicity of CSF glycine. The serine trial also shows that the severity of clinical condition is not directly related to CSF glycine concentration.

The mechanism by which L-serine influences clinical symptoms is unknown. We tried L-serine because it is, like glycine, a one-carbon donor in a reaction in which tetrahydrofolate is involved. Whether the sudden influx of one-carbon units into the depleted one-carbon pool can result in the observed changes is not clear. CSF glycine, however, cannot be the sole contributing factor to the neurological symptoms in non-ketotic hyperglycinaemia, and more subtle mechanisms must be involved.

Although the pathological signs are already present shortly after birth in non-ketotic hyperglycinaemia (Von Wendt *et al.*, 1981), the nature of the symptoms can still be influenced after weeks. We therefore hope that our observation may be of help for further studies in the pathogenesis of the symptoms and investigation of possible therapeutic agents.

REFERENCES

Baumgartner, R., Ando, T. and Nyhan, W. Non-ketotic hyperglycinemia. *J. Pediatr.* 75 (1969) 1022–1025

Hayasaka, K., Tada, K., Fueki, N., Nakamura, Y., Nyhan, W. L., Schmidt, K., Packman, S., Seashore, M., Haan, E., Danks, D. and Schutgens, R. B. H. Non-ketotic hyperglycinemia; analysis of glycine cleavage system in typical and atypical cases. *J. Pediatr.* 110 (1987) 873–877

Krieger, I. and Hart, Z. H. Valine-sensitive non-ketotic hyperglycinemia. *J. Pediatr.* 85 (1974) 43–48

Palmer, T. and Oberholzer, V. G. Amino acid loading tests in a patient with non-ketotic hyperglycinaemia. *J. Inher. Metab. Dis.* 8, Suppl. 2 (1985) 125–126

Tada, K. and Hayasaka, K. Non-ketotic hyperglycinemia: clinical and biochemical aspects. *Eur. J. Pediatr.* 146 (1987) 221–227

Von Wendt, L., Simila, S., Suakkonen, A. L., Koivisto, M. and Kouvalainen, K. Prenatal brain damage in non-ketotic hyperglycinemia. *Am. J. Dis. Child.* 135 (1981) 1072

J. Inher. Metab. Dis. 11 Suppl. 2 (1988) 221–224

Short Communication

The Use of Phenylpropionic Acid as a Loading Test for Medium-Chain Acyl-CoA Dehydrogenase Deficiency

J. W. T. Seakins and G. Rumsby*
Department of Clinical Biochemistry, Hospital for Sick Children and Institute of Child Health, 30 Guilford Street, London, WC1N 1EH, UK

Medium-chain acyl-CoA dehydrogenase (MCAD) deficiency is a disorder of β-oxidation of fatty acids leading to episodes of hypoketotic hypoglycaemia. Typically, excretion of abnormal metabolites, derived from alternative pathways of fatty acid metabolism, only occurs when the patient is unwell (Gregersen *et al.*, 1983). Two approaches have been suggested for investigation of asymptomatic patients and siblings, prolonged fasting or loading test with medium-chain triglycerides. Both tests require the analysis of urine for characteristic metabolites, preferably by gas chromatography–mass spectroscopy. Fasting is not without serious risk (Gregersen *et al.*, 1976) and the latter test may lack specificity (Whyte *et al.*, 1986).

Knoop, in the early 1900s, was the first to use phenyl-labelled fatty acids to elucidate the pathways of fatty acid metabolism. From his experiments and those of others, particularly those by Raper and Wayne (1928) who summarized the earlier studies, the metabolism of fatty acids by β-oxidation was established. The odd chain-length fatty acids such as 3-phenylpropionic or 9-phenylnonanoic acid yield benzoic acid which is excreted as a glycine conjugate. The even chain-length fatty acids like 4-phenylbutyric or 10-phenyldecanoic give rise to phenylacetic acid excreted as phenaceturic acid in dogs (Raper and Wayne, 1928) or as the glutamine conjugate in man.

The molecular weight (150) of 3-phenylpropionic acid (hydrocinnamic acid) is of the same order as those fatty acids which are the substrates for MCAD. Thus the metabolism of 3-phenylpropionic acid would be expected to be impaired in patients with MCAD-deficiency, leading to the accumulation of the corresponding CoA-derivative and the formation of phenylpropionylglycine. By contrast, normal subjects would only excrete hippuric acid.

We describe two simple high performance liquid chromatography (HPLC) methods for the detection of phenylpropionylglycine in urine, one using ion-moderated partition chromatography (IMPC) as described by Rumsby *et al.* (1987), the second a quicker, more selective system, using reverse phase chromatography.

---

* Present address: Department of Chemical Pathology, Royal Marsden Hospital, Fulham Road, London, SW3 6JJ, UK.

Journal of Inherited Metabolic Disease. ISSN 0141–8955. Copyright © SSIEM and MTP Press Limited, Queen Square, Lancaster, UK.

CLINICAL MATERIAL AND INVESTIGATIVE PROCEDURE

Permission for this study was obtained from the Ethical Committee of the Hospital for Sick Children. The clinical material studied consisted of eight children known to have MCAD deficiency, six healthy siblings of known cases and 17 patients initially suspected of MCAD deficiency but where this condition was later excluded on other biochemical evidence. Additionally, three adult obligate carriers and two normal adults were studied.

After a baseline urine specimen had been collected for the measurement of hippuric acid, phenylpropionic acid ($25\,mg\,kg^{-1}$) was given, dissolved in yoghurt, cream, jam or other strongly flavoured food. All urine voided over the next 6 h was collected. The subjects were not fasted, and were encouraged to drink plenty of water to ensure adequate urine flow.

MATERIALS AND METHODS

3-Phenylpropionic acid and 3-phenylpropionyl chloride were purchased from Aldrich Chemical Co Ltd, Gillingham, Dorset, SP8 4JL, UK; HPLC-grade water and solvents from Rathburn Chemicals Ltd., Walkerburn, Peebleshire, Scotland. The glycine conjugate of phenylpropionic acid was synthesized by the Schotten-Baumann reaction starting with the acyl chloride.

For IMPC an Aminex HPX-87H organic acids column ($300 \times 7.8\,mm$) protected by an Aminex-85H guard cartridge ($40 \times 4.6\,mm$) (Bio-Rad Laboratories Ltd., Caxton Way, Watford, WD1 8RP, UK) was used under the conditions described by Rumsby *et al.* (1987).

For reverse phase chromatography an Apex $5\,\mu m$ ODS column ($250 \times 4.5\,mm$) with guard column ($50 \times 4.5\,mm$) containing pellicular ODS (Jones Chromatography Ltd., Tir-Y-Berth Industrial Estate, New Road, Hengoed, CF8 8AU, Wales) was employed. The mobile phase was methanol–acetic–water (21:4:75 v/ v/v), the flow rate $1.5\,mL\,min^{-1}$, the temperature 50°C. The UV detector was set at 260 nm, the absorption maximum for phenylpropionylglycine. Urine samples were centrifuged and $20\,\mu L$ injected directly onto the analytical column.

RESULTS AND DISCUSSION

With IMPC, phenylpropionylglycine and hippuric acid have relative retention times of 1.116 and 0.815, respectively, with reference to the internal standard, 3-(4-hydroxyphenyl)-propionic acid, with absolute retention times of 32 min for hippurate and 43.9 min for phenylpropionylglycine. With reverse phase chromatography the retention times for hippurate and phenylpropionylglycine were 3.88 and 7.08 min, respectively. Both HPLC methods gave very similar results for the measurement of phenylpropionylglycine. IMPC is the method of choice in our laboratory simply because it is in routine use for the screening of urinary organic acids (Rumsby *et al.*, 1987).

In two normal adult subjects and three adult obligate carriers for MCAD deficiency, and the 23 patients investigated, phenylpropionic acid was completely metabolized to hippuric acid. However, in the eight known cases of MCAD deficiency, a substantial amount of the phenylpropionate was excreted conjugated

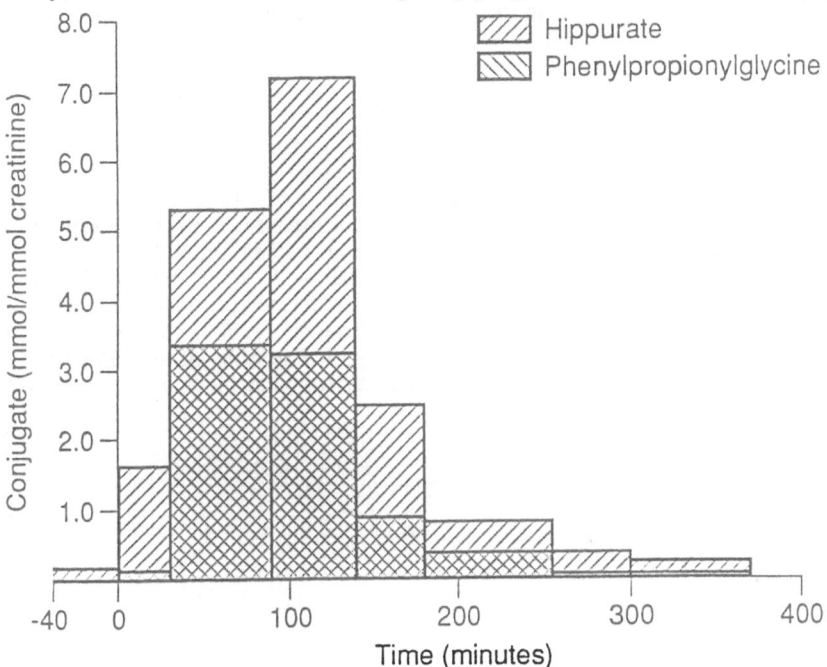

Figure 1 The excretion of phenylpropionylglycine and hippuric acid in a patient with MCAD deficiency following a load of phenylpropionic acid $(25 \, \text{mg} \, \text{kg}^{-1})$

with glycine (Figure 1). The presence of this metabolite was confirmed both by comparison with authentic phenylpropionylglycine and by gas chromatography–mass spectrometry. It is interesting to note that the enzyme block is not complete, for a large part of the phenylpropionic acid was converted to benzoic acid and excreted as hippuric in these subjects. In the patient with MCAD deficiency, illustrated in Figure 1, about two-thirds is excreted as hippuric acid and one-third as phenylpropionylglycine. This is consistent with the clinical features.

The excretion of the glycine conjugates was studied in detail in one patient, the maximum excretion of both conjugates occurred in the period 1.5–3 h and was complete within 6 h (Figure 1). No glucuronide conjugates of benzoic or phenylpropionic acid were detected by HPLC in any post-load urine specimen, indicating that the glycine-conjugating system was not saturated.

The test is simple to administer, does not require fasting and uses readily available HPLC equipment. The test appears to be specific, but its sensitivity is not yet known, and comparisons need to be made with the measurement of octanoylcarnitine which is thought to be a characteristic metabolite of MCAD deficiency (Roe *et al.*, 1985).

ACKNOWLEDGEMENTS

Our thanks to Mr G. Lynes for the mass spectra, to Dr James Leonard for permission to investigate his patients and to Professor Magnus Hjelm for his encouragement.

REFERENCES

Gregersen, N., Lauritzen, R. and Rasmussen, K. Suberylglycine excretion in the urine from a patient with dicarboxylic aciduria. *Clin. Chim. Acta* 70 (1976) 417–425

Gregersen, N., Kølvraa, S., Rasmussen, K., Mortensen, P. B., Divry, P., David, M. and Holboth, N. General (medium chain) acyl-CoA dehydrogenase deficiency (non-ketotic dicarboxylic aciduria): quantitative urinary excretion pattern of 23 biological significant organic acids in three cases. *Clin. Chim. Acta* 132 (1983) 181–191

Raper, H. S. and Wayne, E. J. A quantitative study of the oxidation of phenyl-fatty acids in the animal organism. *Biochem. J.* 22 (1928) 188–197

Roe, C. R., Millington, D. S., Maltby, D. A., Bohan, T. P., Kahler, S. G. and Chalmers, R. A. Diagnostic and therapeutic implications of medium chain acyl-carnitines in the medium-chain acyl-CoA dehydrogenase deficiency. *Pediatr. Res.* 19 (1985) 459–466

Rumsby, G., Belloque, J., Ersser, R. S. and Seakins, J. W. T. Effect of temperature and sample preparation on performance of ion-moderated partition chromatography of organic acids in biological fluids. *Clin. Chim. Acta* 163 (1987) 171–183

Whyte, R. K., Whelan, D., Hill, R. and McClory, S. Excretion of dicarboxylic and $\omega-1$ hydroxy fatty acids by low birth weight infants fed with medium chain triglycerides. *Pediatr. Res.* 20 (1986) 122–125

J. Inher. Metab. Dis. 11 Suppl. 2 (1988) 225–228

Short Communication

Odd-Numbered Long-Chain Fatty Acid Contents in Erythrocyte Membrane Phospholipids in Patients with an Impaired Propionate Utilization

U. Wendel, E. Diekmann and M. D. Laryea

Metabolic Unit, University Children's Hospital, Düsseldorf D-4000 Düsseldorf, Moorenstr. 5, Federal Republic of Germany

Patients with the inherited disorders of propionate and methylmalonate metabolism such as propionic acidaemia (PA; McKusick 23200) and methylmalonic acidurias (MMA; McKusick 25100) accumulate odd-numbered long-chain fatty acids (OLCFA) in lipids of various organs (Hommes *et al.*, 1968; Gompertz, 1970; Gompertz *et al.*, 1970; Kishimoto *et al.*, 1973). The accumulation of OLCFA also occurs in cultured skin fibroblasts of PA and MMA patients (Giudici *et al.*, 1986). This is due to the fact that intracellularly accumulated propionyl-CoA can replace acetyl-CoA as a 'primer' for long-chain fatty acid synthesis.

We studied the fatty acid composition of total lipids and of phosphatidylcholine (PC) and phosphatidylethanolamine (PE) in erythrocyte membranes of patients with PA and MMA. We sought to determine whether there is a relationship between the clinical course of these disorders and the OLCFA contents in erythrocyte phospholipids.

MATERIALS AND METHODS

Blood samples from six patients with propionic acidaemia (age range from 1 month to 22 years) and eight patients with methylmalonic aciduria (age range 2–16 years), and four patients with other organoacidopathies (glutaric aciduria I, $n = 2$; isovaleric acidaemia, $n = 1$; medium chain acyl-CoA dehydrogenase deficiency, $n = 1$) were studied. From six patients, samples were evaluated on two to four occasions at 2 to 6 months intervals.

Venous blood was collected into EDTA tubes and sent to Düsseldorf within 48 h. There, erythrocytes were separated by centrifugation and washed twice with physiological saline containing butylated hydroxytoluene. The packed red cells were stored at −20°C until analysis. The work up of the samples (extraction of membrane lipids, separation into PE and PC by thin-layer chromatography, subsequent base-catalysed transesterification of the fatty acids using sodium methoxide as well as analyses of the fatty acid methyl esters by capillary gas–liquid chromatography) was exactly as we have described previously (Laryea *et al.*, 1988).

225

Journal of Inherited Metabolic Disease. ISSN 0141–8955. Copyright © SSIEM and MTP Press Limited, Queen Square, Lancaster, UK.

Peaks were identified by comparison of retention times with authentic standards and quantified with a Shimadzu C R3A digital integrator.

RESULTS

In subjects with propionic acidaemia and methylmalonic aciduria the proportion of the C 15:0, C 17:0 and C 17:1 fatty acids (expressed as percentage of total fatty acids of chain length 14 to 22) was regularly increased in the phospholipids of erythrocyte membranes. The highest amounts were found in the PC fraction, which usually contains the highest proportion of saturated and mono-unsaturated fatty acids. In all analysed fractions the percentages of these OLCFA were in the order C 17:0>C 15:0>C 17:1. No significant differences between the propionic acidaemia and methylmalonic aciduria patients were observed with respect to the various OLCFA.

In Figure 1a-c the sums of the percentages of the three OLCFA (C 15:0, C 17:0, and C 17:1) are shown for the patients with propionic acidaemia (pa), methylmalonic acidurias (mma) and other organoacidopathies (oa). The values marked by asterisks are from patients with a mild propionic acidaemia variant (Przyrembel *et al.*, 1979) and a cobalamin-responsive methylmalonic aciduria. The values of the oa-patients were within the range of controls.

CONCLUSIONS

The grossly normal OLCFA contents in red-cell membrane phospholipids of patients with unusually mild PA and MMA variants as well as the highly increased and variable OLCFA contents in severely affected and poorly controlled patients might suggest that the OLCFA content is an indicator for the severity of the disorder and the quality of metabolic control. OLCFA synthesis might be proportional to the intramitochondrial/cytoplasmatic propionyl-CoA pool. Thus, the amounts of OLCFA in erythrocyte phospholipids could be a long-term and a more reliable parameter for the evaluation of clinical management in patients with PA and MMA than the sporadic quantification of the organic acids in plasma and urine (in analogy to the glycosylated haemoglobin in diabetics).

Further studies have to elucidate the turnover of the erythrocyte OLCFA as well as the dependence of the OLCFA contents in the different red-cell phospholipid fraction on the plasma lipids.

ACKNOWLEDGEMENT

We would like to thank Dr Böhles, Erlangen; Dr Leonard, London; Dr Leupold, Ulm; Dr Marg, Bremen; Dr Otten, Giessen; Dr Przyrembel, Rotterdam; Dr Schmidt, Heidelberg and Dr Sperl, Innsbruck for providing blood samples from their patients. This work was supported by the Deutsche Forschungsgemeinschaft (We 614/5-1).

total lipids
odd chain fatty acids (%)

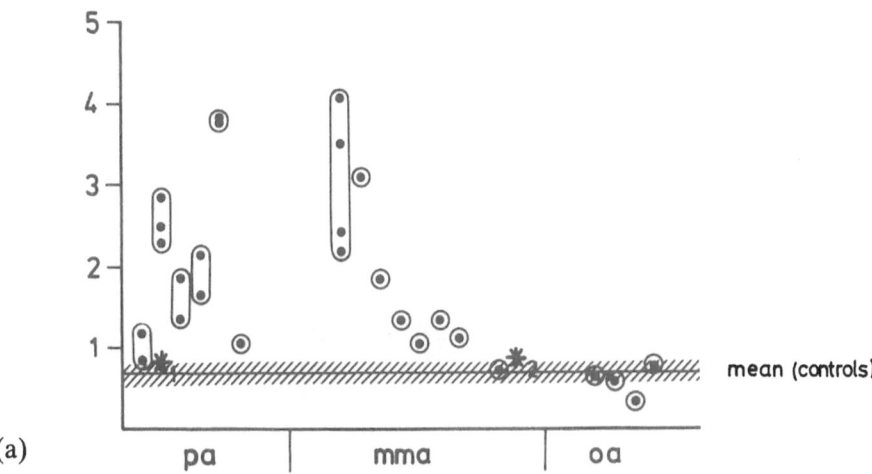

(a)

pe fraction
odd chain fatty acids (%)

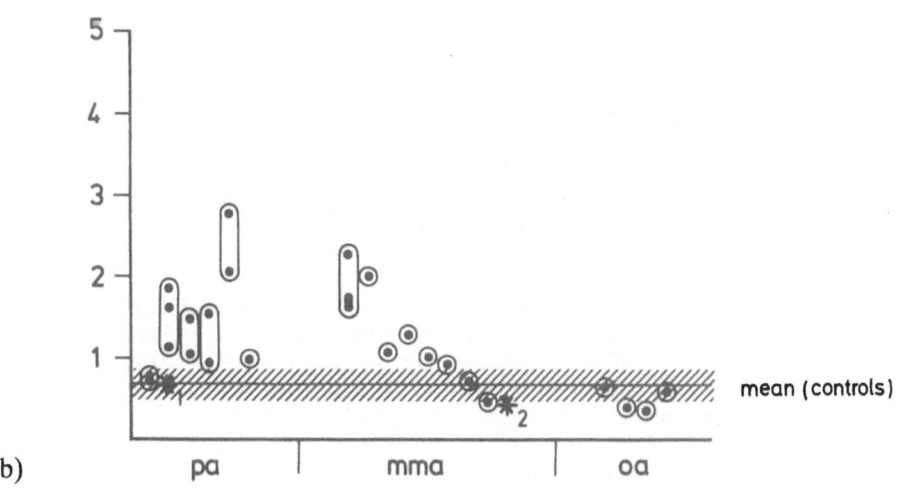

(b)

pc fraction

odd chain fatty acids (%)

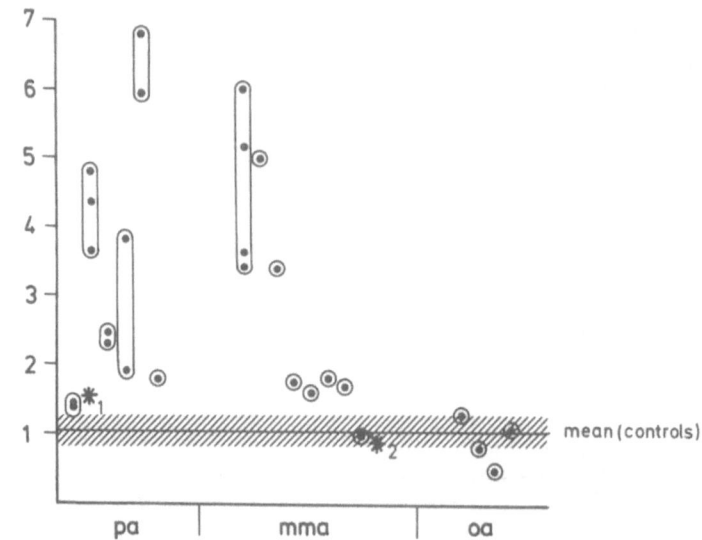

(c)

Figure 1 Sums of the percentages of the OLCFA (C 15:0, C 17:0 and C 17:1) in the phospholipids of erythrocyte membranes for the different patients. (a) Total lipids; (b) PE fractions; and (c) PC fractions of erythrocyte membranes. Pa = propionic acidaemias, mma = methylmalonic acidurias, oa = other organoacidopathies.*1) unusually mild propionic acidaemia variant, 22-years old; *2) cobalamin-responsive methylmalonic aciduria. Vertically arranged marks represent data from one patient on different occasions. Shaded areas: mean±SD for normal controls ($n = 74$)

REFERENCES

Gompertz, D. The distribution of 17-carbon fatty acids in the liver of a child with propionic acidemia. *Lipids* 6 (1970) 576–580

Gompertz, D., Bau, D. C. K., Storrs, C. N., Peters, T. J. and Hughes, E. A. Localisation of enzymic defect in propionic acidemia. *Lancet* 1 (1970) 1140–1143

Giudici, T. A., Chen, R. G., Oizumi, J., Shaw, K. N. F., Ng, W. G. and Donnell, G. N. Methylmalonic and propionic acidemias: lipid profiles of normal and affected human skin fibroblasts incubated with [1–^{14}C] propionate. *Biochem. Med. Metab. Biol.* 35 (1986) 384–398

Hommes, F. A., Kuipers, J. R. G., Elema, J. D., Jansen, J. F. and Jonxis, J. H. P. Propionicacidemia – a new inborn error of metabolism. *Pediatr. Res.* 2 (1968) 519–524

Kishimoto, Y., Williams, M., Moser, H. W., Hignite, Ch. and Biemann, K. Branched-chain and odd-numbered fatty acids and aldehydes in the nervous system of a patient with deranged vitamin B_{12} metabolism. *J. Lipid Res.* 14 (1973) 69–77

Laryea, M. D., Cieslicki, P., Diekmann, E. and Wendel, U. Analysis of the fatty acid composition of erythrocyte phospholipids by a base-catalysed transesterification method – prevention of formation of dimethylacetals. *Clin. Chim. Acta* 171 (1988) 11–18

Przyrembel, H., Bremer, H. J., Duran, M., Bruinvis, L., Ketting, D., Wadman, S. K., Baumgartner, R., Irle, U., and Bachmann, C. Propionyl-CoA carboxylase deficiency with overflow of metabolites of isoleucine catabolism at all levels. *Eur. J. Pediatr.* 130 (1979) 1–14

J. Inher. Metab. Dis. 11 Suppl. 2 (1988) 229–232

Short Communication

Mevalonic Aciduria: Pathobiochemical Effects of Mevalonate Kinase Deficiency on Cholesterol Metabolism in Intact Fibroblasts

G. Hoffmann[1], K. M. Gibson[2], W. L. Nyhan[2] and L. Sweetman[2]

Departments of Pediatrics, [1]University of Göttingen, 3400 Göttingen, Federal Republic of Germany; and [2]University of California, San Diego, La Jolla, CA 92093, USA

Mevalonic aciduria represents the first documented inherited disorder of the pathway for the biosynthesis of cholesterol and non-sterol isoprenes in man (Hoffmann *et al.*, 1986). Two patients have been described (Berger *et al.*, 1985; Hoffmann *et al.*, 1986), whose clinical presentations were very different. A deficiency of mevalonate kinase (ATP: mevalonate-5-phosphotransferase, E.C.2.7.1.36) activity was documented in extracts of cultured fibroblasts and lymphoblasts derived from the index patient and in lymphocytes isolated from whole venous blood (Hoffmann *et al.*, 1986). The level of activity was <5% of controls. In contrast to patients with metabolic defects in catabolic pathways, the index patient with mevalonic aciduria due to mevalonate kinase deficiency did not suffer from recurrent acidosis triggered by excess of dietary nutrients or catabolic crises, but from a chronic multisystemic disease. He produced an enormous amount of mevalonic acid, which was excreted efficiently in the urine.

In mammalian cells, requirements for cholesterol are met through two interacting pathways. The first is the uptake of exogenous cholesterol, which is mediated by the low-density lipoprotein receptor, which is altered in familial hypercholesterolaemia (Brown and Goldstein, 1986). The second and quantitatively more important source of cholesterol is its *de novo* synthesis from acetate via 3-hydroxy-3-methylglutaryl-CoA (HMG-CoA). The synthesis of the first committed intermediate of this pathway, mevalonic acid, is primarily controlled by feedback inhibition by cholesterol of the rate-limiting enzyme HMG-CoA reductase.

METHODS

Monolayers of fibroblasts were cultured in Eagle's minimal essential medium supplemented with 10% fetal bovine serum (FBS). When cells were in late logarithmic growth, the fibroblasts were incubated in Eagle's minimal essential medium supplemented with either 10% FBS or 10% FBS from which lipoproteins and other non-sterol isoprenes had been removed (delipidated FBS) for another 24 h. Then the medium was supplemented with 100 μCi of [2-^{14}C]acetate (1.85 μmol per 10 mL). After 12, 24 and 48 h the cells were harvested by trypsinization and an aliquot

229

Journal of Inherited Metabolic Disease. ISSN 0141–8955. Copyright © SSIEM and MTP Press Limited, Queen Square, Lancaster, UK.

was removed for protein determination by the method of Lowry. [³H]Cholesterol (0.05 μCi) was added as internal standard, and the cells saponified in 75% KOH for 3 h at 80°C. Extraction and separation of the non-saponifiable lipids by thin-layer chromatography was performed as described by Paton and Poulos (1984). Fatty acids were quantified with [9, 10-³H]oleic acid (0.3 μCi) as an internal standard following a modified procedure by Brown *et al.* (1978). Instead of petroleum ether extraction, we prepared chloroform-methanol extracts of the saponifiable lipids (Bligh and Dyer, 1959).

Mevalonic acid was quantified as the lactone or the mono-TMS derivative of the lactone by means of ammonia chemical ionization selected ion monitoring capillary gas chromatography–mass spectroscopy (GC–MS) with [²H₃] mevalonic acid (MSD Isotopes, Pointe-Claire, Dorval, Quebec, Canada) as the internal standard (Hoffmann *et al.*, 1986).

RESULTS AND DISCUSSION

Mevalonate kinase deficient fibroblasts synthesized cholesterol at close to 40% of normal when incubated with [2-¹⁴C]acetate in the presence of FBS (Table 1).

Table 1 Incorporation of [2-¹⁴C] acetate into lipids by intact fibroblasts from a patient with mevalonate kinase deficiency (MK⁻)and normal controls

	Incorporation products	
	Cholesterol	*Fatty acids*
Medium+10% FBS (containing LDL)		
Control A	2.5	167
Control B	2.7	149
MK⁻ cells	1.0	118
MK⁻ as % of control mean	41%	75%
Medium+10% delipidated FBS (no LDL)		
Control A	45	31
Control B	94	39
MK⁻ cells	62	48
MK⁻ as % of control mean	89%	137%

The data shown are the mean calculated from three different experiments at 12, 24 and 48 h of incubation, respectively. All experiments were done in duplicate. Values are dpm h⁻¹ (μg cellular protein)⁻¹

Product formation was linear for up to 48 h of incubation time in all cell lines. FBS is known to supply most of the cholesterol needed by cells in culture. This is reflected in the large increase of cholesterol biosynthesis in all cell lines when incubated in the presence of delipidated FBS (Table 1). Despite the more than 95% reduction in mevalonate kinase activity mevalonate kinase deficient fibroblasts were able to increase their production of cholesterol, when depleted of cholesterol, to levels in the same range as controls. There were no significant differences in rates of fatty-acid synthesis from [2-^{14}C]acetate between the different cell lines under identical culture conditions. The lower fatty-acid synthesis in the presence of delipidated FBS in all cell lines is a result of the reduced cellular growth under these conditions. The delipidation process not only removes low density lipoproteins, but other lipids, probably including growth factors, as well.

HMG-CoA reductase activity in intact cells was estimated by monitoring concentrations of mevalonic acid in culture media by stable isotope dilution GC–MS analysis. Mevalonic acid could not be detected in the culture medium of control cell lines in the presence of FBS nor in the presence of delipidated FBS, to which an excess of 50 μg of non-lipoprotein cholesterol per mL of media was added, and was only barely detectable after incubation in delipidated serum. In contrast, mevalonate kinase deficient cells produced large amounts of mevalonic acid even in the presence of exogenous cholesterol. This could be consistent with a diminution in the feedback inhibition of HMG-CoA reductase resulting from diminished production of the different end products, i.e. cholesterol and non-sterol isoprenes (Brown and Goldstein, 1986). Mevalonate kinase deficient cells may have to produce large amounts of mevalonic acid even in the presence of exogenous cholesterol to insure the production of other non-sterol isoprenes. In addition, accumulation of large amounts of mevalonic acid could overcome a decreased affinity and permit the pathway to function. Mevalonate kinase deficient cells produced 52 fmol of mevalonic acid per cell, when incubated in the presence of FBS, 454 fmol of mevalonic acid per cell in the presence of delipidated FBS, and 57 fmol of mevalonic acid per cell in the presence of delipidated FBS, to which an excess of 50 μg of non-lipoprotein cholesterol per mL of media had been added. The very similar results obtained with FBS and with delipidated FBS, to which non-lipoprotein cholesterol had been added, showed that FBS supplies all of the cholesterol needed by cells in culture. Removal of cholesterol from the culture medium resulted in a further ninefold increase in mevalonic acid production by mevalonate kinase deficient cells, which was reversible by the addition of exogenous cholesterol. Apparently, HMG-CoA reductase is regulated in the deficient cells in a similar fashion as that known from control cell lines by direct enzyme assay (Brown *et al.*, 1978; Brown and Goldstein, 1986).

In summary, even a more than 95% reduction in activity of the enzyme mevalonate kinase only moderately reduces the flow through the entire pathway of cholesterol biosynthesis *in vitro*. The enzyme deficiency appears to be balanced in the cell by the expression of a high activity of HMG-CoA reductase, which markedly increases the concentration of mevalonic acid, the substrate for the deficient enzyme. Excess mevalonic acid then diffuses out of the cell and accumulates in the culture medium and *in vivo* in the patient.

ACKNOWLEDGEMENTS

The authors are grateful to Charles D. Johnson for maintenance of the fibroblast cultures. This work was supported in part by US Public Health Service grant HD04608. Dr Hoffmann is supported by the Deutsche Forschungsgemeinschaft (Ho 966/1-1 and Ho 966/2-1). Dr Gibson is the recipient of a Basil O'Connor Starter Scholar Award No. 5-565 from the March of Dimes Birth Defects Foundation.

REFERENCES

Berger, R., Smit, G. P. A., Schierbeek, H., Bijsterveld, K. and le Coultre, R. Mevalonic aciduria: an inborn error of cholesterol biosynthesis? *Clin. Chim. Acta* 152 (1985) 219–222

Bligh, E. G. and Dyer, W. J. A rapid method of total lipid extraction and purification. *Can. J. Biochem. Physiol.* 37 (1959) 911–917

Brown, M. S., Faust, J. R., Goldstein, J. L., Kanedo, I. and Endo, A. Induction of 3-hydroxy-3-methylglutaryl coenzyme A reductase activity in human fibroblasts incubated with compactin (ML-236B), a competitive inhibitor of the reductase. *J. Biol. Chem.* 253 (1978) 1121–1128

Brown, M. S. and Goldstein, J. L. A receptor-mediated pathway for cholesterol homeostasis. *Science* (NY) 232 (1986) 34–47

Hoffmann, G., Gibson, K. M., Brandt, I. K., Bader, P. I., Wappner, R. S. and Sweetman, L. Mevalonic aciduria: An inborn error of cholesterol and nonsterol isoprene biosynthesis. *N. Engl. J. Med.* 314 (1986) 1610–1614

Paton, B. C. and Poulos, A. Dolichol metabolism in cultured skin fibroblasts from patients with 'neuronal' ceroid lipofuscinosis (Batten's disease). *J. Inher. Metab. Dis.* 7 (1984) 112–116

J. Inher. Metab. Dis. 11 Suppl. 2 (1988) 233–236

Short Communication

A Patient with Mevalonic Aciduria Presenting with Hepatosplenomegaly, Congenital Anaemia, Thrombocytopenia and Leukocytosis

J. B. C. de Klerk, M. Duran, L. Dorland, H. A. A. Brouwers, L. Bruinvis and D. Ketting

University Children's Hospital 'Het Wilhelmina Kinderziekenhuis', Nieuwe Gracht 137, NL-3512 LK Utrecht, The Netherlands

Mevalonic aciduria is an inborn error in cholesterol biosynthesis, due to a deficiency of mevalonate kinase (EC 2.7.1.36), which has recently been discovered. So far two patients have been reported (Berger *et al.*, 1985; Hoffmann *et al.*, 1986).

The patient described by Hoffmann *et al.* (1986) died at the age of 2 years, while the other patient (Berger *et al.*, 1985) is doing rather well. There are some digestive problems due to a low production of bile salts (Koopman *et al.*, 1988). Both patients were reported to have had a rather low serum cholesterol concentration. In this paper we present a third patient with mevalonic aciduria, who died at the age of 5 months. Cholesterol biosynthesis appeared not to be affected.

CASE REPORT

The patient, a girl, was born by caesarean section for decelerating cardiotocogram after a 36 weeks gestation. Her birth weight was 2200 g. Her parents were unrelated. A previous first pregnancy ended at 30 weeks with a stillbirth.

Haematological parameters such as anaemia, severe thrombocytopenia and leukocytosis were suggestive of a congenital infection, which could be excluded, however. Physical examination showed massive hepatosplenomegaly and minor dysmorphic features (micrognathia). Cataracts could not be observed. Echography of the skull revealed enlarged ventricles and signs of ventriculitis. She showed a failure to thrive with gastrointestinal manifestations without positive cultures. On a special formula with casein hydrolysate she managed to reach 2900 g at $3\frac{1}{2}$ months of age. At 4 months she developed a bronchiolitis with overexpansion of both lungs, and needed artificial ventilation. She died suddenly in shock at the age of $4\frac{1}{2}$ months.

Laboratory data: Haemogram at birth: haemoglobin $6.1\,\text{mmol}\,\text{L}^{-1}$ leukocytes $56.4 \times 10^9\,\text{L}^{-1}$; thrombocytes $43 \times 10^9\,\text{L}^{-1}$. There was no metabolic acidosis. Serum cholesterol varied from $1.7–3.4\,\text{mmol}\,\text{L}^{-1}$ (age-matched controls 1.2–4.4). Lathosterol was $1.45\,\mu\text{mol}\,\text{L}^{-1}$. The ratio of plasma lathosterol/cholesterol was 0.71

Journal of Inherited Metabolic Disease. ISSN 0141-8955. Copyright © SSIEM and MTP Press Limited, Queen Square, Lancaster, UK.

(controls 0.70±0.36), an index of normal cholesterol biosynthesis (personal communication, H. J. M. Kempen). Plasma lactate was normal: 1.0–1.9 mmol L^{-1}. Glycoprotein biosynthesis indirectly measured by total sialated protein was normal (1.0 mg mL^{-1}; control 0.54 mg mL^{-1}).

Humoral and cellular immunological parameters were in the normal range for age. There were no autoimmune antibodies against thrombocytes. Chromosomal analysis of bone marrow and peripheral lymphocytes was normal (46; XX).

MATERIALS AND METHODS

Organic acids were extracted from urine, and analysed by gas–liquid chromatography/mass spectrometry as their trimethylsilyl derivatives (Duran *et al.*, 1985). Under these conditions mevalonic acid was converted to its lactone for the major part (Figure 1). Mevalonate kinase activity was determined in lysates of cultured fibroblasts according to the method described by Hoffman *et al.* (1986).

RESULTS AND DISCUSSION

Screening for inherited metabolic disorders in this sick neonate showed a massive mevalonic aciduria. The urinary excretion values ranged from 5.8 to 18 mol (mol creatinine)$^{-1}$ and showed a steady increase during the patient's life. Compared with the previously reported patients, these excretion values may be considered 'intermediate'. Also plasma mevalonate increased during life with a starting value of 70 μmol L^{-1} and a final value of 542 μmol L^{-1} in a perimortal sample. Surprisingly there was rather high residual activity of mevalonate kinase in the patient's fibroblast lysates: 0.27 mmol min^{-1} (mg protein)$^{-1}$ (controls 2.44±0.71). This is in contrast with Berger's patient, who had only 2% residual activity, but produced considerably less mevalonic acid (Gibson *et al.*, 1987). A severely decreased activity of mevalonate kinase does not necessarily lead to a decreased plasma cholesterol concentration; the normal values in our patient were comparable to those observed by Hoffmann *et al.* (1986).

Mevalonate is a precursor for ubiquinone, a component of the respiratory chain, and of dolichol, which is essential for the biosynthesis of glycoproteins. As the present patient did not have lactic acidaemia or a deficiency of glycoproteins, we conclude that mevalonate kinase deficiency does not lead to gross secondary abnormalities in these areas of metabolism. Clinically our patient resembled the patient described by Hoffmann *et al.* (1986) in several aspects, i.e. they both had a marked anaemia and appeared to have multiple infections without a causing agent ever being found. Our patient neither developed cataracts nor did she have obvious microcephaly. Gastrointestinal problems occurred in all three patients; this could be related to a reduction of bile acid production and thus be related to the basic defect.

It is tempting to speculate that abnormalities of cell membranes will be the consequence of this defect. So far preliminary investigations have not shown

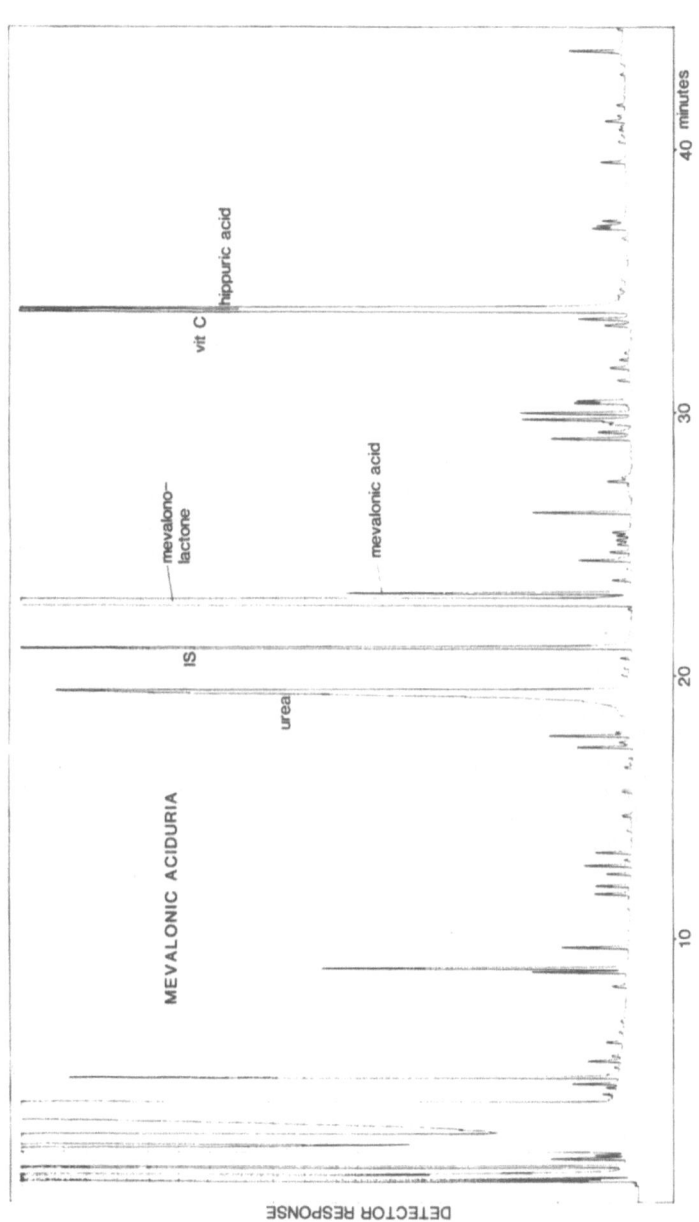

Figure 1 Gas chromatogram of urinary organic acids as TMS-derivatives. IS: internal standard 3-phenylbutyric acid; vitamin C is routinely added as an antioxidant

unequivocal results (Hübner *et al.*, 1986). Thus further investigations in this direction will be needed.

ACKNOWLEDGEMENTS

We want to express our thanks to the following persons: Dr K. M. Gibson (University of California, San Diego) for the assay of mevalonate kinase; Dr W. J. Kleijer (Erasmus University, Rotterdam) for his help in culturing fibroblasts; Dr H. J. M. Kempen (Gaubius Instituut TNO, Leiden) for measuring the lathosterol content, and Professor Dr D. van den Eijnden (Free University, Amsterdam) for his assay of the glycoprotein.

REFERENCES

Berger, R., Smit, G. P. A., Schierbeek, H., Bijsterveld, K. and le Coultre, R. Mevalonic aciduria: an inborn error of cholesterol biosynthesis? *Clin. Chim. Acta* 152 (1985) 219–222

Duran, M., Ketting, D., van Vossen, R., Beckeringh, T. E., Dorland, L., Bruinvis, L. and Wadman, S. K. Octanoylglucuronide excretion in patients with a defective oxidation of medium-chain fatty acids. *Clin. Chim. Acta* 152 (1985) 253–260

Gibson, K. M., Hoffmann, G., Nyhan, W. L., Sweetman, L., Brandt, I. K., Wappner, R. S. and Bader, P. I. Mevalonic aciduria: family studies in mevalonate kinase deficiency, an inborn error of cholesterol biosynthesis. *J. Inher. Metab. Dis.* 10 Suppl. 2 (1987) 282–285

Hoffmann, G., Gibson, K. M., Brandt, I. K., Bader, P. I., Wappner, R. S. and Sweetman, L. Mevalonic aciduria – An inborn error of cholesterol and nonsterol isoprene biosynthesis. *N. Engl. J. Med.* 314 (1986) 1610–1614

Hübner, C., Hoffmann, G., Kohlschütter, A., Hermanussen, M., Gibson, K. M. and Sweetman, L. Mevalonic aciduria. Membrane composition and fluidity in an inborn error of cholesterol synthesis. *Pediatr. Res.* 20 (1986) 1044

Koopman, B. J., Kuipers, F., Bijleveld, C. M. A., van der Molen, J. C., Nagel, G. T., Vonk, R. J. and Wolthers, B. G. Determination of cholic acid and chenodeoxycholic acid pool sizes and fractional turnover rates by means of stable isotope dilution technique, making use of deuterated cholic acid and chenodeoxycholic acid. *Clin. Chim. Acta* (1988) (in press)

J. Inher. Metab. Dis. 11 Suppl. 2 (1988) 237–239

Short Communication

A Closer Look at the Eye in Homocystinuria: A Screened Population

J. P. BURKE[1], M. O'KEEFE[1], R. BOWELL[1] and E. R. NAUGHTEN[2]
[1]*Department of Ophthalmology and* [2]*The Metabolic Unit, The Children's Hospital, Temple Street, Dublin 1, Ireland*

Homocystinuria (HCU) due to cystathionine-β-synthetase deficiency (McKusick 23620) was initially detected in 1962 (Carson and Neill) and the underlying defect was defined in 1964 (Mudd *et al.*, 1985). The incidence in Ireland is 1 in 52 000 live births. Ectopia lentis occurs in 97% of untreated cases by age 39 years (Mudd *et al.*, 1985). Early detection by neonatal screening programmes with dietary and/or vitamin therapies have improved the prognosis. We describe the ocular findings in 19 patients and correlate these with age at commencement of therapy and with biochemical control.

MATERIALS AND METHODS

Nineteen patients (10 males, nine females) with cystathionine-β-synthetase deficiency were studied. Fourteen patients (Group 1) were on treatment from the first 6 weeks of life and five patients (Group 2) commenced treatment in childhood. The Guthrie microbiological inhibition assay for methionine was used for screening and the diagnosis was confirmed by estimation of plasma homocyst(e)ine, methionine and cyst(e)ine levels (Spackman *et al.*, 1958). Each patient had a detailed slit-lamp ophthalmic examination. Historical information was obtained from the medical and ophthalmic records. Control was defined as poor, fair or good according to the plasma homocyst(e)ine and cyst(e)ine levels (Figure 1).

RESULTS

The mean age of commencement of treatment in Group 1 (14 patients) was 22.7 days (range 7–42 days) and the mean follow-up period was 8.2 years (range 3.5–15.1 years). Three patients were poorly controlled with mean plasma homocyst(e)ine $>15\,\mu\text{mol}\,\text{L}^{-1}$ (Figure 1). Two of these patients aged 8 and 15.2 years, respectively, had high myopia. The third patient, aged 3.8 years, had a spontaneously-resolving episode of idiopathic bilateral optic-disc oedema. The remaining 11 patients did not have myopia nor ectopia lentis. Two of these were partially pyridoxine responsive.

The mean age of commencement of treatment in Group 2 (five patients) was 3.7 years (range 1.5–7.0 years). The mean age of follow-up was 9 years. Three patients aged 2.9, 4.8 and 7 years, respectively, presented with ectopia lentis. The last

237

Journal of Inherited Metabolic Disease. ISSN 0141–8955. Copyright © SSIEM and MTP Press Limited, Queen Square, Lancaster, UK.

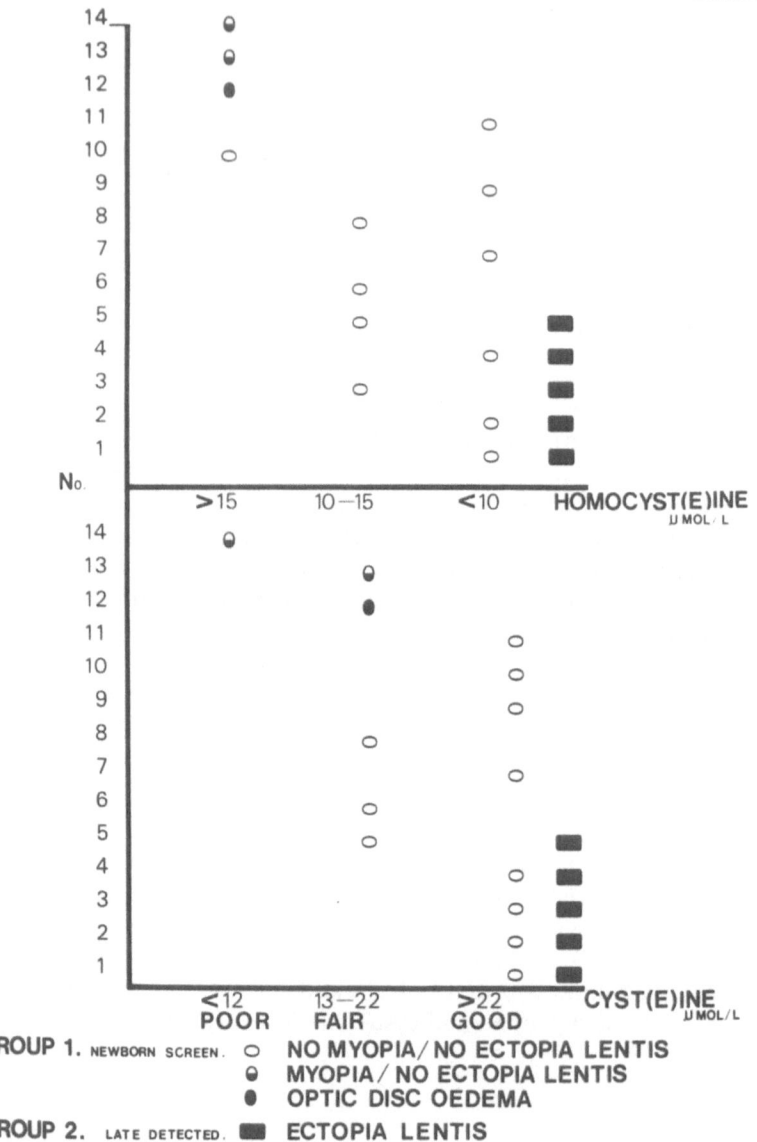

Figure 1 Eye findings related to biochemical control in homocystinuria

patient was pyridoxine responsive and does not require diet. One patient presented at 1.5 years with an intracranial haemorrhage and developed ectopia lentis within 6 months. The fifth patient was diagnosed at 2.4 years following identification of HCU in a sibling on neonatal screening. This patient developed unilateral ectopia lentis within 1 year of diagnosis. Ectopia lentis progressed or developed in all patients after diagnosis despite good biochemical control (mean plasma homocyst(e)ine $<10\,\mu$mol L^{-1}; mean plasma cyst(e)ine $>22\,\mu$mol L^{-1}).

DISCUSSION

Ectopia lentis preceded by lenticular myopia is the ocular hallmark of untreated homocystinuria. Currently, elevated homocyst(e)ine or low cyst(e)ine (Nelson and Maumenee, 1982) are thought to be responsible for zonular disintegration leading to myopia and lens subluxation. Our findings suggest that ectopia lentis, once present, progresses inexorably despite good biochemical control. This agrees with the findings of other authors (Mudd *et al.*, 1985).

The outcome in late-detected cases where treatment is initiated prior to development of ectopia lentis is unclear. The two patients in group 2 who fall into this category subsequently developed ectopia lentis despite good biochemical control (Figure 1). Boers (1985), however, has followed five such patients for periods of 6 months to 15 years and all have normal lenses.

The ophthalmic benefit of treatment from the newborn period is uncertain because of the small numbers of patients studied to date. The probability of lens dislocation in pyridoxine-unresponsive patients who are untreated for 8 years is said to be 75% (Mudd *et al.*, 1985). It is also suggested that treatment may only delay rather than prevent lens dislocation.

To date, none of the 14 patients in our series detected by screening have developed ectopia lentis (mean follow-up 8.2 years). Three of our patients in group 1 were poorly controlled (Figure 1) and two of these, whose treatment started at 42 days, have recently become myopic. Long-term follow-up on this group of patients will help clarify the history of the treated *versus* the untreated patient.

ACKNOWLEDGEMENT

We wish to thank Dr S. Cahalane and his staff for their excellent laboratory service and I. Saul for her help with this study.

REFERENCES

Boers, G. H. H., *Homocystinuria, homozygosity versus heterozygosity*. MD Thesis DRUK, LCG Printing BV, Dordrecht, 1985

Carson, N. A. J. and Neill, D. N. Metabolic abnormalities detected in a survey of mentally backward individuals in Northern Ireland. *Arch. Dis. Child.* 37 (1962) 505–513

Mudd, S. H., Finkelstein, J. D., Irreverre, F. and Laster, L. Homocystinuria. An enzymatic defect. *Science* 143 (1964) 1443–1445

Mudd, S. H., Skovby, F., Levy, H. L., Pettigrew, K. D., Wilcken, B., Pyeritz, R. E., Andria, G., Boers, G. H. J., Bromberg, I. L., Cerone, R., Fowler, B., Gröbe, H., Schmidt, H. and Schweitzer, L. The natural history of homocystinuria due to cystathio-nine-β-synthase deficiency. *Am. J. Hum. Genet.* 37 (1985) 1–31

Nelson, L. B. and Maumenee, I. H. Ectopia lentis, *Survey Ophthalmol.* 27 (1982) 143–160

Spackman, D. H., Stein, W. H. and Moore, S. Automatic recording apparatus for use in the chromatography of amino acids, *Anal. Chem.*, 30 (1958) 1190–1206

J. Inher. Metab. Dis. 11 Suppl. 2 (1988) 240–242

Short Communication

Peptiduria Presumably Caused by Aminopeptidase-P Deficiency. A New Inborn Error of Metabolism

N. Blau[1], A. Niederwieser[1]† and D. H. Shmerling[2]
[1]Divisions of Clinical Chemistry and [2]Gastroenterology, Department of Pediatrics, University of Zürich, 8032 Zürich, Switzerland

Aminopeptidase-P (APP; EC 3.4.11.9) is an exopeptidase highly specific in cleaving the *N*-terminal imino bond in oligopeptides of an Xaa-Pro-sequence (Yaron and Berger, 1970). Active participation of APP in collagen metabolism, as well as a role in inactivation of biologically active peptides, has been proposed (Fleminger and Yaron, 1983); however its real biological function is still unknown. We report on a 5-month-old boy with epilepsy, and his healthy 5-year-old brother, both of whom were found to excrete massive quantities of imino-oligopeptides, amongst which Gly-Pro-4Hyp-Gly (GPOG) was the most prominent. With the exception of the tetrapeptide, all other peptides can also be detected in lower concentrations in the urines of healthy children. Specific GPOG-aminopeptidase activity in a small-intestinal biopsy of the healthy boy was absent or markedly decreased as compared with controls, while saccharase activity in the same biopsy was normal. An exogenous source of the tetrapeptide was excluded by a gelatin loading test. Because the enzyme system found in dog and human small intestine seems to be identical with APP we propose an APP deficiency in these brothers with and without clinical symptoms. This is, beside the defect of prolidase (EC 3.4.13.9) deficiency (Buist *et al.*, 1972), the second known peptidase deficiency.

CASE REPORT

The patient (M.F.), a male infant born at term after an uncomplicated pregnancy, is the third child of healthy and unrelated parents. He came to our clinic at the age of 5 months because of epilepsy, severe developmental retardation, microcephaly, and episodes of eye rolling. His EEG was pathological, and the striking biochemical finding was intense yellow spots on the high-voltage electropherogram of urinary amino acids (Steiner and Niederwieser, 1979). The healthy 5-years-older brother (M.L.), but not the 9-years-older sister or parents, shows the same biochemical abnormality.

MATERIALS AND METHODS

Urinary amino acids were analysed by high-voltage electrophoresis and ion-exchange chromatography. Hydroxyproline (Hyp) was determined according to the

Journal of Inherited Metabolic Disease. ISSN 0141–8955. Copyright © SSIEM and MTP Press Limited, Queen Square, Lancaster, UK.

method of Bergmann and Loxley (1970). High performance liquid chromatography (HPLC) of peptides was performed on a Lichrosorb RP8 column (4.6 × 250 mm) using 10 mmol L^{-1} sodium perchlorate in 0.1% (v/v) phosphoric acid containing 0.3% (v/v) acetonitrile as the mobile phase (Blau, 1984). Gas chromatography–mass spectrometry of peptides was performed with trifluoroacetyl-trifluoroethyl esters using a 20 m OV-1 glass capillary column in a Finnigan gas chromatograph coupled with a Finnigan 3200 mass spectrometer as described elsewhere (Blau, 1984). GPOG was isolated from patients' urines by column chromatography on Dowex 50W, preparative high-voltage electrophoresis and HPLC. Determination of saccharase activity in small-intestinal biopsies was performed according to Auricchio *et al.* (1963). Assay of GPOG specific aminopeptidase in small-intestinal biopsies was performed using GPOG as a substrate by HPLC monitoring of substrate cleavage and the formation of glycine and Pro-4Hyp-Gly (Blau, 1984). Peroral biopsies from the ligament of Treitz or from the first jejunal loop were homogenized in 30 volumes of ice-cold 0.9% (w/v) sodium chloride. The incubation mixture contained 17 mmol L^{-1} sodium tetraborate (pH 8.0), 8 mmol L^{-1} manganese chloride, 0.117 mmol L^{-1} GPOG, and 0.1 mL dialysed homogenate in a total volume of 0.25 mL. After incubation for 3 h at 37°C the reaction was stopped by heating the sample at 95°C for 5 min. After centrifugation, 10 μL of clear supernatant were analysed by HPLC.

RESULTS AND DISCUSSION

The following peptides were found in the urine of the brothers M.F. and M.L. with suspected peptiduria: Gly-Pro-4Hyp-Gly, Gly-Pro-4Hyp, Gly-3Hyp-4Hyp, Pro-4Hyp and Gly-Pro. With the exception of the tetrapeptide GPOG, these peptides were also detected in urine from the parents and the 9-years-older sister in, however, much lower concentrations. The structure of GPOG was derived from amino acid analysis of the hydrolysate and the products of Edman degradation, as well as from fast atom bombardment mass spectra of the isolated tetrapeptide. Excretion of GPOG in patient M.F. seemed to be age-dependent, decreasing from 145 mmol (mol creatinine)$^{-1}$ at the age of 7 months to 17 mmol (mol creatinine)$^{-1}$ at the age of 7 years. To determine the exogenous source of tetrapeptide, GPOG and bound Hyp were measured in the urine of M.F. following loading with 20 g gelatin (18.4 mmol peptide-bound Hyp). The excretion of GPOG remained unchanged at 12–24 mmol (mol creatinine)$^{-1}$, whereas the excretion of peptide-bound Hyp increased from 150 to 976 mmol (mol creatinine)$^{-1}$. Patient M.F., as well as age-matched controls, excreted 3–4% of the total Hyp ingested. This is about 15 times less than expected in the case of prolidase deficiency. From these data, and from measurement of prolidase activity in erythrocytes, we excluded a prolidase deficiency and concluded that GPOG was from an endogenous source.

The enzyme system found in the brush border of both dog and human small intestine splits the tetrapeptide GPOG into glycine and the tripeptide Pro-4Hyp-Gly. It was found that membrane-bound enzyme required manganese cations and a pH of 8.0 for optimal activity and seemed to be identical to APP. No such activity was detected in leukocytes or erythrocytes. APP activity in the small-intestinal biopsies of the healthy boy, M.L., was markedly decreased or absent as compared

Table 1 Aminopeptidase-P and saccharase activities in small-intestinal biopsies of M.L. and healthy controls

Sample	Age	Aminopeptidase-P	Saccharase
Patient M.L.	15	<1	30
Controls	1–17	49	70
(n = 17)		30–68	11–250

Aminopeptidase-P units of activity are μmol GPOG cleaved h^{-1} (g protein)$^{-1}$
Saccharase units are μmol saccharose degraded min^{-1} (g protein)$^{-1}$

with controls (Table 1), while saccharase activity in the same biopsy was normal. The normal values for APP in the small intestine were established in 17 controls and were found to correlate well with intestinal saccharase activity.

These findings indicate an APP deficiency in the brothers with and without clinical symptoms. The finding of this new inborn error of metabolism will have no consequence for the therapy of the epileptic child. However, the perception that a highly specific peptidase is required for the digestion of a simple peptide like GPOG will, we hope, stimulate further investigations into the role of APP in collagen and endorphin degradation.

ACKNOWLEDGEMENTS

We are grateful to Dr Peter Royce for reading the manuscript. This work was supported by the Swiss National Science Foundation, project no. 3.266-0.82.

REFERENCES

Auricchio, S., Rubino, A., Tosi, R., Semenza, S., Landolt, M., Kistler, H. and Prader, A. Disaccharidase activities in human intestinal mucosa. *Enzymol. Biol. Clin.* 3 (1963) 193–208

Bergman, I. and Loxley, R. The determination of hydroxyproline in urine hydrolysates. *Clin. Chim. Acta* 27 (1970) 347–349

Blau, N. Zwei neue angeborene Stoffwechseldefekte: Gly-Pro-Hyp-Gly-Aminopeptidase-und GTP-Cyclohydrolase I-Mangel. Doctoral Thesis, University of Zürich, 1984

Buist, N. R. M., Strandholm, J. J., Bellinger, J. F. and Kennaway, N. G. Further studies on a patient with iminopeptiduria: a probable case of prolidase deficiency. *Metabolism* 21 (1972) 1113–1123

Fleminger, G. and Yaron, A. Sequential hydrolysis of proline-containing peptides with immobilized aminopeptidase. *Biochim. Biophys. Acta* 743 (1983) 437–446

Steiner, W. and Niederwieser, A. Peptide analysis as amino alcohols by gas chromatography–mass spectrometry. Application to hyperoligopeptiduria. Detection of Gly-3Hyp-4Hyp and Gly-Pro-4Hyp-Gly. *Clin. Chim. Acta* 92 (1979) 431–441

Yaron, A. and Berger, A. Aminopeptidase-P. In Perlmann, G. E. and Lorand, L. (eds.) *Methods in Enzymology*, Academic Press, New York, 1970, pp. 521–534

J. Inher. Metab. Dis. 11 Suppl. 2 (1988) 243–245

Short Communication

Early Morning Urine Galactitol Levels in Relation to Galactose Intake: A Possible Method of Monitoring the Diet in Galactokinase Deficiency

J. T. ALLEN[1], J. B. HOLTON[1], A. C. LENNOX[1] and I. C. HODGES[2]
[1]Clinical Chemistry Department, Southmead Hospital, Bristol, BS10 5NB;
[2]Paediatric Department, East Glamorgan Hospital, Church Village, nr Pontypridd, CF38 1AB, UK

Galactokinase (EC 2.7.1.6) deficiency is a rare disorder of galactose metabolism affecting the conversion of galactose to galactose-1-phosphate (McKusick 23020). This causes an almost total inability to metabolize galactose by the main pathway and, consequently, it accumulates in body fluids and tissues, as does galactitol, a reduction product of galactose. The concentration of the latter compound within the eye lens is thought to produce the nuclear cataracts which are usually noticed in early infancy and are the only clinical manifestation of the disorder.

Treatment for galactokinase deficiency is by a galactose-restricted diet. This should prevent progression of the disease and may produce a regression of cataracts already present. As no method of monitoring the diet has been described, we have explored the use of urinary galactitol measurement for this. Some investigations of urinary galactitol levels in classical galactosaemia have been reported (Roe et al., 1973), but they seemed to give only an insensitive and short-term indication of galactose intake.

CASE HISTORIES

The subjects of this study were R.O. and her younger sister. R.O. was born to unrelated parents after an uncomplicated pregnancy, labour and delivery. She was bottle fed on a standard, lactose-containing, formula and there were no apparent neonatal problems until urine analysis, as part of a screening programme, revealed abnormal galactose levels. However, galactose-1-phosphate uridyl transferase deficiency screening on blood spots was negative. R.O. was first seen in the paediatric clinic at 6 weeks when she was found to be healthy apart from the presence of bilateral cataracts. Erythrocyte galactokinase activity was found to be negligible and she was placed immediately on Formula S Baby Milk (soya-based and lactose-free). After remaining on the galactose-free diet for 21 months the lens opacities were still present although much smaller than they were when seen originally and her vision and overall development appeared normal.

Journal of Inherited Metabolic Disease. ISSN 0141–8955. Copyright © SSIEM and MTP Press Limited, Queen Square, Lancaster, UK.

The pregnancy and birth of the younger sister were also uneventful and she was placed immediately on a lactose-free diet. Galactokinase deficiency was confirmed, using erythrocytes, at 1 week of age.

METHODS AND RESULTS

Galactitol was measured in aliquots of urine which were dried and to which perseitol had been added as internal standard (Pettit *et al.*, 1980). The hexa-acetate derivatives of the hexitols were separated by gas chromatography on a fused silica capillary column (25 m × 0.22 mm i.d.) coated with Cp Sil 5 (Chrompack, Middleburg, Netherlands). The identity of the galactitol peak was confirmed by gas chromatography–mass spectrometry. Urine creatinine was measured on a Technicon RA1000.

Random urines were collected from R.O. during two consecutive 24-h periods in order to assess diurnal variation in galactitol excretion on a galactose-restricted diet. In two further studies a supplement of 1 g of galactose was added to three separate feeds throughout a day and random urines were collected during the loading day and subsequent days. Inspection of the data suggested that the galactitol/creatinine ratio was the most useful measure of galactitol excretion. In the first

Table 1   Galactitol excretion in first morning urines

Period	No. of urine	Galactitol[a] Mean	Galactitol[a] Range
Galactose-free diet	10	303	279–333
Post 3 g galactose load			
Day 1	2	564	560–567
Day 2	1	490	
Day 3	1	436	
Day 4	1	405	

[a] Values are mmol (mol creatinine)$^{-1}$

morning this ratio was very constant on a galactose-restricted diet. Table 1 shows the range found for this parameter and the significant increases occurring in the ratio of the first morning urine for up to 4 days following the galactose-loading day.

At 1 week of age the affected sibling gave a first morning urine galactitol/creatinine ratio of 127 mmol mol^{-1}. This was much higher than values found in normal infants but lower than the range established in her sister.

CONCLUSION

The results suggest that the galactitol:creatinine ratio in a first morning urine gives a sensitive and retrospective indication of galactose intake in galactokinase deficient subjects and this could be useful in monitoring their diets. More work is needed to determine whether an acceptable range could be established generally applicable to all patients, but it seems probable that this would be age-dependent.

REFERENCES

Pettit, B. R., King, G. S. and Blau, K. The analysis of hexitols in biological fluid by selected ion monitoring. *Biomed. Mass Spectrom.* 7 (1980) 309–313

Roe, T. F., Ng, W. G., Bergren, W. R. and Donnell, G. N. Urinary galactitol in galactosemic patients. *Biochem. Med.* 7 (1973) 266–273

J. Inher. Metab. Dis. 11 Suppl. 2 (1988) 246–248

Short Communication

Cataracts in Children with Classical Galactosaemia and in their Parents

J. P. BURKE[1], M. O'KEEFE[1], R. BOWELL[1] and E. R. NAUGHTEN[2]
[1]*Department of Ophthalmology and* [2]*The Metabolic Unit, The Children's Hospital, Temple Street, Dublin 1, Ireland*

Classical galactosaemia (McKusick 23040) is an autosomal recessive multisystem disorder caused by a deficiency of galactose-1-phosphate uridyl transferase (GPUT) (Kalckar *et al.*, 1956). The incidence in Ireland is 1:30000 live births (Cahalane and Naughten, 1983). Early detection and treatment may prevent the development of its major complications: cataracts, jaundice, hepatomegaly and mental retardation (Donnell and Bergren, 1975). The rationale for newborn screening is based on this experience. The reports of an association between GPUT heterozygosity and an increased risk of visually significant pre-senile cataracts are conflicting (Wilson and Donnell, 1958; Skalka and Prchal, 1978) in a group of patients in whom therapeutic implications may be important. We describe the ophthalmic findings in a group of children with classical galactosaemia (homozygous) and their parents (obligate heterozygotes).

MATERIALS AND METHODS

Eighteen children with classical galactosaemia (11 females, seven males) and 23 (obligate heterozygote) parents were studied. The *E. coli* metabolite inhibition assay was used for screening. The Beutler erythrocyte transferase assay (Beutler and Baluda, 1966), plasma galactose (normal = 0–240 μmol L^{-1}) and erythrocyte galactose–1–phosphate levels (normal = 5–10 μg (mL packed cells)$^{-1}$; homozygote = 30–45 μg (mL packed cells)$^{-1}$) are used for confirmation and in high-risk screening. Each case underwent a detailed ophthalmic examination. The results were correlated with age at commencement of treatment and with biochemical control. Historical information was obtained from the medical and ophthalmic records.

RESULTS

Children

Eleven had no lens opacities detected using slit-lamp biomicroscopy (Group 1). Seven patients had lens opacities (Group 2, Table 1). The mean age at commencement of dietary treatment in Group 1 was 5.8 days (range 1–11 days) and the mean period of follow up was 6.7 years (range 1–15.3 years). At diagnosis, six cases were asymptomatic and five had systemic complications (vomiting, hepatitis and/or jaundice) but none had cataracts. Serial biochemical control is good (plasma

Journal of Inherited Metabolic Disease. ISSN 0141–8955. Copyright © SSIEM and MTP Press Limited, Queen Square, Lancaster, UK.

Table 1 Lens opacities in galactosaemia

Case no.	Age at diagnosis (days)	Follow-up period (years)	Medical problems At diagnosis	Now (1987)	Lens opacities At diagnosis	Now[b]	Bio-chemical control[c]
1	42	1.9	Failure to thrive	Nil	Lamellar cataract	Nil	Very good
2	19	6.9	Septicaemia	Nil	Foetal and lamellar	i.s.q.	Very good
3	12 years	2.8	Mental retardation	i.s.q.	Foetal and lamellar with riders	i.s.q.	Very good
4	1	7.1	Nil	Nil	Nil	Peripheral opacities	Fair
5	1	8.3	Nil	Nil	Nil	Nil[d]	Fair
6	6	7.7	Jaundice	Nil	Antr. axial embryonic cataract	i.s.q.	Very good
7	3	4.8	Nil	Nil	Nil[a]	Pulverant	Very good

[a]Direct ophthalmoscope
[b]Slit lamp
[c]Erythrocyte galactose-1-phosphate: very good <50; good 50 to 70; fair 70 to 100 μg (mL packed cells)$^{-1}$
[d]Peripheral opacities documented during period off diet, which regressed
i.s.q. means unchanged

galactose-1-phosphate levels <50 μg (mL packed cells)$^{-1}$) in eight patients and fair (50–70 μg mL^{-1}) in three patients. At present, one patient has an intelligence quotient in the mild mental handicap range but the others are within normal limits. All 11 Group 1 patients have normal visual acuities

In six of seven cases in Group 2 the mean age at commencement of dietary treatment was 12.7 days (range 1–46 days) and the mean follow-up was 6.1 years (range 1.9–7.7 years). The seventh patient commenced treatment aged 12 and is now 14.8 years. At initial diagnosis, three were asymptomatic, and three presented with jaundice, failure to thrive and septicaemia, respectively. The seventh patient presented with mental retardation (Table 1). Lens opacities were observed in four patients at diagnosis and regressed completely in Case 1 within 6 months of treatment. Two patients (Cases 4 and 5) who were asymptomatic at birth developed lens opacities during a 3-month period off diet. These regressed in one patient within 9 months of improved dietary control. One patient (Case 7) had a stationary pulverant cataract identical to that of her mother, which was not seen with a direct ophthalmoscope at initial diagnosis but was noted when examined with a slit-lamp aged 2 years.

Parents

A detailed history was taken which ruled out other causes for cataract, and an ocular examination was performed in 23 of 24 parents (mean age 34.9 years; range 23–54 years). Five cases (mean age 37.1 years) had mild bilateral lens opacities (cortical – 3, subcapsular – 1, pulverant – 1) which did not impair visual acuity.

DISCUSSION

Some authors suggest that cataracts may develop in approximately 75% of individuals with untreated classical galactosaemia (Duke-Elder, 1966). The severity of lens involvement has been related to the severity of the galactosaemia, the age at diagnosis and the age at commencement of treatment. The lens may become totally opaque before 3 months of age in untreated cases. Cases 4 and 5 developed lens opacities visible only with a slit-lamp while off diet. These regressed with improvement in dietary control. The appearence of opacities may be the most sensitive initial index of inadequate biochemical control.

All cataracts in galactosaemia patients may not be due to galactosaemia. An anterior axial embryonic cataract was noted in Case 6. This type of opacity is frequently visible in normal lenses. Case 7 clearly demonstrates how thin lens opacities may be overlooked by direct ophthalmoscopy when it has been impossible or difficult to use the slit-lamp. Lens opacities do not regress in all cases despite early commencement of diet (Cases 1 and 2). Dietary treatment in cases with cataract prevented major visual disability however.

There are conflicting reports concerning the association of visually significant pre-senile cataracts and GPUT heterozygosity. We found no case of visually significant cataract among 23 obligate heterozygote parents. Nevertheless, four out of 23 had early cortical or sub-capsular lens opacities (mean age 37.1 years) which is unexpectedly high at this age. Prospective studies are urgently required to resolve this question as the therapeutic implications may be great.

We would recommend that slit-lamp examination be performed regularly as part of the monitoring process in the individual with galactosaemia, particularly in centres where there may be a delay in obtaining biochemical results.

ACKNOWLEDGEMENTS

We wish to thank Dr S. Cahalane and his laboratory staff for their excellent service and I. Saul for her help with this study.

REFERENCES

Beutler, E. R. and Baluda, M. C. A simple spot screening test for galactosaemia. *J. Lab. Clin. Med.* 68 (1966) 137–141

Cahalane, S. F. and Naughten, E. R. Total newborn population and high risk screening for galactosaemia in Ireland 1972–1982. In Naruse, H. (ed.) *Neonatal Screening* Excerpta Medica, Japan, 1983, pp. 250–251

Donnell, G. N. and Bergren, W. R. The galactosaemias. In Raine, D. N. (ed.) *The Treatment of Inherited Metabolic Disease*, Medical and Technical Publishing Co. Ltd., Lancaster (1975) 91–114

Duke-Elder, S. Galactosaemic cataract. In *System of Ophthalmology*, Vol. 11, Henry Kimpton, London (1966) 172

Kalkar, H. M., Anderson, E. P. and Isselbacher, K. J. Galactosaemia. A congenital defect in nucleotide transferase. *Biochim. Biophys. Acta* 20 (1956) 262–268

Skalka, H. N. and Prchal, J. T. Presenile cataract formation and decreased activity of galactosaemic enzymes. *Arch. Ophthalmol.* 98 (1980) 269–273

Wilson, W. A. and Donnell, G. N. Cataracts in galactosaemia. *Arch. Ophthalmol.* 60 (1958) 215–222

J. Inher. Metab. Dis. 11 Suppl. 2 (1988) 249–251

Short Communication

A Patient with Severe Type of Epimerase Deficiency Galactosaemia

I. B. Sardharwalla[1], J. E. Wraith[1], C. Bridge[1], B. Fowler[1] and S. A. Roberts[2]

[1]Willink Biochemical Genetics Unit, Royal Manchester Children's Hospital, Manchester, M27 1HA, [2]Duchess of York Hospital, Nell Lane, Manchester 20, UK

UDP-galactose-4-epimerase (EC 5.1.3.2) is a key enzyme in the metabolic route by which man metabolises galactose to produce glucose. Its absence will impair the continual action of galactose-1-phosphate uridyl transferase as this enzyme depends upon a constant regeneration of UDP-glucose catalysed by the action of epimerase. In addition a lack of endogenous synthesis of UDP-galactose will prevent the normal production of galactocerebroside.

A deficiency of epimerase activity (McKusick 23035) was first demonstrated during a mass newborn screening programme aimed at the detection of galactosaemia (Gitzelman, 1972). In the patients so identified, the enzyme deficiency was limited to red blood cells alone and there was no evidence of clinical disease in them. However, a severe type, caused by *generalized* deficiency of epimerase, was later described (Holton *et al.*, 1981) in an infant who presented with clinical manifestations very similar to those seen in neonates with galactose-1-phosphate uridyl transferase deficiency galactosaemia. This paper reports a further case of the severe type of epimerase deficiency galactosaemia.

CASE REPORT

On the newborn amino acid screening programme using one dimensional paper chromatography of plasma, a female infant was found to have a marked elevation of plasma methionine. Further investigation showed evidence of liver disease associated with galactosuria and amino aciduria. It was subsequently demonstrated that the infant's red blood cell galactose-1-phosphate level was markedly elevated, $(1738 \mu g \, mL^{-1}$ packed red cells). In the red blood cells and cultured skin fibroblasts the activities of galactokinase and galactose-1-phosphate uridyl transferase were normal but there was a marked deficiency of UDP galactose-4-epimerase activity (erythrocytes, $0.023 \mu mol \, h^{-1} \, mL^{-1}$, controls range 0.75–3.05, mean 1.69, $n = 30$; cultured fibroblasts none detected, controls 0.27, 0.34 $\mu mol \, h^{-1} \, mg^{-1}$).

The infant was the fourth child of Asian Muslim parents who were first cousins. During the first week of life the child had become unwell with poor feeding and vomiting. In addition, she was mildly jaundiced, had hepatomegaly measuring 3 cm

Journal of Inherited Metabolic Disease. ISSN 0141–8955. Copyright © SSIEM and MTP Press Limited, Queen Square, Lancaster, UK.

below the subcostal margin and generalised hypotonia. Blood cultures grew a staphylococcal organism and she was treated with the appropriate antibiotics. Biochemical investigations confirmed the presence of liver disease as well as demonstrating abnormal thyroid function with a low T_4 and T_3 but normal TSH. The latter returned to normal without specific treatment.

A galactose-*restricted* diet was started at the age of 21 days. The child responded quickly. The hepatomegaly regressed, the liver function tests returned to normal and the cataracts which were present on slit lamp examination resolved. The concentrations of red cell galactose-1-phosphate and UDP-galactose decreased and were maintained at low levels on an intake of 30 mL of cow's milk or 1 g of galactose per day. A direct correlation was demonstrated between red cell galactose-1-

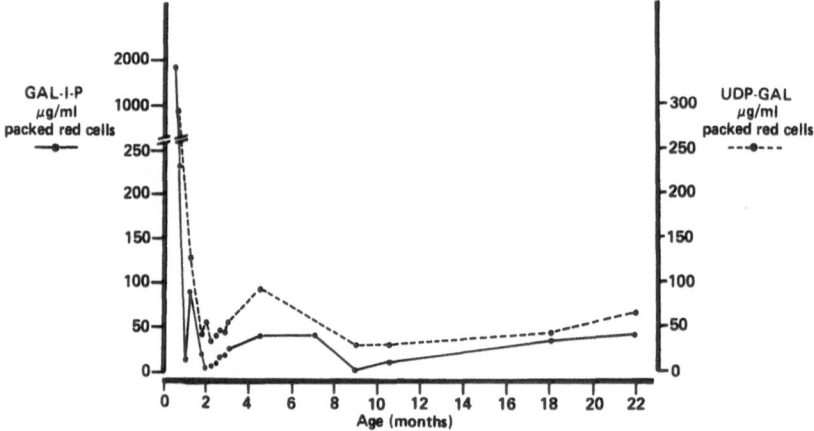

Figure 1 Relationship between red cell galactose-1-phosphate and UDP-galactose before and during treatment

phosphate and UDP-galactose levels (Figure 1). However, despite early treatment and satisfactory biochemical control, clinical assessment at the age of 2 years 9 months showed that the child was severely mentally retarded with an IQ score of 55. In addition she had profound sensori-neuronal deafness. A CT scan of the brain was reported to be normal.

DISCUSSION

Galactose is an essential constituent of many cerebral structural components such as sphingolipids. Incorporation of galactose into these compounds takes place via UDP-galactose. In normal individuals and patients with transferase or galactokinase deficiency galactosaemia, UDP-galactose can be produced endogenously via a reversal of the pyrophosphorylase pathway. However, a patient with epimerase deficiency galactosaemia is unable to utilise this pathway and is therefore dependent on exogenous galactose to produce these essential compounds. For this reason a galactose-*restricted* rather than a galactose-free diet is recommended in management.

Despite early diagnosis and treatment and apparently satisfactory levels of red cell UDP-galactose our patient has considerable mental retardation. Two possible explanations may be offered. First, UDP-galactose may be far more toxic than galactose-1-phosphate so that even a short term exposure of the brain to it or in combination with galactose-1-phosphate may cause severe impairment of neurological function. Second, a lack of endogenous synthesis of galactocerebrosides during intrauterine life may be the important factor that determines the ultimate intellectual development in these individuals. In either case, the prognosis in terms of mental development in patients with generalised epimerase deficiency galactosaemia may prove to be worse than in those with the 'transferase' deficiency galactosaemia.

REFERENCES

Gitzelman, R. Deficiency of uridine diphosphate galactose 4-epimerase in blood cells of an apparently healthy infant. *Helvet. Pediatr. Acta* 27 (1972) 125–130

Holton, J. B., Gillett, M. G., MacFaul, R. and Young, R. Galactosaemia: A new severe variant due to uridine diphosphate galactose-4-epimerase deficiency. *Arch. Dis. Child.* 56 (1981) 885–887

J. Inher. Metab. Dis. 11 Suppl. 2 (1988) 252–254

Short Communication

Branching Enzyme in Erythrocytes. Detection of Type IV Glycogenosis Homozygotes and Heterozygotes

Y. S. Shin[1], H. Steigüber[1], P. Klemm[1], W. Endres[1], O. Schwab[2] and G. Wolff[3]

[1]*Children's Hospital, University of Munich,* [2]*Children's Hospital, University of Würzburg and* [3]*Institute of Human Genetics, University of Freiburg, Federal Republic of Germany*

Type IV glycogen storage disease (Andersen's disease: McKusick 23250) is a rare congenital disorder characterized by hepatosplenomegaly and growth retardation. The onset and the progress of the disease as well as the severity of the clinical manifestation are greatly variable (Andersen, 1956; Guerra *et al.*, 1986; Landing *el al.*, 1968; McMaster *et al.*, 1979). The activity of α-1, 4-glucan: α-1, 4-glucan 6-transglucosylase (branching enzyme, E.C. 2.4.1.18) was found to be deficient in fibroblasts (Brown and Brown, 1966; Landing *et al.*, 1968) and in leukocytes (Fernandes and Huijing, 1968) of the patients with type IV glycogenosis. In this study we report the branching enzyme activity in various human tissues, especially in erythrocytes where the enzyme activity is not only readily measurable but also clearly distinguishable between various phenotypes.

PATIENTS

Patient 1: An 18-month-old German boy with hepatosplenomegaly and muscular hypotonia. Dr B. Brown determined the branching enzyme activity in fibroblasts.

Patient 2: A 19-year-old German male with occasional febrile seizures. One of the younger brothers died at 3 years of age (Schaub, Hübner and Shin, unpublished).

Patient 3: The second child of a German family who died at birth. The first child of the family also died at birth of respiratory distress syndrome.

MATERIALS AND METHODS

Samples: Heparinized blood was sent by mail. The cells were separated by centrifugation, washed with saline three times and frozen at −25°C until used. The chorionic villous samplings were analysed directly and amniotic fluid cells and skin fibroblasts were cultured until confluency.

Enzyme assay: The branching enzyme activity has been determined by tracing the

Journal of Inherited Metabolic Disease. ISSN 0141–8955. Copyright © SSIEM and MTP Press Limited, Queen Square, Lancaster, UK.

glycogen synthesis from two different substrates and exogenous enzymes. For method 1, rabbit muscle phosphorylase a and glucose-1-phosphate were used and the release of inorganic phosphates was determined (Brown and Brown, 1966). The second method was basically the same as that of Brown and Brown (1983). The reaction mixture contained $1\,\mu$mol glycylglycine buffer, pH 7.2, $5\,\mu$mol sodium fluoride, $0.8\,\mu$mol dithiothreitol, $0.56\,\mu$mol UDP-glucose, $0.1\,\mu$Ci UDP-[^{14}C]glucose, $1\,\mu$mol glucose-6-phosphate and the extracts in a total volume of $100\,\mu$L. UDP-[^{14}C]glucose and [^{14}C]glycogen were separated on a Sephadex column (Shin *et al.*, 1984).

RESULTS AND DISCUSSION

The enzyme activity in erythrocytes was readily determinable by the spectrophotometric method. Less than $5\,\mu$L blood was necessary for the assay. For the second method slightly larger samples were required for the activity measurement

Table 1 Branching enzyme activity in human erythrocytes and other tissues

Sample	Method 1	Method 2
Erythrocytes[a]		
Patient 1	0.018–0.40	n.d.
Mother 1	3.58	0.093
Patient 2	1.66	n.d.
Patient 3	0	0
Mother 3	3.95	0.078
Father 3	3.19	0.064
Controls[b]	7.3±2.6 (75)	0.29±0.09 (36)
Other tissues[c]		
Liver	0.45±0.15 (7)	0.017, 0.02
Fibroblasts	0.85±0.23 (8)	0.007, 0.016
Chorionic villi	0.32±0.09 (6)	n.d.
Amniotic cells	0.19±0.05 (10)	n.d.

n.d. = not determined
[a]Expressed as μmol (g haemoglobin)$^{-1}$min^{-1}
[b]Means ± SD, *n* in parentheses
[c]Expressed as μmol (mg protein)$^{-1}$min^{-1}, *n* in parentheses

(Table 1). These results indicate that the branching enzyme assay in erythrocytes offers an easy means for the diagnosis of type IV glycogenosis and the carriers. The enzyme in heparinized whole blood was stable as long as the blood samples were not haemolysed and was stable in washed erythrocytes at -25°C for several months. This signifies that the diagnosis can be done by whole blood sent by mail as well as by stored blood samples. The possible detection of heterozygotes by blood samples is valuable in genetic counselling. It is also interesting that all three patients showed different clinical symptoms and had correspondingly a different degree of residual enzyme activity in erythrocytes. Although possibilities exist for the prenatal diagnosis of the disease using cultivated amniotic fluid cells, the relatively abundant activity in uncultured chorionic villi suggests a potential use of the villi for this purpose.

REFERENCES

Andersen, D. H. Familial cirrhosis of the liver with storage of abnormal glycogen. *Lab. Invest.* 5 (1956) 11–20

Brown, B. I. and Brown, D. H. Lack of an α-1,4-glucan: α-1,4-glucan 6-transglucosylase in a case of type IV glycogenosis. *Proc. Natl. Acad. Sci. USA* 56 (1966) 725–729

Brown, D. H. and Brown, B. I. Studies of the residual glycogen branching enzyme activity present in human skin fibroblasts from patients with type IV glycogen storage disease. *Biochem. Biophys. Res. Commun.* 111 (1983) 636–643

Guerra, A. S., van Diggelen, O. P., Carneiro, F., Tsou, R. M., Simoes, S. and Santos, N. T. A juvenile variant of glycogenosis IV (Andersen disease). *Eur. J. Pediatr.* 145 (1986) 179–181

Fernandes, J. and Huijing, F. Branching enzyme deficiency glycogenosis: Studies in therapy. *Arch. Dis. Child.* 43 (1968) 437–452

Landing, B. H., Reed, G. B., Dixon, J. F. P., Neustein, H. B. and Donnell, G. N. Patient with absence of a branching enzyme α-1,4-glucan: α-1,4-glucan 6-glucosyl transferase. *Lab. Invest.* 19 (1968) 546–547

McMaster, K. R., Powers, J. M., Henninger, G. R. Jr, Wohltmann, H. and Farr, G. H. Jr. Nervous system involvement in type IV glycogenosis. *Arch. Pathol. Lab. Med.* 103 (1979) 105–111

Shin, Y. S., Rieth, M., Ungar, R. and Endres, W. A simple method for amylo-1,6-glucosidase assay. Detection of heterozygotes for type III glycogenosis in erythrocytes. *Clin. Chem.* 30 (1984) 1717–1718

J. Inher. Metab. Dis. 11 Suppl. 2 (1988) 255–258

Short Communication

β-Mannosidosis in Two Brothers with Hearing Loss

L. DORLAND[1], M. DURAN[1], F. E. T. HOEFNAGELS[1], J. N. BREG[2],
H. FABERY DE JONGE[1], K. CRANSBERG[1], F. J. VAN SPRANG[1] and O. P. VAN
DIGGELEN[3]
[1]*University Children's Hospital 'Het Wilhelmina Kinderziekenhuis', Nieuwe
Gracht 137, NL-3512 LK Utrecht, The Netherlands;* [2]*Department of Bio-Organic
Chemistry, University of Utrecht, Utrecht, The Netherlands;* [3]*Department of
Clinical Genetics, Erasmus University, Rotterdam, The Netherlands*

β-Mannosidosis is a lysosomal storage disorder caused by a deficiency of the enzyme β-mannosidase (EC 3.2.1.25). This inborn error of glycoprotein catabolism has been described for goats (Jones and Dawson, 1981). Recently, two independent papers have appeared on human β-mannosidase deficiency (Cooper *et al.*, 1986; Wenger *et al.*, 1986). We have diagnosed this disorder in two brothers with hearing problems by analysing urinary oligosaccharides and by measuring the enzyme activity in leukocytes and plasma.

PATIENTS

Patient A.B., born in 1978, is the first child of consanguineous Turkish parents (first cousins). At the age of 3 weeks he started to have feeding difficulties and diarrhoea; later on he suffered from recurring respiratory infections with high fever. A.B. was a troublesome child. Speech retardation was noticed at the age of 4 years. There was a 40 dB hearing loss observed in BERA. His motor development was normal: sitting at 7 months, standing at 10 months and walking at 11 months. Physical examination revealed no abnormalities.

The brother H.B. was born in 1980. Just like his older brother, this boy had severe feeding difficulties and recurring respiratory infections. At the age of 3 years his speech development appeared to be retarded. There was a hearing-loss of 50 dB. The motor development was normal. Physical examination revealed only a slight hypotonia. When the boys were 8 and 6 years old, respectively, they were screened for inherited metabolic disorders because of their abnormal behaviour.

METHODS

Thin-layer chromatography of urinary oligosaccharides was performed on alumina sheets coated with silica using the solvent system *n*–butanol–acetic acid–water (2:1:1 v/v) as published earlier (Dorland *et al.*, 1986). Alternatively the solvent

Journal of Inherited Metabolic Disease. ISSN 0141-8955. Copyright © SSIEM and MTP Press Limited, Queen Square, Lancaster, UK.

system n–propanol–acetic acid–water (85:1:15 v/v) was used. Urinary mucopolysaccharide excretion was examined with the Ames/MPS spot test paper (Biodex Ltd., Jerusalem, Israel), with two-dimensional electrophoresis of the cetylpyridinium chloride precipitate of the urine (Abeling et al., 1974), and by measuring the total uronic acid excretion (Di Ferrante and Rich, 1956).

Urinary oligosaccharides were isolated on a large scale by a sequence of cation-exchange chromatography, anion-exchange chromatography and gel filtration on BioGel P-2. Finally the material was purified on a silicic acid column according to Wenger et al. (1986). Carbohydrate composition analysis was performed by gas chromatography, after methanolysis, re-N-acetylation and trimethylsilylation (Kamerling et al., 1975).

Gas chromatography combined with mass spectrometry of trimethylsilyl derivatives was performed on a Ribermag R10-10 C (Rueil Malmaison, France) quadrupole mass spectrometer in the electron-impact mode (ionising voltage 70 eV). Desorption chemical ionization mass spectrometry of the isolated substance was carried out at the same apparatus using ammonia as the reactant gas at a pressure of 13.3 Pa. The probe filament current was 250 mA.

500-MHz ^{1}H-NMR spectroscopy was performed on a Bruker AM-500 spectrometer, operating in the Fourier-transform mode and equipped with a Bruker Aspect-3000 computer. The spectra were obtained with quadrature phase-detection using a 90° pulse. Natural-abundance ^{13}C-NMR spectra were obtained at 50.76 MHz on a Bruker WP-200 spectrometer.

β-Mannosidase activity in leukocytes and plasma was measured by the method of Panday et al. (1984), α-mannosidase activity according to Galjaard (1980).

RESULTS AND DISCUSSION

One-dimensional thin-layer chromatography (TLC) of oligosaccharides in the urines of both patients in n-butanol–acetic acid–water (2:1:1 v/v) revealed a strong band in a position virtually coinciding with that of lactose. In the additional one-dimensional TLC system n-propanol–acetic acid–water (85:1:15 v/v) the unknown compound was clearly separated from lactose. Isolation from 25 mL of urine by ion-exchange, gel-filtration and silica chromatography yielded a purified fraction with the compound of interest. The substance (18 mg) appeared to be pure in both one-dimensional TLC systems.

Sugar composition analysis of the isolated compound revealed mannose and N-acetylglucosamine in equal amounts. The EI mass spectrum of the pertrimethylsilyl derivative of this compound showed m/z 872 as the highest observed mass, in accordance with [M-15]$^+$ of a pertrimethylsilyl disaccharide, containing a hexose residue and an N-acetylhexosamine residue. It could be deduced from this spectrum that an N-acetyl-amino sugar (N-acetylglucosamine) is present in the reducing position. Desorption chemical ionisation mass spectrometry with ammonia as reactant gas of the underivatized substance showed a quasi-molecular ion at m/z 384 [M+H]$^+$, also indicative of a disaccharide consisting of a hexose and an N-acetylhexosamine.

500-MHz ^{1}H-NMR spectroscopy in deuterium oxide of this disaccharide showed that the configuration of the non-reducing mannose (Man) residue is β (δH1 = 4.769 ppm; $J_{1,2} \leq$ 1Hz). The (1→4) type of bonding between mannose and *N*-acetylglucosamine was unambiguously proven by ^{13}C-NMR spectroscopy.

On the basis of these data this urinary compound was identified as the disaccharide mannosyl-β(1→4)-*N*-acetylglucosamine.

β-Mannosidase activities in leukocytes and plasma of the patients were undetectable (Table 1). The parents had decreased plasma β-mannosidase activities and low–normal activities in leukocytes, consistent with heterozygosity. The activities of α-mannosidase were normal for all family members.

Table 1 *β*-**Mannosidase and** *α*-**mannosidase activities in leukocytes and plasma**

	β-Mannosidase		*α-Mannosidase*	
	Leukocytes[a]	*Plasma*[b]	*Leukocytes*[a]	*Plasma*[b]
A.B.	0.0	0.0	171	32
H.B.	0.0	0.0	219	28
Father	43	58	137	41
Mother	41	58	189	32
Normal range	45–150	80–200	110–320	20–120

Values are [a]nmol h^{-1}(mg protein)$^{-1}$; [b]nmol h^{-1} mL^{-1}

The original patient (Wenger *et al.*, 1986), but not the patients of Cooper *et al.* (1986), also had a heparan sulphamidase deficiency (Sanfilippo A disease; MPS IIIA). We checked the possibility of an abnormal mucopolysacchariduria in both patients. Their MPS spot tests were negative, neither of them excreted heparan sulphate and their uronic acid excretions expressed as mg (g creatinine)$^{-1}$ were normal. This suggests that our patients do not have the combined enzyme deficiencies. Retrospectively we have checked the urines of nine Sanfilippo A patients for the presence of the disaccharide mannosyl-β(1→4)-*N*-acetylglucosamine. It was absent in all investigated urine samples. The hearing loss was also observed in Wenger's patient (Wenger *et al.*, 1986), but it was not reported for the two patients reported by Cooper *et al.* (1986).

Screening for β-mannosidosis can easily be performed by one-dimensional TLC of untreated urine. However, in the standard TLC system for oligosaccharides *n*-butanol–acetic acid–water (2:1:1 v/v) the β-mannoside cannot be distinguished from lactose. Therefore, in selected cases chromatography should also be carried out in the solvent system *n*-propanol–acetic acid–water (85:1:15 v/v), which separates the disaccharide mannosyl-β(1→4)-*N*-acetylglucosamine from lactose.

REFERENCES

Abeling, N. G. G. M., Wadman, S. K. and Van Gennip, A. H. Two-dimensional electrophoresis of urinary mucopolysaccharides on cellulose acetate after *N*-cetylpyridiniumchloride (CPC) precipitation: a method suitable for the routine laboratory. *Clin. Chim. Acta* 56 (1974) 297–303

Cooper, A., Sardharwalla, I. B. and Roberts, M. M. Human β-mannosidase deficiency. *N. Engl. J. Med.* 315 (1986) 1231

Di Ferrante, N. and Rich, C. The mucopolysaccharide of normal human urine. *Clin. Chim. Acta* 1 (1956) 519–524

Dorland, L., Wadman, S. K., Fabery de Jonge, H. and Ketting, D. 1,6-Anhydro-β-D-glucopyranose (β-glucosan), a constituent of human urine. *Clin. Chim. Acta* 159 (1986) 11-16

Galjaard, H. *Genetic Metabolic Diseases. Early Diagnosis and Prenatal Analysis* (Monograph) Elsevier-North Holland, Amsterdam (1980) p. 818

Jones, M. Z. and Dawson, G. Caprine β-mannosidosis: inherited deficiency of β-D-mannosidase. *J. Biol. Chem.* 256 (1981) 5185–5188

Kamerling, J. P., Gerwig, G. J., Vliegenthart, J. F. G. and Clamp, J. R. Characterization by gas–liquid chromatography–mass spectrometry and proton-magnetic resonance spectroscopy of pertrimethylsilyl methyl glycosides obtained in the methanolysis of glycoprotein and glycopeptides. *Biochem. J.* 151 (1975) 491–495

Panday, R. S., Van Diggelen, O. P., Kleijer, W. J. and Niermeijer, M. F. β-Mannosidase in human leukocytes and fibroblasts. *J. Inher. Metab. Dis.* 7 (1984) 155–156

Wenger, D. A., Sujansky, E., Fennessey, P. V. and Thompson, J. N. Human β-mannosidase deficiency. *N. Engl. J. Med.* 315 (1986) 1201–1205

J. Inher. Metab. Dis. 11 Suppl. 2 (1988) 259–262

Short Communication

Infantile Sialic Acid Storage Disease in Two Siblings

A. Cooper[1], I. B. Sardharwalla[1], M. Thornley[1] and K. P. Ward[2]
[1]*Willink Biochemical Genetics Unit, Royal Manchester Children's Hospital, Pendlebury, Manchester, M27 1HA, UK and* [2]*Airedale General Hospital, Steeton, Keighley, West Yorkshire, UK*

Infantile sialic acid storage disease (ISSD: McKusick 26992) is a rare disorder related to Salla disease (McKusick 26874) and sialuria (McKusick 26992). The disorder is caused by progressive accumulation of free sialic acid within lysosomes due to defective efflux (Renlund *et al.*, 1986). This is probably caused by defective transport across the lysosomal membrane (Mancini *et al.*, 1986). Free sialic acid content of tissues is elevated and there is hypersialuria (Stevenson *et al.*, 1983).

Few cases have been reported to date. Presentation, in the neonatal period, includes coarse facies, blond hair, hepatosplenomegaly, cardiomegaly, increased muscle tone, delayed psychomotor development, moderate radiological changes, recurrent infection and vacuolation of peripheral lymphocytes (Hancock *et al.*, 1982; Tondeur *et al.*, 1982; Stevenson *et al.*, 1983; Gillan *et al.*, 1984; Paschke *et al.*, 1986).

We report clinical and biochemical findings in two patients with this disorder.

CASE REPORTS

Case 1

At age 3 months this female infant presented with coarse dysmorphic features, delayed psychomotor development and hepatosplenomegaly. At 6 months the liver edge was 6 cm below the costal margin and the spleen was 2 cm enlarged. An echocardiogram showed increased ventricular size and ECG indicated left ventricular hypertrophy. Skeletal survey showed no abnormality and a computed tomography (CT) brain scan was normal. Muscle tone and reflexes were unaffected, the tongue was not enlarged but she had a short neck. Mental and motor development was grossly retarded. Electron microscopy of peripheral and bone-marrow lymphocytes showed marked cytoplasmic vacuolation; skin fibroblast and epithelial cells also showed evidence of lysosomal storage.

Case 2

The male sibling of Case 1 died aged 18 months and diagnosis was achieved retrospectively on stored urine and cultured skin fibroblasts. The patient's clinical progress was similar to that of his sister. Facial coarsening was present at birth

Journal of Inherited Metabolic Disease. ISSN 0141–8955. Copyright © SSIEM and MTP Press Limited, Queen Square, Lancaster, UK.

and hepatosplenomegaly had developed by 3 months. Head circumference was enlarged, psychomotor development was markedly delayed and there were repeated episodes of respiratory tract infection. In addition there was beaking of lumbar vertebrae.

METHODS

Lysosomal enzymes were assayed and urinary mucopolysaccharide and oligosaccharide excretion investigated as previously described (Cooper et al., 1988). Sialic acid-containing compounds of urine were detected with resorcinol spray following chromatography on plastic-backed silica TLC plates (Merck). Chromatograms were developed for 10 cm: once in butanol-acetic acid-water (10:5:5) followed by two further developments in nitromethane-propanol-water (4:5:3). Skin fibroblasts were cultured, leukocytes harvested and lysosomal enzymes assayed as previously described (Cooper et al., 1988) as were patterns of urinary mucopolysaccharide and orcinol-positive oligosaccharides. Free and bound sialic acid was quantitatively determined in urine, fibroblasts, white cells and cell-free amniotic fluid by a scaled down version of the method of Warren (1959).

RESULTS

Lysosomal hydrolase activities of leukocytes and fibroblasts were normal in both patients and urinary mucopolysaccharide and orcinol-positive oligosaccharide excretion showed a normal pattern. Resorcinol staining of urinary oligosaccharide chromatograms revealed intense bands of free sialic acid in both cases but sialyloligosaccharides were not increased (data not shown).

Quantitative determination showed a 10-fold elevation of free sialic acid in the patients' urines whilst in cultured skin fibroblasts a 25-fold increase was observed. In leukocytes of Case 1, free sialic acid was 15 times the mean control value. In all tissues from both patients the level of bound sialic acid did not differ from controls (Table 1).

Table 1   Urine, leukocyte and fibroblast sialic acid content

	Urine[a]		Leukocytes[b]		Fibroblasts[b]	
	Free	*Bound*	*Free*	*Bound*	*Free*	*Bound*
Case 1	1240	32.0	99.0	25	54.2	10.6
Case 2	1497	ND	ND	ND	56.6	8.1
Sialidosis	89	1196	ND	ND	4.6	68.8
Controls (n)	146±64(8)	203±106(8)	6.8±2.9(8)	36.2±4.9(8)	2.2±3.3(6)	24.6±6.1(6)

ND = not determined
[a] Expressed in nmol (μg creatinine)$^{-1}$; [b] nmol (mg protein)$^{-1}$

Amniotic fluid was obtained from the pregnancy of Case 1 (gestational age 15 weeks) for an unrelated procedure. At this stage the diagnosis of ISSD had not been established. When analysed after storage at −20°C for 12 months the free sialic acid concentration was slightly elevated at 15.4 nmol mL^{-1} (controls 3.2–13.9, mean 7.5, $n = 6$).

DISCUSSION

Elevated free sialic acid in the urine of Case 1, demonstrated by TLC, together with early onset of symptoms, suggested a diagnosis of ISSD. An identical result was observed retrospectively in urine from a deceased sibling (Case 2) presenting with similar clinical findings. The diagnosis in both cases was confirmed by quantitative demonstration of elevated free sialic acid in urine and cultured fibroblasts (Table 1). The elevation in cultured cells (and leukocytes from Case 1) is presumably due to defective transport across the lysosomal membrane as demonstrated by Mancini *et al.* (1986).

Prenatal diagnosis of ISSD has been reported (Vamos *et al.*, 1986). A seven-fold elevation of free sialic acid was described in cell-free amniotic fluid from an affected fetus (gestational age 17 weeks), whilst in cultured amniocytes levels were approximately 100 times greater than controls. Free sialic acid was only slightly elevated in amniotic fluid from Case 1. Whilst this may be a reflection of the lower gestational age, it would appear that cultured amniocytes are the tissue of choice for reliable prenatal diagnosis.

Conventional screening techniques for mucopolysaccharide and oligosaccharide storage disorders do not detect free sialic acid. As ISSD is clinically similar to these diseases it is likely that cases have been missed in the past (as Case 2 originally was). A specific test for free sialic acid should be included in biochemical investigations of all patients presenting with coarse facies, organomegaly or marked psychomotor delay. If this is done routinely it is possible that many more cases will be diagnosed in the future. The most reliable initial screening test would be oligosaccharide chromatography stained with resorcinol.

REFERENCES

Cooper, A., Hatton, C., Thornley, M. and Sardharwalla, I. B. Human β-mannosidase deficiency: Biochemical findings in plasma, fibroblasts, white cells and urine. *J. Inher. Metab. Dis.* 11 (1988) 17–29

Gillan, J. E., Lowden, A. J., Gaskin, K. and Cutz, E. Congenital ascites as a presenting sign of lysosomal storage disease. *J. Pediatr.* 104 (1984) 225–231

Hancock, L. W., Tahler, M. M., Horwitz, A. L. and Dawson, G. Generalised N-acetylneuraminic acid storage disease: Quantitation and identification of the monosaccharide accumulating in brain and other tissues. *J. Neurochem.* 38 (1982) 803–809

Mancini, G. M. S., Verheijen, F. W. and Galjaard, H. Free N-acetylneuraminic acid (NANA) storage disorders: evidence for defective NANA transport across the lysosomal membrane. *Hum. Genet.* 73 (1986) 214–217

Paschke, E., Trinkl, G., Erwa, W., Pavelka, M., Mutz, I. and Roscher, A. Infantile type of sialic acid storage disease with sialuria. *Clin. Genet.* 29 (1986) 417–424

Renlund, M., Tietze, F. and Gahl, W. A. Defective sialic acid egress from isolated fibroblast lysosomes of patients with Salla disease. *Science (New York)* 232 (1986) 759–762

Stevenson, R. E., Lubinsky, M., Taylor, H. A., Wenger, D. A., Schroer, R. J. and Olmstead, P. M. Sialic acid storage disease with sialuria: clinical and biochemical features in the severe infantile type. *Pediatrics* 72 (1983) 441–449

Tondeur. M., Libert, J., Vamos, E., Van Hoof, F., Thomas, G. H. and Strecker, G. Infantile sialic acid storage disease: Clinical, ultrastructural and biochemical studies in two siblings. *Eur. J. Paediatr.* 139 (1982) 142–147

Vamos, E., Libert, J., Elkhazen, N., Jauniaux, E., Hustin, J., Wilkin, P., Baumkotter, J., Mendla, K., Cantz, M. and Strecker, G. Prenatal diagnosis and confirmation of infantile sialic acid storage disease. *Prenat. Diagn.* 6 (1986) 437–446

Warren, L. The thiobarbituric acid assay of sialic acids. *J. Biol. Chem.* 234 (1959) 1971-1975

J. Inher. Metab. Dis. 11 Suppl. 2 (1988) 263–266

Short Communication

Evaluation of Lysosomal Enzymes in Uncultured and Cultured Chorionic Villi and Amniocytes

G. BARTALINI, M. A. MARGOLLICCI, P. BALESTRI and A. FOIS
Institute of Clinical Pediatrics, University of Siena, Via P. A. Mattioli, 10 – 53100 Siena, Italy

Chorionic villi obtained between the 8th and 12th weeks of gestation can be utilized for prenatal diagnosis of fetal sex (Gosden *et al.*, 1982), chromosomal abnormalities (Simoni *et al.*, 1983), enzyme defects (Poenaru *et al.*, 1984) and DNA analysis (Old *et al.*, 1982).

Reports on lysosomal enzyme activities in chorionic villi are scarce and discordant. Therefore reference values must be obtained from each laboratory engaged in this activity. For this reason we have evaluated the activities of 13 lysosomal enzymes in uncultured and cultured villi obtained from 30 women who had requested a voluntary termination of pregnancy. The same determinations were carried out in eight cultured amniotic fluid samples obtained at 16–20 weeks gestation from pregnancies at risk for chromosomal abnormalities.

MATERIALS AND METHODS

Abortive material obtained at 8–12 weeks gestation was collected into a Petri dish with sterile saline solution. This material was dissected and cleaned from any non-villous tissues under the inverted microscope. The selected villi were washed twice with saline solution and approximately 15–50 mg of tissue were homogenized, sonicated and utilized for enzymatic analysis. From the 30 samples it was possible to obtain chorionic tissue suitable for culture in 15. This material was cultured in minimum Eagle's medium (MEM) with 25% fetal calf serum, L-glutamine and antibiotics. Enzymatic assays were performed on confluent subcultures, obtained after 20 to 30 days.

The eight amniotic fluid samples were centrifuged gently and the cell pellets suspended in MEM supplemented with 25% fetal calf serum, subsequently transferred to culture dishes. The enzymatic activities were determined on the fibroblast-like cells after 20–30 days of culture.

Activities of β-hexosaminidase (EC 3.2.1.30), α-mannosidase (EC 3.2.1.24), β-mannosidase, β-galactosidase (EC 3.2.1.23), α-fucosidase (EC 3.2.1.51), β-glucosidase (EC 3.2.1.21), β-glucuronidase (EC 3.2.1.31), α-L-iduronidase (EC 3.2.1.76), arylsulphatase C (EC 3.1.6.2) and α-neuraminidase (EC 3.2.1.18) were determined with fluorimetric methods using 4-methylumbelliferyl derivatives as substrate,

Journal of Inherited Metabolic Disease. ISSN 0141–8955. Copyright © SSIEM and MTP Press Limited, Queen Square, Lancaster, UK.

but α-neuraminidase activity was assayed on fresh samples without sonication (Galjaard, 1980). For the determination of arylsulphatases A (EC 3.1.6.8) and B (EC 3.1.6.12) and *N*-acetyl-α-D-glucosaminidase a colorimetric method utilizing *p*-nitrocatechol or *p*-nitrophenyl derivatives as substrate was used (Galjaard, 1980). The protein concentration of each sample was measured by the method of Lowry and colleagues.

RESULTS

Table 1 shows the mean value and the standard deviation of lysosomal enzyme activities in homogenized chorionic villi, in cultured villi and in amniocytes. The activities of the 13 enzymes tested in cultured villi and amniocytes were similar. The activity of β-hexosaminidase, α-mannosidase, *N*-acetyl-α-D-glucosaminidase and β-glucosidase in homogenated villi were also similar to those in cultured villi or amniocytes. Activity of β-glucuronidase, α-fucosidase, arylsulphatase B and C in homogenated villi were higher than in cultured villi or amniocytes but the activity of arylsulphatase A, β-mannosidase and β-galactosidase was lower. The α-neuraminidase and α-L-iduronidase assays indicated a much lower level in respect to the same enzymatic activities assayed in cultured villi or amniocytes.

DISCUSSION

According to Gatti *et al.* (1985), Kleijer (1986) and Poenaru *et al.* (1984) activities of *N*-acetyl-α-D-glucosaminidase, heparan sulphaminidase, β-glucuronidase, α- and β-glucosidase, iduronate sulphatase, α-mannosidase, α-fucosidase, β-hexosaminidase, cerebroside β-galactosidase, α- and β-galactosidase, arylsulphatases B and C in homogenized chorionic villi and cultured villi and amniocytes seem to be in general juxtaposable. In contrast, the activities of α-neuraminidase, α-L-iduronidase, sphingomyelinase and arylsulphatase A cannot be detected in homogenized chorionic villi since in this material they are much lower.

Our findings confirm that chorionic material can be used for the biochemical prenatal diagnosis of lysosomal diseases caused by a defect of the enzymes that we have tested and also that first trimester diagnosis using uncultured chorionic villi is reliable for GM2 gangliosidosis, mannosidosis (α-mannosidase deficiency), Gaucher's, Sanfilippo B, Maroteaux–Lamy (MPS VI) and Sly (MPS VII) diseases, fucosidosis and X-linked ichthyosis. The same material cannot be utilized for prenatal diagnosis of Hurler's disease or sialidosis because the activities of α-L-iduronidase and α-neuraminidase are very low. Since the activities of β-mannosidase, β-galactosidase and arylsulphatase A in villous homogenates appear to be low in comparison to cultured villi and amniocytes, in pregnancies at risk for β-mannosidase deficiency, GM1 gangliosidosis and metachromatic leukodystrophy, the enzymatic assay on this material can be considered indicative only when a total enzyme deficiency is demonstrated.

In conclusion, our data confirm the reliability of the lysosomal hydrolase determinations in cultured and uncultured villous material. However the choice of the

Table 1 Activities of 13 lysosomal enzymes in homogenized chorionic villi, in cultured villi and in amniocytes

Enzyme	Homogenized chorionic villi			Cultured villi			Amniocytes		
	n	Mean	±SD	n	Mean	±SD	n	Mean	±SD
N-Acetyl-α-D-glucosaminidase	22	4.17	1.23	13	4.84	1.71	7	4.80	1.56
β-Hexosaminidase	30	6441.70	2752.50	15	6151.20	3386.00	8	6694.75	2926.40
β-Galactosidase	30	227.71	72.06	15	510.03	221.20	7	494.40	243.90
β-Glucuronidase	30	79.50	37.20	15	38.90	15.80	7	47.10	12.70
α-Mannosidase	30	171.40	68.20	15	159.40	113.50	7	144.90	69.80
β-Mannosidase	30	62.60	31.20	15	169.00	71.40	7	136.80	51.20
β-Glucosidase	30	244.80	82.12	15	264.30	78.20	8	268.50	113.00
α-L-Iduronidase	21	24.20	7.90	15	152.10	53.70	8	164.70	73.00
α-Neuraminidase	21	15.30	6.80	13	79.70	25.90	7	102.10	52.10
α-Fucosidase	30	769.15	309.90	15	49.31	16.60	7	74.00	36.90
Arylsulphatase A	30	207.97	82.40	15	309.90	54.80	7	272.80	95.10
Arylsulphatase B	28	323.50	195.00	15	185.30	65.60	7	203.40	87.30
Arylsulphatase C	28	843.40	408.90	12	3.89	2.30	7	2.98	1.40

Values are expressed in $\text{nmol h}^{-1}(\text{mg protein})^{-1}$; n = number of samples tested procedure (uncultured or cultured villi) must be made according to the possibility of demonstrating sufficient enzyme levels. Moreover, it seems important for each laboratory to establish its own reference values on a sufficient number of normal samples.

REFERENCES

Galjaard, H., *Genetic Metabolic Disease. Early Diagnosis and Prenatal Analysis*. Elsevier, North-Holland Biomedical Press, Amsterdam, New York, Oxford, 1980

Gatti, R., Lombardo, C., Filocamo, M., Borrone, C. and Porro, E. Comparative study of 15 lysosomal enzymes in chorionic villi and cultured amniotic fluid cells. *Prenat. Diagn.* 5 (1985) 329–336

Gosden, J. R., Mitchell, A. R. and Gosden, C. M. Direct vision chorion biopsy and chromosome-specific DNA probes for determination of fetal sex in first-trimester prenatal diagnosis. *Lancet* 2 (1982) 1416–1419

Kleijer, W. J. First-trimester diagnosis of genetic metabolic disorders. *Contrib. Gynecol. Obstet.* 15 (1986) 80–89

Old, J. M., Ward, R. H. T., Petrou, M., Karagözlü, F., Modell, B. and Weatherall, D. J. First-trimester fetal diagnosis for haemoglobinopathies: three cases. *Lancet* 2 (1982) 1413–1416

Poenaru, L., Kaplan, L., Dumez, J. and Dreyfus, J. C. Evaluation of possible first trimester prenatal diagnosis in lysosomal diseases by trophoblast biopsy. *Pediatr. Res.* 18 (1984) 1032–1034

Simoni, G., Brambati, B., Danesino, C., Rossella, F., Terzoli, G., Ferrari, M. and Fraccaro, M. Efficient direct chromosome analyses and enzyme determinations from chorionic villi samples in the first trimester of pregnancy. *Hum. Genet.* 63 (1983) 349–357

SOCIETY FOR THE STUDY OF
INBORN ERRORS OF METABOLISM

The SSIEM was founded in 1963 by a small group in the North of England but now has more than 70% of its members outside the UK. The aim of the Society is to promote the exchange of ideas between professional workers in different disciplines who are interested in inherited metabolic disorders. This aim is pursued in scientific meetings and publications.

The Society holds an annual symposium concentrating on different topics each year with facilities for poster presentations. There is always a clinical aspect as well as a laboratory component. The meeting is organized so that there is ample time for informal discussion; this feature has allowed the formation of a network of contacts throughout the world. The international and multidisciplinary approach is also reflected in the *Journal of Inherited Metabolic Disease.*

If you are interested in joining the SSIEM then contact the Treasurer: Dr. I. B. Sardharwalla, Willink Biochemical Genetics Unit, Royal Manchester Children's Hospital, Pendlebury, Manchester M27 1HA, UK. The current subscription is £25 per year payable January 1st each year. This subscription includes 4 issues and 2 supplements of the *Journal of Inherited Metabolic Disease* as well as the regular circulation of a newsletter.

Contents (*continued from inside back cover*)

Odd-numbered long-chain fatty acid contents in erythrocyte membrane phospholipids in patients with an impaired propionate utilization
U. Wendel, E. Diekmann, M. D. Laryea — 225

Mevalonic aciduria: pathobiochemical effects of mevalonate kinase deficiency on cholesterol metabolism in intact fibroblasts
G. Hoffmann, K. M. Gibson, W. L. Nyhan, L. Sweetman — 229

A patient with mevalonic aciduria presenting with hepatosplenomegaly, congenital anaemia, thrombocytopaenia and leukocytosis
J. B. C. de Klerk, M. Duran, L. Dorland, H. A. A. Brouwers, L. Bruinvis, D. Ketting — 233

A closer look at the eye in homocystinuria – a screened population
J. P. Burke, M. O'Keefe, R. Bowell, E. R. Naughten — 237

Peptiduria presumably caused by aminopeptidase-P deficiency. A new inborn error of metabolism
N. Blau, A. Niederwieser, D. H. Shmerling — 240

Early morning urine galactitol levels in relation to galactose intake: a possible method of monitoring the diet in galactokinase deficiency
J. T. Allen, J. B. Holton, A. C. Lennox, I. C. Hodges — 243

Cataracts in children with classical galactosaemia and in their parents
J. P. Burke, M. O'Keefe, R. Bowell, E. R. Naughten — 246

A patient with severe type of epimerase deficiency galactosaemia
I. B. Sardharwalla, J. E. Wraith, C. Bridge, B. Fowler, S. A. Roberts — 249

Branching enzyme in erythrocytes. Detection of type IV glycogenosis homozygotes and heterozygotes
Y. S. Shin, H. Steiguber, P. Klemm, W. Endres, O. Schwab, G. Wolff — 252

β-Mannosidosis in two brothers with hearing loss
L. Dorland, M. Duran, F. E. T. Hoefnagels, J. N. Breg, H. Fabery de Jonge, K. Cransberg, F. J. van Sprang, O. P. van Diggelen — 255

Infantile sialic acid storage disease in two siblings
A. Cooper, I. B. Sardharwalla, M. Thornley, K. P. Ward — 259

Evaluation of lysosomal enzymes in uncultured and cultured chorionic villi and amniocytes
G. Bartalini, M. A. Margollicci, P. Balestri, A. Fois — 263